METHODS IN MOLECULAR BIOLOGY

Series Editor
John M. Walker
School of Life and Medical Sciences
University of Hertfordshire
Hatfield, Hertfordshire, AL10 9AB, UK

For further volumes:
http://www.springer.com/series/7651

Modeling Peptide-Protein Interactions

Methods and Protocols

Edited by

Ora Schueler-Furman

Microbiology and Molecular Genetics, Hebrew University Hadassah Medical School, Jerusalem, Israel

Nir London

Organic Chemistry, The Weizmann Institute of Science, Rehovot, Israel

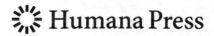 Humana Press

Editors
Ora Schueler-Furman
Microbiology and Molecular Genetics
Hebrew University Hadassah Medical School
Jerusalem, Israel

Nir London
Organic Chemistry
The Weizmann Institute of Science
Rehovot, Israel

ISSN 1064-3745   ISSN 1940-6029   (electronic)
Methods in Molecular Biology
ISBN 978-1-4939-8302-5   ISBN 978-1-4939-6798-8   (eBook)
DOI 10.1007/978-1-4939-6798-8

Cover Caption: Peptide docking: Programs such as GalaxyPepDock can generate a refined model (dark green) that is very similar to the native structure (yellow), starting from an initial similar model (green), or even farther. *Image courtesy of Hasup Lee and Chaok Seok.*

Printed on acid-free paper

This Humana Press imprint is published by Springer Nature
The registered company is Springer Science+Business Media LLC
The registered company address is: 233 Spring Street, New York, NY 10013, U.S.A.

Preface

Peptide-mediated interactions play prominent roles in many cellular processes: their weak, transient character, and their easy manipulation by targeted changes such as post-translational modifications, makes them especially amenable to versatile regulation.

The structure of a peptide-protein complex allows its detailed characterization and fine-tuned manipulation and provides important leads for targeted inhibitor design. It is therefore not surprising that much effort has been put into the development of tailored tools for the modeling of peptide-protein complex structures. However, until not long ago, such approaches were significantly limited, mainly due to challenges of sampling (peptides predominantly do not adopt a defined conformation prior to binding, so that peptide docking may be seen as a "fold-and-dock" challenge), but also scoring (peptide-protein interactions are often transient and weak, and modeling of solvation can be particularly challenging for these small interfaces).

The last few years have witnessed an unprecedented interest, and consequently advance, in our abilities to model and manipulate peptide-mediated interactions. This started with the development of dedicated protocols for local peptide-protein docking that apply a range of different algorithms to tackle the sampling problem. These now generate on a regular basis accurate, near-atom resolution models. It did not take long for the development of a second wave of approaches that extend and complement these tools towards full blind docking, without prior knowledge of the binding site, or an approximate starting conformation for the peptide. Such global docking may be accomplished either by combining binding site prediction with subsequent peptide docking or, alternatively, by performing both together.

Another area of fruitful advance has been our improved ability to predict not only the structure but also the binding affinity and specificity of peptide-protein interactions. These come together with dramatic improvement in the design of inhibitory peptides for the fine-tuned manipulation of protein interactions. Such advances bring us closer to be able to perform peptide-protein modeling on proteomic scale.

It is truly impressive how, in a short time, peptide-protein modeling has risen from a challenged side topic to an ever improving, buzzing field! Key to this improvement has been benchmarks, in the form of curated datasets of peptide-protein complex structures (such as PeptiDB), and last but not least, the CAPRI challenge for the assessment of the modeling of protein interactions: CAPRI has enthusiastically embraced peptide docking and included several peptide-protein docking targets over the past few years. This has further spurred the development of peptide docking protocols; many of them have been discussed in detail at the latest CAPRI evaluation meeting in 2016 in Tel Aviv (www.cs.tau.ac.il/conferences/**CAPRI2016**/).

In this book we have collected a series of chapters from the leading figures in the field of peptide-protein docking. The chapters are bundled into four inter-related parts, including (1) peptide binding site prediction; (2) peptide-protein docking; (3) prediction and design of peptide binding specificity; and (4) the design of inhibitory peptides. In their combination in this book, the chapters provide a diverse and unified state-of-the-art overview of this rapidly advancing field of major interest and applicability.

We look forward to seeing the many applications that will result from applying the methodologies described in this book.

Jerusalem, Israel   *Ora Schueler-Furman*
Rehovot, Israel   *Nir London*

Contents

Preface. . . . . *v*
Contributors. . . . . *ix*

PART I PREDICTION OF PEPTIDE BINDING SITES

1 The Usage of ACCLUSTER for Peptide Binding Site Prediction. 3
   Chengfei Yan, Xianjin Xu, and Xiaoqin Zou

2 Detection of Peptide-Binding Sites on Protein Surfaces
   Using the Peptimap Server 11
   Tanggis Bohnuud, George Jones, Ora Schueler-Furman,
   and Dima Kozakov

3 Peptide Suboptimal Conformation Sampling for the Prediction
   of Protein-Peptide Interactions 21
   Alexis Lamiable, Pierre Thévenet, Stephanie Eustache, Adrien Saladin,
   Gautier Moroy, and Pierre Tuffery

PART II PEPTIDE DOCKING

4 Template-Based Prediction of Protein-Peptide Interactions
   by Using GalaxyPepDock 37
   Hasup Lee and Chaok Seok

5 Application of the ATTRACT Coarse-Grained Docking and Atomistic
   Refinement for Predicting Peptide-Protein Interactions. 49
   Christina Schindler and Martin Zacharias

6 Highly Flexible Protein-Peptide Docking Using CABS-Dock 69
   Maciej Paweł Ciemny, Mateusz Kurcinski, Konrad Jakub Kozak,
   Andrzej Kolinski, and Sebastian Kmiecik

7 AnchorDock for Blind Flexible Docking of Peptides to Proteins 95
   Michal Slutzki, Avraham Ben-Shimon, and Masha Y. Niv

8 Information-Driven, Ensemble Flexible Peptide Docking
   Using HADDOCK. 109
   Cunliang Geng, Siddarth Narasimhan, João P.G.L.M. Rodrigues,
   and Alexandre M.J.J. Bonvin

9 Modeling Peptide-Protein Structure and Binding Using Monte
   Carlo Sampling Approaches: Rosetta FlexPepDock and FlexPepBind. 139
   Nawsad Alam and Ora Schueler-Furman

PART III PREDICTION AND DESIGN OF PEPTIDE BINDING SPECIFICITY

10 Flexible Backbone Methods for Predicting and Designing Peptide Specificity 173
    Noah Ollikainen

11   Simplifying the Design of Protein-Peptide Interaction Specificity
with Sequence-Based Representations of Atomistic Models   189
Fan Zheng and Gevorg Grigoryan

12   Binding Specificity Profiles from Computational Peptide Screening   201
Stefan Wallin

13   Enriching Peptide Libraries for Binding Affinity and Specificity Through
Computationally Directed Library Design   213
*Glenna Wink Foight, T. Scott Chen, Daniel Richman,
and Amy E. Keating*

PART IV   DESIGN OF INHIBITORY PEPTIDES

14   Investigating Protein-Peptide Interactions Using the Schrödinger
Computational Suite   235
Jas Bhachoo and Thijs Beuming

15   Identifying Loop-Mediated Protein-Protein Interactions using LoopFinder   255
Timothy R. Siegert, Michael Bird, and Joshua A. Kritzer

16   Protein-Peptide Interaction Design: PepCrawler and PinaColada   279
Daniel Zaidman and Haim J. Wolfson

17   Modeling and Design of Peptidomimetics to Modulate
Protein–Protein Interactions   291
Andrew M. Watkins, Richard Bonneau, and Paramjit S. Arora

Index *309*

Contributors

NAWSAD ALAM • *Department of Microbiology and Molecular Genetics, Institute for Medical Research Israel-Canada, Faculty of Medicine, Hadassah Medical School, The Hebrew University of Jerusalem, Jerusalem, Israel*

PARAMJIT S. ARORA • *Department of Chemistry, New York University, New York, NY, USA*

AVRAHAM BEN-SHIMON • *The Institute of Biochemistry, Food Science and Nutrition, The Robert H Smith Faculty of Agriculture, Food and Environment, and the Fritz Haber Center for Molecular Dynamics, The Hebrew University, Jerusalem, Israel*

THIJS BEUMING • *Schrödinger, Inc., New York, NY, USA*

JAS BHACHOO • *Schrödinger, Inc., New York, NY, USA*

MICHAEL BIRD • *Department of Chemistry, Tufts University, Medford, MA, USA*

TANGGIS BOHNUUD • *Department of Biomedical Engineering, Boston University, Boston, MA, USA*

RICHARD BONNEAU • *Department of Biology, Center for Genomics and Systems Biology, New York University, New York, NY, USA; Computer Science Department, Courant Institute of Mathematical Sciences, New York University, New York, NY, USA*

ALEXANDRE M.J.J. BONVIN • *Computational Structural Biology Group, Bijvoet Center for Biomolecular Research, Faculty of Science—Chemistry, Utrecht University, Utrecht, The Netherlands*

T. SCOTT CHEN • *Department of Biology, Massachusetts Institute of Technology, Cambridge, MA, USA; Google Inc., Mountain View, CA, USA*

MACIEJ PAWEŁ CIEMNY • *Faculty of Chemistry, University of Warsaw, Warsaw, Poland*

STEPHANIE EUSTACHE • *Université Paris Diderot, Sorbonne Paris Cité, Molécules Thérapeutiques In Silico, Inserm UMR-S 973, Paris, France*

GLENNA WINK FOIGHT • *Department of Biology, Massachusetts Institute of Technology, Cambridge, MA, USA; Department of Chemistry, University of Washington, Seattle, WA, USA*

CUNLIANG GENG • *Computational Structural Biology Group, Bijvoet Center for Biomolecular Research, Faculty of Science—Chemistry, Utrecht University, Utrecht, The Netherlands*

GEVORG GRIGORYAN • *Department of Biological Science, Dartmouth College, Hanover, NH, USA; Department of Computer Sciences, Dartmouth College, Hanover, NH, USA*

GEORGE JONES • *Department of Applied Mathematics and Statistics, Stony Brook University, New York, NY, USA*

AMY E. KEATING • *Department of Biology, Massachusetts Institute of Technology, Cambridge, MA, USA; Department of Biological Engineering, Massachusetts Institute of Technology, Cambridge, MA, USA*

SEBASTIAN KMIECIK • *Faculty of Chemistry, University of Warsaw, Warsaw, Poland*

ANDRZEJ KOLINSKI • *Faculty of Chemistry, University of Warsaw, Warsaw, Poland*

KONRAD JAKUB KOZAK • *Faculty of Chemistry, University of Warsaw, Warsaw, Poland*

DIMA KOZAKOV • *Department of Biomedical Engineering, Boston University, Boston, MA, USA; Department of Applied Mathematics and Statistics, Stony Brook University, New York, NY, USA*

JOSHUA A. KRITZER • *Department of Chemistry, Tufts University, Medford, MA, USA*

MATEUSZ KURCINSKI • *Faculty of Chemistry, University of Warsaw, Warsaw, Poland*

ALEXIS LAMIABLE • *Uniersité Paris Diderot, Sorbonne Paris Cité, Molécules Thérapeutiques In Silico, Inserm UMR-S 973, Paris, France*

HASUP LEE • *Department of Chemistry, Seoul National University, Seoul, Republic of Korea*

GAUTIER MOROY • *Université Paris Diderot, Sorbonne Paris Cité, Molécules Thérapeutiques In Silico, Inserm UMR-S 973, Paris, France*

SIDDARTH NARASIMHAN • *Computational Structural Biology Group, Bijvoet Center for Biomolecular Research, Faculty of Science—Chemistry, Utrecht University, Utrecht, The Netherlands*

MASHA Y. NIV • *The Institute of Biochemistry, Food Science and Nutrition, The Robert H Smith Faculty of Agriculture, Food and Environment, and the Fritz Haber Center for Molecular Dynamics, The Hebrew University, Jerusalem, Israel*

NOAH OLLIKAINEN • *Division of Biology and Biological Engineering, California Institute of Technology, Pasadena, CA, USA*

DANIEL RICHMAN • *Department of Biology (T38), Massachusetts Institute of Technology, Cambridge, MA, USA*

JOÃO P.G.L.M. RODRIGUES • *Computational Structural Biology Group, Bijvoet Center for Biomolecular Research, Faculty of Science—Chemistry, Utrecht University, Utrecht, The Netherlands; Department of Structural Biology, Stanford University School of Medicine, Stanford, CA, USA*

ADRIEN SALADIN • *Université Paris Diderot, Sorbonne Paris Cité, Molécules Thérapeutiques In Silico, Inserm UMR-S 973, Paris, France*

CHRISTINA SCHINDLER • *Physics Department (T38), Center for Integrated Protein Science Munich (CIPSM), Technical University of Munich, Garching, Germany*

ORA SCHUELER-FURMAN • *Department of Microbiology and Molecular Genetics, Institute for Medical Research Israel-Canada, Hadassah Medical School, The Hebrew University of Jerusalem, Jerusalem, Israel*

CHAOK SEOK • *Department of Chemistry, Seoul National University, Seoul, Republic of Korea*

TIMOTHY R. SIEGERT • *Department of Chemistry, Tufts University, Medford, MA, USA*

MICHAL SLUTZKI • *The Institute of Biochemistry, Food Science and Nutrition, The Robert H Smith Faculty of Agriculture, Food and Environment, and the Fritz Haber Center for Molecular Dynamics, The Hebrew University, Jerusalem, Israel*

PIERRE THÉVENET • *Université Paris Diderot, Sorbonne Paris Cité, Molécules Thérapeutiques In Silico, Inserm UMR-S 973, Paris, France*

PIERRE TUFFERY • *Université Paris Diderot, Sorbonne Paris Cité, Molécules Thérapeutiques In Silico, Inserm UMR-S 973, Paris, France*

STEFAN WALLIN • *Department of Physics and Physical Oceanography, Memorial University, St. John's, NF, Canada*

ANDREW M. WATKINS • *Department of Chemistry, New York University, New York, NY, USA*

HAIM J. WOLFSON • *Blavatnik School of Computer Science, Tel Aviv University, Tel Aviv, Israel*

XIANJIN XU • *Department of Physics and Astronomy, Department of Biochemistry, Dalton Cardiovascular Research Center, Informatics Institute, University of Missouri, Columbia, MO, USA*

CHENGFEI YAN • *Department of Physics and Astronomy, Department of Biochemistry, Dalton Cardiovascular Research Center, Informatics Institute, University of Missouri, Columbia, MO, USA*

MARTIN ZACHARIAS • *Physics Department, Center for Integrated Protein Science Munich (CIPSM), Technical University of Munich, Garching, Germany*

DANIEL ZAIDMAN • *Blavatnik School of Computer Science, Tel Aviv University, Tel Aviv, Israel*

FAN ZHENG • *Department of Biological Sciences, Dartmouth College, Hanover, NH, USA*

XIAOQIN ZOU • *Department of Physics and Astronomy, Department of Biochemistry, Dalton Cardiovascular Research Center, Informatics Institute, University of Missouri, Columbia, MO, USA*

Part I

Prediction of Peptide Binding Sites

The Usage of ACCLUSTER for Peptide Binding Site Prediction

Chengfei Yan, Xianjin Xu, and Xiaoqin Zou

Abstract

Peptides mediate up to 40 % of protein–protein interactions in a variety of cellular processes and are also attractive drug candidates. Thus, predicting peptide binding sites on the given protein structure is of great importance for mechanistic investigation of protein–peptide interactions and peptide therapeutics development. In this chapter, we describe the usage of our web server, referred to as ACCLUSTER, for peptide binding site prediction for a given protein structure. ACCLUSTER is freely available for users without registration at http://zougrouptoolkit.missouri.edu/accluster.

Key words Binding site prediction, Molecular docking, Protein–peptide interaction

1 Introduction

Peptides mediate up to 40 % of protein–protein interactions involved in signal transduction, immune responses, transcriptional regulation, and other cellular processes [1]. The development of peptide therapeutics has also become an attractive direction for cancer therapy [2].

However, due to the difficulties and cost for resolving the protein–peptide complex structures, the current number of released protein–peptide complexes in Protein Data Bank (PDB) [3] is quite limited in comparison with the number of protein monomeric structures. The lack of structures significantly hinders our understanding of the mechanisms underlying protein–peptide interactions and impedes the peptide-based therapeutic development. Efficient and effective in silico methods to complement the experimental techniques are urgently needed. The prediction of peptide binding site is generally the first step towards the more challenging protein–peptide complex structure prediction.

ACCLUSTER is a web server for the application of our previously developed computational method for predicting peptide binding sites [4]. Given a protein structure, ACCLUSTER employs

Ora Schueler-Furman and Nir London (eds.), *Modeling Peptide-Protein Interactions: Methods and Protocols*, Methods in Molecular Biology, vol. 1561, DOI 10.1007/978-1-4939-6798-8_1, © Springer Science+Business Media LLC 2017

amino acid residue probes to globally detect the protein surface to identify the regions allowing for forming good chemical interactions with these probes, which are considered as putative peptide binding sites. The binding sites predicted by ACCLUSTER can be used for guiding experimental mutant design to find critical residues and also providing inputs to molecular docking tools, which usually require the binding site information.

In this chapter, we describe the usage of ACCLUSTER for peptide binding site prediction.

2   Materials

2.1   An Overview of ACCLUSTER Methodology

ACCLUSTER is a prediction method for identifying the peptide binding sites on a given protein. The method was developed based on the assumption that the peptide binding site is generally the region on the protein surface with which amino acids can form good chemical interactions [4].

Specifically, given the protein structure, the 20 standard amino acids (or the residues in the bound peptide if the sequence of the peptide is known) are used as the probes to globally scan the protein surface with ZDOCK3.02 [5, 6]. The modes sampled by ZDOCK are then re-scored and optimized with our knowledge-based scoring function, ITScorePP [7]. Only modes that form good chemical interactions with the protein according to the ITScorePP scores are kept and then clustered by the Density-Based Spatial Clustering Applications with Noise (DBSCAN) method [8]. The produced clusters are ranked based on the cluster sizes; the top cluster is considered as the predicted peptide binding site (*see* Fig. 1).

2.2   The Website of ACCLUSTER

ACCLUSTER can be reached through http://zougrouptoolkit. missouri.edu/accluster. The submission page is shown in Fig. 2. The details are described as follows.

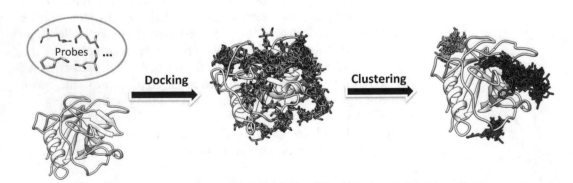

Fig. 1 The schematic diagram of the ACCLUSTER method

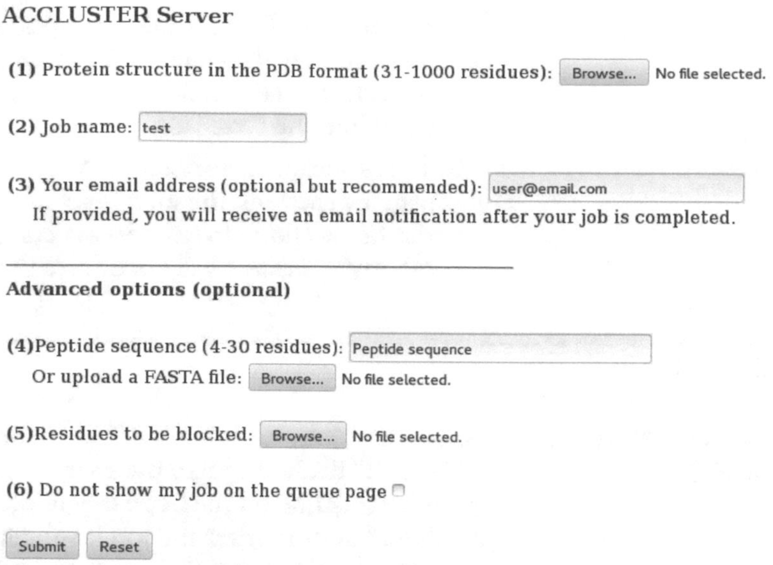

ACCLUSTER Server

(1) Protein structure in the PDB format (31-1000 residues): [Browse...] No file selected.

(2) Job name: [test]

(3) Your email address (optional but recommended): [user@email.com]
If provided, you will receive an email notification after your job is completed.

Advanced options (optional)

(4) Peptide sequence (4-30 residues): [Peptide sequence]
Or upload a FASTA file: [Browse...] No file selected.

(5) Residues to be blocked: [Browse...] No file selected.

(6) Do not show my job on the queue page ☐

[Submit]   [Reset]

Fig. 2 The submission page of the ACCLUSTER server

2.3 Input

2.3.1 Basic Input

1. ACCLUSTER requires only the Brookhaven PDB file of the protein structure as the input. The PDB file is allowed to contain either one protein chain or multiple protein chains. Currently, the total chain length is limited to a range between 31 and 1000. Although the uploaded pdb will be cleaned such as the deletion of solvates, ligands, and hydrogens), uploading a pre-checked structure which contains only standard amino acid residues is strongly recommended.

2. The user is encouraged to type a name to specify the job. If not, the default job name is set as "test."

3. An email address for receiving the notification after the job is completed is strongly recommended. Otherwise, the user will have to manually check the job status in the queue page.

2.3.2 Advanced Options

4. ACCLUSTER allows the user to provide the sequence of the bound peptide (if known) to assist the prediction. The peptide sequence can be input either by entering the single-letter amino acid codes or by uploading a FASTA file. The allowed sequence length is between 4 and 30. According to our previous systematic tests, the inclusion of the peptide sequence only slightly improves the performance of ACCLUSTER. However, the main advantage of the use of the peptide sequence is that this approach significantly reduces the computational time, because only the residues occurring in the sequence of the bound peptide (usually much fewer than 20 amino acids) are used to scan the protein surface [4].

5. ACCLUSTER allows the user to block the protein residues that are known to be remote from the binding site, by uploading

a ".txt" file which specifies the residue names. Specifically, the user can copy the columns 18 to 26 of the atom lines that correspond to these residues in the protein pdb file, and paste them into the ".txt" file.

6. The user may choose not to show the job information in the queue page by checking the box. If so, an email address for receiving the notification after the job is completed must be provided; otherwise, the user will not be able to reach the result page.

3   Methods

3.1   Submitting Jobs to ACCLUSTER

Submitting jobs to ACCLUSTER is easy. After navigating to the ACCLUSTER homepage, the user can upload the pdb file of the protein structure, fill the necessary information, and then click the "Submit" button. After the job is submitted, the job status can be monitored by checking the queue page. If an email address is provided, a notification with the link to the result page will be sent after the job is completed.

3.2   The Result Page

Each job can usually be completed in 30 min. One example of the result page is shown in Fig. 3. On the result page, up to three predicted binding sites are displayed via a JavaScript library 3Dmol.js [9]. The top prediction is colored red, the second prediction is colored cyan, and the third prediction is colored magenta. The pdb files representing the predicted binding sites (accluster_site1. pdb accluster_site2.pdb accluster_site3.pdb) are stored in a compressed file (Result.tar), which can be downloaded and then analyzed locally using a molecular visualization program such as UCSF Chimera [12].

4   A Case Study

Here, we use the ubiquitin-specific protease, USP7, as an example. USP7 plays a key role in the p53 pathway. The PDB entry, 2FOJ, contains the N-terminal domain of USP7 bound with p53 peptides [10]. We use the unbound structure of the N-terminal domain of USP7, 2F1W [11] for binding site prediction to mimic the real practice in which the bound structure of the protein is usually not available. We test two different conditions, either use or not use the sequence of the bound peptide. In both cases, ACCLUSTER successfully identified the native peptide binding site on the unbound protein structure as the top prediction, as shown in Fig. 4. However, ACCLUSTER took about 17 min for the prediction without being given the sequence of the bound peptide, and took about only 6 min with the use of the sequence of the bound peptide.

Job information:
Job id: S00071
Job name: test
Job starting time: 2016-03-02 16:49:59
Job ending time: 2016-03-02 17:06:39
Protein receptor structure: rec_clean.pdb
Peptide sequence: N/A
Blocked residues: N/A

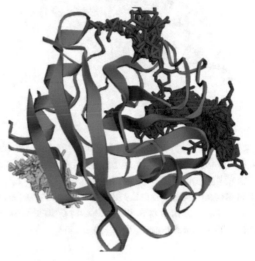

Download
Result: Result.tar

Fig. 3 An example of the result page of ACCLUSTER. The three predicted binding sites are displayed via 3Dmol.js. The top prediction is colored *red*. The second prediction is colored *cyan*. The third prediction is colored *magenta*. The protein is shown in a *gray-colored ribbon* diagram

5 Notes

1. The only required input for ACCLUSTER is a pdb file which contains the protein structure. Although the pdb file will be cleaned by deleting the hydrogens, solvates, and ligands, the pre-checking of the pdb file to mutate non-standard residues is strongly encouraged.

2. ACCLUSTER allows the user to provide the sequence of the bound peptide to assist the prediction. According to the previous test, the inclusion of the sequence of the bound peptide can significantly improve the computational efficiency of ACCLUSTER.

3. Protein residues that are known to be remote from the peptide binding site can be blocked by uploading a ".txt" file that specifies these residues by copying the columns 18 to 26 of the corresponding atom lines in the protein pdb file.

4. By default, the job submitted to ACCLUSTER will be shown in the queue page, which is accessible to the public. On the

A   **B**

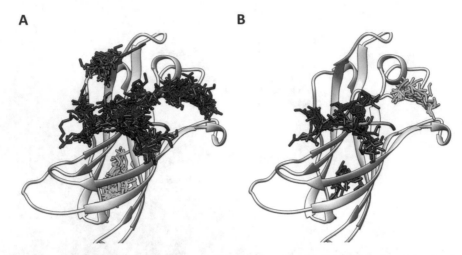

Fig. 4 ACCLUSTER successfully identifies the peptide binding site on the unbound protein structure of the N-terminal domain of USP7 when the sequence of the bound peptide is not provided (Panel (**a**)) and when the sequence of the bound peptide is provided (Panel (**b**)), respectively. The top prediction is colored *red*, the second prediction is colored *cyan*, and the third prediction is colored *magenta*. The bound peptide after the superimposition of the complex structure onto the unbound protein structure is plotted in a *rainbow-colored ribbon* diagram. The protein structure is shown as a *gray ribbon*

other hand, the user can also choose not to show the job in the queue page for the sake of privacy. In this case, a valid email address must be provided for receiving the notification when the job is completed.

5. The details are provided in the Tutorial posted on the server site.

Acknowledgements

This work is supported by NSF CAREER Award DBI-0953839 and the NIH R01GM109980 (Xiaoqin Zou). The computations were performed on the high performance computing infrastructure supported by NSF CNS-1429294 (PI: Chi-Ren Shyu) and the HPC resources supported by the University of Missouri Bioinformatics Consortium (UMBC).

References

1. Petsalaki E, Russell RB (2008) Peptide-mediated interactions in biological systems: new discoveries and applications. Curr Opin Biotechnol 19:344–350

2. Wells JA, McClendon CL (2007) Reaching for high-hanging fruit in drug discovery at protein-protein interfaces. Nature 450:1001–1009

3. Berman HM, Westbrook J, Feng Z, Gilliland G, Bhat T, Weissig H, Shindyalov IN, Bourne PE (2000) The Protein Data Bank. Nucleic Acids Res 28:235–242

4. Yan C, Zou X (2015) Predicting peptide binding sites on protein surfaces by clustering chemical interactions. J Comput Chem 36:49–61

5. Chen R, Li L, Weng, Z (2003) ZDOCK: an initial-stage protein-docking algorithm. Proteins 52:80–87

6. Pierce BG, Hourai Y, Weng, Z (2011) Accelerating protein docking in ZDOCK using an advanced 3D convolution library. PLoS One 6:e24657

7. Huang S-Y, Zou X (2008) An iterative knowledge-based scoring function for protein-protein recognition. Proteins 72:557–579

8. Ester M, Kriegel HP, Sander J, Xu, X (1996) A density-based algorithm for discovering clusters in large spatial databases with noise. In: Proceedings of KDD, vol 96, pp 226–231

9. Rego N, Koes, D (2014) 3Dmol.js: molecular visualization with WebGL. Bioinformatics 31:1322–1324. doi:10.1093/bioinformatics/btu829

10. Sheng Y, Saridakis V, Sarkari F, Duan S, Wu T, Arrowsmith CH, Frappier, L (2006) Molecular recognition of p53 and MDM2 by USP7/HAUSP. Nat Struct Mol Biol 13:285–291

11. Hu M, Gu L, Li M, Jeffrey PD, Gu W, Shi, Y (2006) Structural basis of competitive recognition of p53 and MDM2 by HAUSP/USP7: implications for the regulation of the p53–MDM2 pathway. PLoS Biol 4:e27

12. Pettersen EF, Goddard TD, Huang CC, Couch GS, Greenblatt DM, Meng EC, Ferrin TE (2004) UCSF chimera–a visualization system for exploratory research and analysis. J Comput Chem 25:1605–1612

Chapter 2

Detection of Peptide-Binding Sites on Protein Surfaces Using the Peptimap Server

Tanggis Bohnuud, George Jones, Ora Schueler-Furman, and Dima Kozakov

Abstract

Peptide-mediated interactions are of primordial importance to the cell, and the structure of such interaction provides an important starting point for their further characterization. In many cases, the structure of the peptide-protein complex has not been solved by experiment, and modeling tools need to be applied to generate structural models of the interaction. PeptiMap is a protocol that identifies the peptide-binding site when only the structure of the receptor is known, but no information about where the peptide binds is available. This is achieved by mapping the surface for solvents to identify ligand-binding sites, similar in approach to ANCHORMAP in which amino acids are mapped. Peptimap is a free open access web-based server. It can be accessed at http://peptimap.cluspro.org.

Key words Peptide-protein interactions, Binding site prediction, Solvent mapping, Peptide mapping

1 Introduction

Peptide-mediated interactions are of primordial importance to the cell. It has been estimated that up to half of known protein-protein reactions are mediated by peptides in higher Eukaryotes (e.g., [1]). It is believed that peptides play an essential role in the modulation of these protein-protein interactions. As such, they are important drug targets, and it is therefore crucial to improve our understanding of these important players. The structures of such interactions provide good starting points for their further characterization. However, in many cases, such a structure has not been solved by experiment, and in silico modeling tools need to be developed and applied to fill the gap. These methods can be broken down into two main steps: the identification of possible binding spots on the protein surface, and the prediction of the peptide sequence and peptide pose in the binding spots. If the binding site is known, modeling protocols such as FlexPepDock [2, 3], peptide Haddock [4], pepCrawler [5], and others can be used to generate the full

Ora Schueler-Furman and Nir London (eds.), *Modeling Peptide-Protein Interactions: Methods and Protocols*, Methods in Molecular Biology, vol. 1561, DOI 10.1007/978-1-4939-6798-8_2, © Springer Science+Business Media LLC 2017

atom model of the interaction (reviewed in [6], and covered by other chapters in this volume).

PeptiMap is a protocol that identifies peptide-binding sites when only the structure of the receptor is known. The approach is based on the experimental observation, in both crystal soaking and NMR experiments, that small organic molecules of varying size and polarity tend to accumulate on the protein surface in regions that bind larger ligands and proteins [7, 8]. FTmap is a Fast Fourier Transform (FFT)-based method for computational mapping of the surface for solvents [9] to identify ligand-binding sites [10, 11]. The approach is similar to ANCHORMAP in which amino acids are mapped [12]. In PeptiMap, the FTmap is modified to specify the search for peptide-binding sites. This chapter will introduce the reader to Peptimap's functionality and the means to use the freely accessible Peptimap server at http://peptimap.cluspro.org.

2 Materials

The following are needed to run PeptiMap on the server (at *peptimap.cluspro.org*; Fig. 1a):

1. An input structure of the protein receptor on which we would like to identify peptide-binding site(s) (in Protein Data Bank format, *see* www.pdb.org [13]). It can be uploaded, or provided as PDB id.

2. Chain(s) of the protein receptor to be mapped.

The output of the mapping will be available for visual inspection on the server (Fig. 1c), as well as a PyMOL session file for download and inspection using the molecular viewer PyMOL (PyMOL can be obtained at www.pymol.org; Schroedinger LLC).

3 Methods

We first describe the setup of the PeptiMap Server, and detail the input options. We then present case studies to demonstrate the utility of PeptiMap. Finally, we provide more details about the underlying algorithms, as implemented in the server. Further details are also available in a previous publication on PeptiMap [14].

3.1 Input Interface

Peptimap provides a simple interface with which to search for peptide-binding sites on a given receptor protein surface (*see* Fig. 1a). The following information is required/accepted as input:

1. PDB file to map, provided either in the form of a PDB id, or uploaded onto the server.

2. Chains to be investigated: indicate simply the chain(s) name, separated by spaces in case of multiple chains.

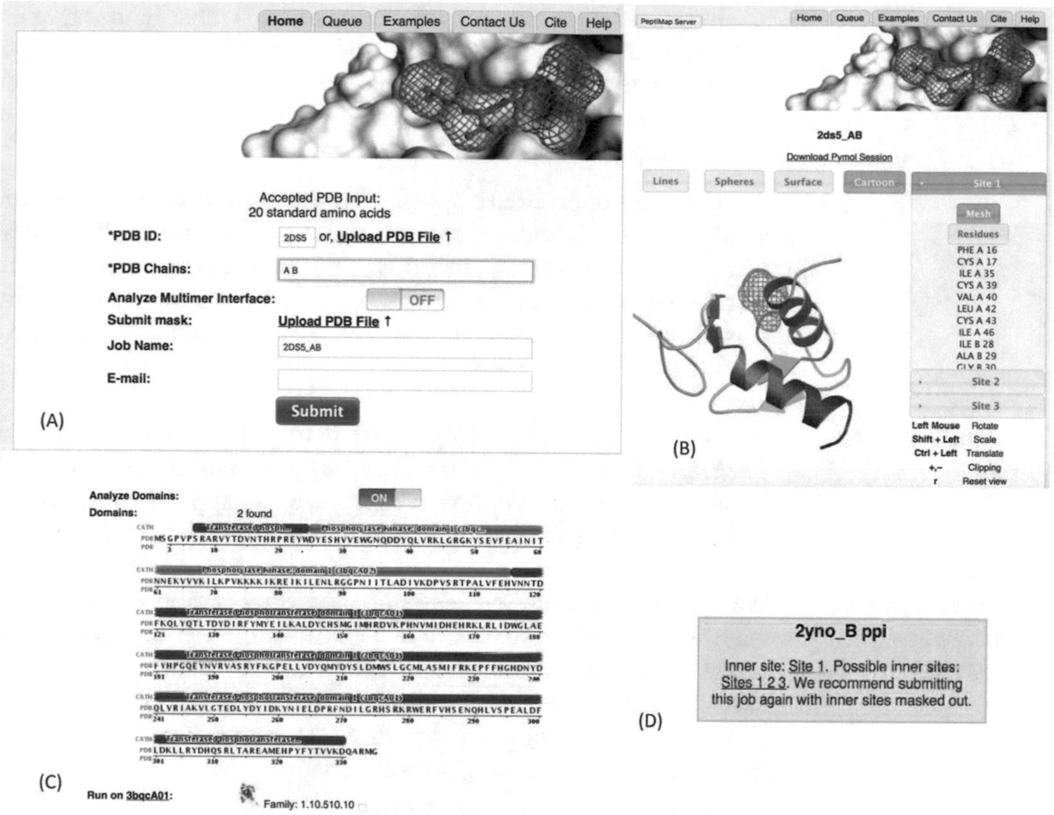

Fig. 1 Overview of the PeptiMap server pages. (**a**) The main page where the job is submitted. (**b**) Results page for 2DS5 (*see* also Fig. 2): For each site, buttons allow to toggle the display of the predicted site (as mesh), as well as the binding residues on the protein receptor (shown as *sticks* and displayed as *list*). A Java applet allows changing the view of the protein, by scaling, rotating, and translating. In addition, the receptor can be shown as *cartoon* (default option), *sticks*, *spheres*, or surface representation. (**c**) Output of domain mapping option for 3BQC (*see* also Fig. 3). (**d**) Example for Output that detects inaccessible pockets for 2YNO, and suggests repeating the mapping using the provided PDB file for masking (download the PDB file and upload it in your new submission)

3. For multi-chain structures: Map single or multiple chains (*see* **Note 1** for guidelines on what unit to map, and Subheading 3.3.2 on how these guidelines are implemented).

4. For single-chain structures: Map specific domains or full receptor chain (currently the PDB must be provided as PDB ID for invoking this option). You can search the provided protein to identify (multiple) domains in the provided structure (*see* Subheading 3.3.3 and Fig. 1c). If the structure contains multiple domains, Peptimap provides the option to search the entire structure, or alternatively to focus on specified domains (*see* **Note 2** for guidelines/suggestions of what part of the structure to map).

5. Mask option—removes from the search on the protein surface the atoms present in a provided mask PDB file. Such a file can

be provided, or alternatively, Peptimap will identify sites that are too small for peptides to bind, and suggest filtering them out, by automatically providing the mask file (*see* Subheading 3.2.3 and Fig. 1d). The criteria used to define such sites inaccessible to peptides are defined in Subheading 3.3.4.

Once the appropriate options for the desired search have been selected, an email address and job name can be entered. The email address allows Peptimap to inform the submitting party that the search has been completed. The results of the search will be displayed on the Peptimap site (*see* Fig. 1b), and can also be downloaded as a Pymol session.

3.2 PeptiMap Case Studies

In the following three examples, we demonstrate how the *Analyze Multimer Interface, Analyze Domain,* and *Submit Mask* options implemented in the PeptiMap server can be applied to focus and improve binding site prediction.

3.2.1 The Analyze Multimer Interface Option, Applied to 2DS5

We will use the *Analyze Multimer Interface* option to determine whether or not the structure is a homo-multimer and if so, how many monomers should be included in the mapping unit. The example used is the Zinc Binding Domain of ClpX (PDB id 2DS5 [15]). In the following, we describe each step involved for the prediction of peptide-binding sites in detail.

1. Go to the site *peptimap.bu.edu*.

2. Locate the **PDB ID** field and type 2DS5.

3. In the **PDB Chains** field type A B.

4. Now click on the **Analyze Multimer Interface** button function. After a few seconds, a message should appear specifying the buried surface area between the individual chains, in this case:

    A B are tight multimers (Buried SA: 42.6% of monomer surface, >25%). We recommend submitting the multimer.

5. Once you have finished selecting options, specify an email where the notifications will be sent, and submit the job. This job took approximately 30 min to finish.

The result of this search revealed six binding hotspots on the protein surface (the top site is shown on the server page, and two additional sites can be displayed using the buttons to the right, Fig. 1b, while details of the top-6 sites are included in the PyMOL session). Figure 2 shows the corresponding protein bound to two copies of an SSPB tail peptide (2DS8 [15]), superimposed over the results generated by Peptimap. One peptide interacts with hotspots ranked 1 and 4, while a second peptide interacts with the predicted hotspot ranked 2. Thus, Peptimap was able to identify the binding sites of the two peptides.

(A)   (B)

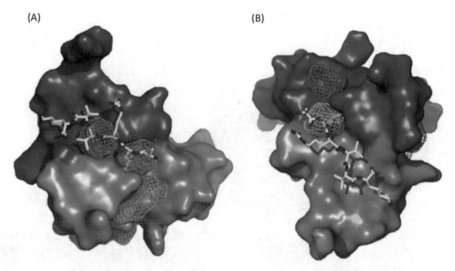

Fig. 2 Mapping on a homodimer: the predicted binding site on homodimeric Zinc Binding Domain of ClpX (2DS5), overlapped with the structure of ClpX complexed with SspB-tail peptides ALRVVK (2DS8). The two domains are shown as surface, in different *shades of gray*. Two views are shown, to capture the two peptides (*yellow stick* representation): (**a**) In the first peptide, residues L2 and V4 overlap with the predicted binding hotspots ranked 1 and 4 (*red* and *pink*), respectively. (**b**) In the second peptide, residue L2 overlaps with the predicted site ranked 2 (*green*). This and the following figures were generated from the PyMOL sessions provided by the server. The binding sites are colored according to their ranking (ranks 1–5 *red-green-blue-pink-yellow* meshes)

3.2.2 The Analyze Domain Function, Applied to 3BQC

We will analyze the domain composition of the catalytic subunit of protein kinase CK2 to identify the domain to map for peptide-binding sites. The structure of this protein is available as PDB ID 3BQC, chain A [16]. Go to the site *peptimap.bu.edu*.

1. Locate the PDB ID field and type 3BQC.

2. In the PDB Chain field type A. This automatically defines the job name to 3bqc_a.

3. Now click on the Analyze Domain button function. The domain composition according to CATH [17] will be displayed (Fig. 1c). For our case, this includes (a) the Transferase Phosphorylase family (CATH id 1.10.510.10) and (b) the Phosphorylase Kinase Family (CATH id 3.30.200.20).

4. In this example, we will look at one of the domains, the Phosphorylase Kinase Family domain. Click on the box next to this family: 3.30.200.20.

5. Once you have finished selecting options, specify an email where the notifications will be sent, and submit the job. This job took approximately 40 min to finish.

The result of this search revealed six binding hotspots on the protein surface. Figure 3 shows the bound protein-peptide complex (PDB ID 4IB5, human protein kinase CK2 catalytic subunit in a

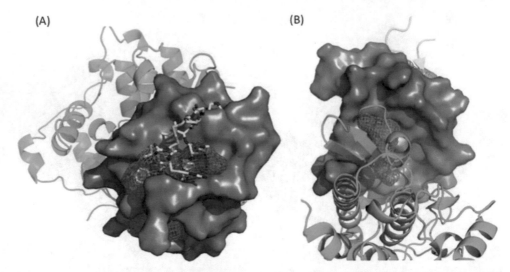

Fig. 3 Single domain mapping: shown are the predicted binding sites on protein kinase CK2 (3BQC), overlapped with the peptide-bound structure (to CK2β-competitive cyclic peptide GCRLYGFKIHGC; 4IB5). The mapped domain is shown as *gray* surface, the second domain as green cartoon. (**a**) View of the peptide with residues F7 and Y5 overlapping sites ranked 3 and 5 (*blue* and *yellow* meshes, respectively). (**b**) View of the domain interface, covering the top-ranking site (*red*), that could be masked

complex with a CK2β-competitive cyclic peptide [16]), superimposed onto the results generated by Peptimap. The bound peptide is clearly located within the predicted binding hotspots ranked 3 and 5 (Fig. 3a). Examination of the structure reveals the best-ranking site is inaccessible within the context of the second domain (Fig. 3b). (Note that to identify only sites outside this domain interface, a second search could be performed using the *Submit mask* option, *see* next Subheading).

3.2.3 The Submit mask Option, Applied to 1EJ1 and 2YNO

We will use the *Submit mask* function to remove from the search parts of the protein that will be inaccessible due to a ligand bound, on the example of the mRNA Cap-Binding interface of eIF4E bound to M7G (PDB id 1EJ1 [18]).

1. Go to the site *peptimap.bu.edu*.

2. Locate the *PDB ID* field and type 1EJ1.

3. In the *specify chains* field type A.

4. Using a text processor, remove from the PDB file all residues that do not contact the ligand M7G (i.e., keep only residues with at least one atom within 10 Å of M7G; remove also non-amino acid entities). Save the resulting PDB file.

5. Click on the *Submit mask* option and select the file that you just have generated.

6. Once you have finished selecting options, specify an email where the notifications will be sent, and submit the job. This job took approximately 32 min to finish.

(A)   (B)

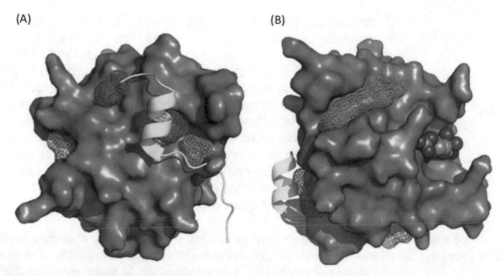

Fig. 4 Using a mask to exclude a ligand binding site: the predicted peptide-binding site on Eukaryotic Translation Initiation Factor 4E, eIF4E (1EJ1), overlapped with the structure of eIF4E complexed with the 4E-BP1 peptide (1WKW). (**a**) The peptide overlaps sites ranked 2, 3, and 5 (*green, blue,* and *yellow,* respectively). (**b**) Rotated view, showing the masked ligand binding site

The result of this search revealed six binding hotspots on the protein surface. Figure 4 shows the bound protein (PDB id 1WKW [19]) superimposed over the results generated by Peptimap. The peptide 4EBP1 interacts with hot spots ranked 2, 3, and 5.

Masking may also be used to skip deep binding pockets inaccessible to peptides. In the example of 2YNO, the propeller WD40 repeat domain of COP1 [20], such sites are automatically identified by the server, and a PDB file is provided for submission as masking file (Fig. 1d; *see* Subheading 3.3.4 for details how these sites are identified).

3.3   Details of Implementation

We provide here a short overview of the protocol as implemented in the PeptiMap server. More details can be found in our previous publication of the PeptiMap algorithm [14].

3.3.1   Mapping of a Receptor Surface for Peptide-Binding Sites

Peptimap uses Fast Fourier Transform (FFT)-based grid-sampling to search the receptor surface for binding sites [14]. Water molecules and other ligands are removed prior to calculations. Sixteen small molecules are used as probes, as done previously for ligand mapping (*see*, e.g., Fig. 1b in [11]). These small molecules are varied enough to cover all the functional groups found in amino acids. The probes are allowed to adopt different conformations, which are all subjected to an FFT search to select favorable positions and conformations on the receptor surface. The 2000 best poses are then clustered using a greedy algorithm. This selects the lowest energy pose and incorporates all poses within 4 Å into the cluster (*see* [14]). The poses in the cluster are removed from the general bin and the procedure is repeated with the lowest energy pose remaining. If a cluster contains less than ten probes it is removed. The remaining clusters are then minimized with the CHARMM potential, using the Analytic Continuum Electrostatic (ACE) model [21]. The minimized results are re-clustered using 4 Å full atom RMSD as the clustering restraint, and clusters with less than ten members are discarded. The remaining clusters are re-ranked based upon Boltzmann-averaged energy, and the six lowest energy poses for each type of probe are retained. Hotspots are generated by grouping neighboring clusters together. If there are multiple neighbors to a cluster (neighbor being defined as a cluster falling within 4 Å), then they are considered a better hotspot.

3.3.2   Definition of Tight Homo-Multimer Interaction

Homo-multimers are identified based on their sequence similarity—two monomers are considered homomers only if they are exactly the same. To define if monomers in these homo-multimers are tightly or loosely associated, the ratio of buried surface area is calculated: if >25 % of the surface area is buried in the homo-multimer, the multimer is considered to be obligatory and it is suggested to consider it as one unit to map (*see* also **Note 1**).

3.3.3   Definition of Domains in Receptor Protein

Peptimap uses the CATH classification [17] to define domains on the receptor structure. If no classification is available, sequence alignment is used to find the CATH domain closest to the provided structure.

3.3.4   Identification of Binding Pockets Inaccessible to Peptides

The use of small molecules as a search method allows for identification of sites that are not necessarily accessible to peptides, such as those in the core of a protein. In order to remove these sites, a sphere is generated at the center of the site and one hundred rays are projected outward. If 80 % or more of the rays contact the protein (i.e., they pass within 2 Å of an atom), the site is considered internal and it is suggested to the reader to discard it.

4   Notes

1. Criteria for deciding on mapping monomers or homo-multimers. Detection of peptide-binding sites depends on defining the structure that the peptide will encounter on its search. For obligate homo-multimers, it can be assumed that the individual monomer structure is never encountered (if it exists as stable, separate structure at all). In contrast, transient homo-multimers might also expose single monomers. Therefore, the following rule is suggested:

   Rule #1: If the interaction between the individual monomers is strong, map the homo-multimer, else map individual monomer structures. Strength of interaction is estimated based on the buried surface area of the interface (*see* Subheading 3.3.2 for details on buried surface area criteria).

2. Criteria for deciding on which region to map in a multi-domain protein. Multi-domain protein structures solved by x-ray crystallography might display pockets at the domain interface that could fit a peptide well. However, these pockets are not necessarily stable, as the two domains might not be oriented in a fixed orientation shown in the crystal, but rather show a more flexible orientation, which will affect the stability of the binding pocket. In calibrating PeptiMap, we have therefore devised simple rules to distinguish true binding sites at domain interfaces from unstable false-positive hits, for protein chains that contain more than one structured domain (*see* Subheading 3.3.3 for details of how domains are mapped).

Rule #2: If the mapped domains are identical, it is recommended to map the full structure.

Rule #3: If the domains are different (hetero-domain structure), it is suggested to map single specific domains of interest. In that case, we assume that even if a binding site is at a domain interface, one of the domains will be the predominant contributor to binding, and consequently that binding site is also identified on single domains.

References

1. Neduva V, Linding R, Su-Angrand I, Stark A, de Masi F, Gibson TJ, Lewis J, Serrano L, Russell RB (2005) Systematic discovery of new recognition peptides mediating protein interaction networks. PLoS Biol 3:e405

2. Raveh B, London N, Schueler-Furman O (2010) Sub-angstrom modeling of complexes between flexible peptides and globular proteins. Proteins 78:2029–2040

3. Raveh B, London N, Zimmerman L, Schueler-Furman O (2011) Rosetta FlexPepDock ab-initio: simultaneous folding, docking and refinement of peptides onto their receptors. PLoS One 6, e18934

4. Trellet M, Melquiond AS, Bonvin AM (2013) A unified conformational selection and induced fit approach to protein-peptide docking. PLoS One 8, e58769

5. Donsky E, Wolfson HJ (2011) PepCrawler: a fast RRT-based algorithm for high-resolution refinement and binding affinity estimation of peptide inhibitors. Bioinformatics 27:2836–2842

6. London N, Raveh B, Schueler-Furman O (2013) Peptide docking and structure-based characterization of peptide binding: from knowledge to know-how. Curr Opin Struct Biol 23:894–902

7. Mattos C, Ringe D (1996) Locating and characterizing binding sites on proteins. Nat Biotechnol 14:595–599

8. Shuker SB, Hajduk PJ, Meadows RP, Fesik SW (1996) Discovering high-affinity ligands for proteins: SAR by NMR. Science 274:1531–1534

9. Brenke R, Kozakov D, Chuang GY, Beglov D, Hall D, Landon MR, Mattos C, Vajda S (2009) Fragment-based identification of druggable 'hot spots' of proteins using Fourier domain correlation techniques. Bioinformatics 25:621–627

10. Ngan CH, Hall DR, Zerbe B, Grove LE, Kozakov D, Vajda S (2012) FTSite: high accuracy detection of ligand binding sites on unbound protein structures. Bioinformatics 28:286–287

11. Kozakov D, Grove LE, Hall DR, Bohnuud T, Mottarella SE, Luo L, Xia B, Beglov D, Vajda S (2015) The FTMap family of web servers for determining and characterizing ligand-binding hot spots of proteins. Nat Protoc 10:733–755

12. Ben-Shimon A, Eisenstein M (2010) Computational mapping of anchoring spots on protein surfaces. J Mol Biol 402:259–277

13. Berman HM, Westbrook J, Feng Z, Gilliland G, Bhat TN, Weissig H, Shindyalov IN, Bourne PE (2000) The Protein Data Bank. Nucleic Acids Res 28:235–242

14. Lavi A, Ngan CH, Movshovitz-Attias D, Bohnuud T, Yueh C, Beglov D, Schueler-Furman O, Kozakov D (2013) Detection of peptide-binding sites on protein surfaces: the first step toward the modeling and targeting of peptide-mediated interactions. Proteins 81:2096–2105

15. Park EY, Lee BG, Hong SB, Kim HW, Jeon H, Song HK (2007) Structural basis of SspB-tail recognition by the zinc binding domain of ClpX. J Mol Biol 367:514–526

16. Raaf J, Klopffleisch K, Issinger OG, Niefind K (2008) The catalytic subunit of human protein kinase CK2 structurally deviates from its maize homologue in complex with the nucleotide competitive inhibitor emodin. J Mol Biol 377:1–8

17. Sillitoe I, Lewis TE, Cuff A, Das S, Ashford P, Dawson NL, Furnham N, Laskowski RA, Lee D, Lees JG, Lehtinen S, Studer RA, Thornton J, Orengo CA (2015) CATH: comprehensive structural and functional annotations for genome sequences. Nucleic Acids Res 43: D376–D381

18. Marcotrigiano J, Gingras AC, Sonenberg N, Burley SK (1997) Cocrystal structure of the messenger RNA 5′ cap-binding protein (eIF4E) bound to 7-methyl-GDP. Cell 89: 951–961

19. Tomoo K, Matsushita Y, Fujisaki H, Abiko F, Shen X, Taniguchi T, Miyagawa H, Kitamura K, Miura K, Ishida T (2005) Structural basis for mRNA Cap-Binding regulation of eukaryotic initiation factor 4E by 4E-binding protein, studied by spectroscopic, X-ray crystal structural, and molecular dynamics simulation methods. Biochim Biophys Acta 1753:191–208

20. Jackson LP, Lewis M, Kent HM, Edeling MA, Evans PR, Duden R, Owen DJ (2012) Molecular basis for recognition of dilysine trafficking motifs by COPI. Dev Cell 23: 1255–1262

21. Brooks BR III, Jr CLB, Nilsson ADM, Petrella L, Roux RJ, Won B, Archontis Y, Bartels G, Boresch C, Caflisch S, Caves A, Cui L, Dinner Q, Feig AR, Fischer M, Gao S, Hodoscek J, Im M, Kuczera W, Lazaridis K, Ma T, Ovchinnikov J, Paci V, Pastor E, Post RW, Pu CB, Schaefer JZ, Tidor M, Venable B, Woodcock RM, Wu HL, Yang X, York WDM, Karplus M (2009) CHARMM: the biomolecular simulation program. J Comput Chem 30:1545–1614

Chapter 3

Peptide Suboptimal Conformation Sampling for the Prediction of Protein-Peptide Interactions

Alexis Lamiable, Pierre Thévenet, Stephanie Eustache, Adrien Saladin, Gautier Moroy, and Pierre Tuffery

Abstract

The blind identification of candidate patches of interaction on the protein surface is a difficult task that can hardly be accomplished without a heuristic or the use of simplified representations to speed up the search. The PEP-SiteFinder protocol performs a systematic blind search on the protein surface using a rigid docking procedure applied to a limited set of peptide suboptimal conformations expected to approximate satisfactorily the conformation of the peptide in interaction. All steps rely on a coarse-grained representation of the protein and the peptide. While simple, such a protocol can help to infer useful information, assuming a critical analysis of the results. Moreover, such a protocol can be extended to a semi-flexible protocol where the suboptimal conformations are directly folded in the vicinity of the receptor.

Key words Peptide-protein interactions, Blind docking, Rigid docking, Peptide conformational sampling, PEP-FOLD, PEP-SiteFinder

1 Introduction

Protein peptide interactions have recently been the object of many studies, owing to several reasons. For one part, peptides have gained attraction as candidate therapeutics, as the consequence of a series of progress in better control of their bioavailability, biodelivery, and manufacturing, among others [1–3]. New classes of natural peptides have focused attention, such as the antimicrobial peptides [4], or venom peptides [5], not talking about non-ribosomal peptides [6]. For another part, peptides that correspond to proteins fragments have also raised interest, following the emerging promise that targeting protein-protein interactions could become a new Eldorado for drug development [7].

Protein-peptide interactions have become a new field for in silico approaches, which comes with a series of new specific challenges. Compared with small compounds for which in silico docking protocols are now routinely used, peptides are in general

Ora Schueler-Furman and Nir London (eds.), *Modeling Peptide-Protein Interactions: Methods and Protocols*, Methods in Molecular Biology, vol. 1561, DOI 10.1007/978-1-4939-6798-8_3, © Springer Science+Business Media LLC 2017

more flexible which limits the application of approaches set for chemicals to very short peptides [8]. The tuning of protocols to characterize and model protein-peptide complexes, with questions in terms of the conformations adopted by the peptide alone or in complex, remains a challenge to which answers start to be brought.

Here, we investigate a model in which the conformation of the protein receptor is kept rigid, and the ligand peptide is represented using a limited collection of suboptimal conformations, also rigid in a first approximation. We show how such a model can be used for the identification of candidate sites of interaction given the structure of the protein receptor and the sequence of the candidate peptide. This model can be extended to the de novo modeling of peptides in contact with the protein receptor.

1.1 Peptide-Protein Complexes: Moderate Conformational Changes for Proteins, Possibly Larger for Peptides

PeptiDB [9] is a collection of protein-peptide complexes obtained at a resolution of the crystal structure of less than 2 Å for which protein sequence identity is less than 70% and peptide size comprised between 5 and 15 residues. From the initial collection of 103 structures of protein-peptide complexes (holo conformation), the authors could identify for 78 cases the structure of the protein solved without the peptide (apo conformation). It is striking that for most cases (67 over 78) the authors reported a very small difference between the apo and holo conformations of the protein (average Cα RMSD of 0.83 Å). This observation can lead to postulate that most of peptides could adapt their conformation to allow their interaction with proteins.

Checking this assertion turns however to be difficult due to data sparsity. Despite the number of entries of the Protein Data Bank (PDB) [10] being presently more than 100,000 structures, the number of systems for which both the conformation of the peptide alone in solution and the conformation of the same peptide in complex with a protein is known is very small. Scanning the PDB entries that have only one chain and a short size—e.g., between 4 and 19 amino acids, and looking for protein-peptide complexes in which one chain has the exact same amino acid sequence than that of the free peptide, only 29 such cases could be identified (November 2015). Sixteen of those correspond to peptides that either contain un-natural amino acids, have a structure stabilized by ions or appear to be homo-multimers and are thus not suitable for a direct analysis. For six of the 13 cases, the apo and holo conformations look very similar—backbone Root Mean Square Deviation (RMSD) below 1 Å. These conformations appear stabilized by the existence of a secondary structure: two α helices, 3 β-hairpins, and one helix-turn-helix motif. Such structuration tends to rigidify peptide conformation, and is also favorable for interactions since it is associated with a reduction of the loss of conformational entropy upon binding and consequently facilitates the interaction with the protein. For three other peptides, the

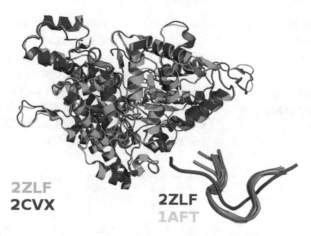

Fig. 1 Peptide adaptation upon the binding. Example of the Eukaryotic ribonucleotide reductase in interaction with the FTLDADF peptide. The conformations of the holo protein, in *green* (PDB ID: 2ZLF), and the apo protein, in *blue* (PDB ID: 2CVX), are very close (backbone RMSD = 0.27 Å); the bound peptide of sequence FTLDADF is in *magenta. Bottom right inset*: the conformation of the bound peptide, in *magenta*, has a backbone RMSD of 2.75 Å with the free peptide in *orange* (PDB ID: 1AFT)

structure of the complex does not include all the amino acids, suggesting that the missing residues are too flexible to determine their structure. The four remaining peptides adopt a random coil conformation for both bound and unbound conformations and have a backbone RMSD of more than 1.0 Å—from 1.37 to 2.75 Å. One such example is depicted in Fig. 1.

An outcome of this analysis is that for seven cases of 13 (54%), the peptide tends to undergo large conformational changes. It is tempting to transpose to peptides observations made for protein-protein interactions. The conformational plasticity of proteins plays a role in most binding events. Among the conformations in equilibrium, only some of them are able to bind a specific partner whereas others are not [11–13]. This is the well-accepted concept of conformational selection for small ligand binding. Another scenario, called induced-fit hypothesis, proposes that the presence of the partner causes changes of the active site to allow its binding. For large ligands, such as peptides, proteins, or nucleic acids, the conformational selection and the induced-fit mechanisms can be intricately coupled to enable the interaction with specific proteins [14–16]. Present observations, together with previous ones [9], suggest that the generally more preserved conformation of proteins compared to that of peptides may drive a conformational selection of unbound peptides among their conformations in equilibrium, for peptides not stabilized by secondary structure or disulfide bridges.

A first approximation of peptide protein interactions is thus a model in which the protein conformation is kept rigid, and the peptide conformation should include some flexibility.

1.2 The Holo Conformation of Peptides Can Be Approached by Suboptimal Sampling of the Apo Peptide

In this section, only the peptide is considered. Sampling peptide conformations can be achieved using a modified version of PEP-FOLD [17, 18] which returns not just a few optimal conformations, but series of suboptimal conformations (in preparation). Results are presented for a subset of 41 peptides of the peptiDB that corresponds to the proteins in different folds, according to CATH classification [19] and with little conformational changes between holo and apo forms. Figure 2a depicts how, on average,

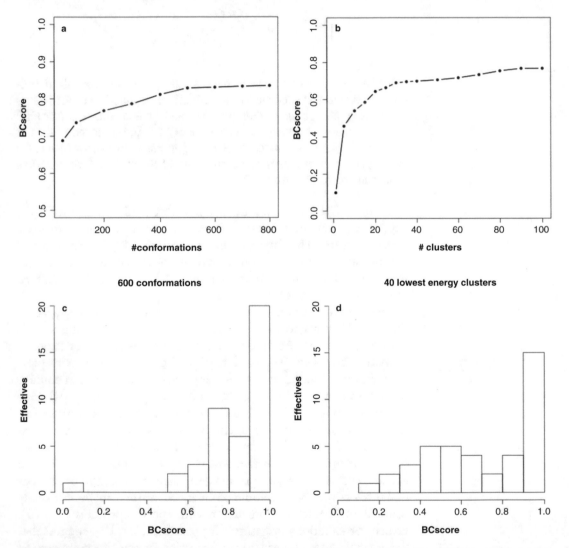

Fig. 2 Identification of peptide holo conformation in the apo suboptimal conformation. (**a**) Best peptide conformation generated as a function of the number of suboptimal conformations drawn. Conformation similarity is measured using the BCscore (*see* the text). A value of 1.0 corresponds to perfect similarity, and a value of 0.8 to a cutoff above which conformations can be considered native like. (**b**) BCscore of the best representative conformation identified after clustering, as a function of the number of clusters. (**c**) Distribution of the BCscores for the best conformation generated for each of the 41 peptides of the collection. (**d**) Distribution of the BCscores for the best cluster representative for a number of clusters limited to up to 40

suboptimal sampling is able to approximate the holo peptide conformation as a function of the number of conformations generated. Conformation similarity is measured using the BCscore [20], a score more selective than the widely used Root Mean Square Deviation (RMSD), and independent of the size. Briefly, a BCscore value of 1 corresponds to perfect structural identity (−1 to mirror conformation), and values of more than 0.8 can be considered close enough to the experimental conformation, values above 0.6 can be considered similar, and values below irrelevant conformations. One observes that for a number of conformations of more than 400, it is possible, on average, to generate a conformation close to the experimental conformation in the complex. Note that being able to generate such conformations does not mean their identification is feasible. Since such a large number of conformations can rapidly become incompatible with systematic docking owing to a high computational cost, a possible approach to limit the number of conformations is to cluster the conformations generated and to consider one representative per cluster. Here, we use as a distance $d = 1 - BCscore$, and a complete linkage approach with a cutoff value of 0.8. Figure 2b shows the average BCscore of the best cluster representative as a function of the number of clusters considered. Two major observations can be drawn. First, rapid increase in quality is reached up to 20 clusters, and the gain in average quality becomes small for a number of clusters of more than 40, which gives an upper limit to the number of conformations to consider. Second, the average quality, while not reaching 0.8 on average, i.e., being degraded compared to Fig. 2a, remains in the range of values where the conformations still have a significant similarity with that of the peptide holo conformation. Thus, it seems possible to identify a limited set of suboptimal conformations in which some are similar to the peptide holo conformation. Of course, large variations can be observed depending on the peptides. Figure 2c shows the distribution of the BCscores observed for the 41 peptides. For only three peptides, no conformation significantly similar to that of the peptide in complex could be generated among 600. Figure 2d shows the distribution of the BC score after selecting 40 representatives only. One also observes that the distribution of the similarity of cluster representatives tends to be bimodal, with in 15 cases conformations very close to the experimental one, the other cases being distributed around a BCscore value of 0.55. It is important to note that BC score values around 0.5, while implying dissimilarity with the reference conformation, do not imply a total dissemblance, fully unrelated conformations being for values close to 0. Some examples of how similar the conformations can be to the experimental one are depicted in Fig. 3. The best BCscores over 600 conformations are of 0.97, 0.89, 0.62, and 0.11 for 1OU8, 1MFG, 1SFI, and 1LVM, respectively. Selecting the representatives of only the 40 lowest energy clusters, the corresponding scores are of 0.58, 0.73, 0.56,

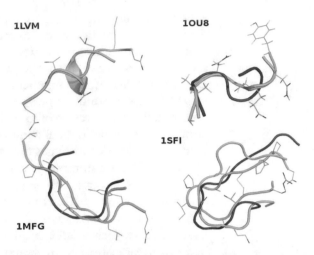

Fig. 3 Examples of conformations generated using PEP-FOLD and their similarity with the peptide holo conformation (*green*). *Cyan*: best conformation generated among 600. The BCscores are of 0.97, 0.89, 0.62, and 0.11 for 1OU8, 1MFG, 1SFI, and 1LVM, respectively. *Magenta*: best cluster representative among the 40 clusters of lowest energy. The corresponding scores are of 0.58, 0.73, 0.56, and 0.11

and 0.11. Despite a general decrease in similarity, note that except for 1LVM where a helical conformation is preferred to a more extended one, global conformational trends tend to be preserved for BCscore around 0.5.

1.3 PEP-SiteFinder: Blind Rigid Docking of Ensembles of Suboptimal Peptide Conformations to Identify Candidate Binding Sites

Given that the peptide bound conformation can be approached by sampling the conformational space of the peptide in isolation, one can expect that the blind rigid docking of a limited number of suboptimal peptide conformations onto the protein receptor should make it possible to infer information about the putative peptide binding site. PEP-SiteFinder is a four-step protocol that starts with the peptide amino acid sequence, the structure of the receptor, and returns information about the candidate-binding site.

2 Methods

2.1 The PEP-SiteFinder Protocol

PEP-SiteFinder consists of rigid blind rigid docking starting from a collection of representative conformations of the peptide to infer information about the candidate peptide-binding site. Figure 4 shows the four main steps of the protocol used by PEP-SiteFinder as available online at http://bioserv.rpbs.univ-paris-diderot.fr/services/PEP-SiteFinder.

2.1.1 Generation of a Large Number of Peptide Conformations

Achieved using a modified version of PEP-FOLD, as discussed in Subheading 1.2. In order to limit the calculation cost, the number of suboptimal conformation generated is limited to 200.

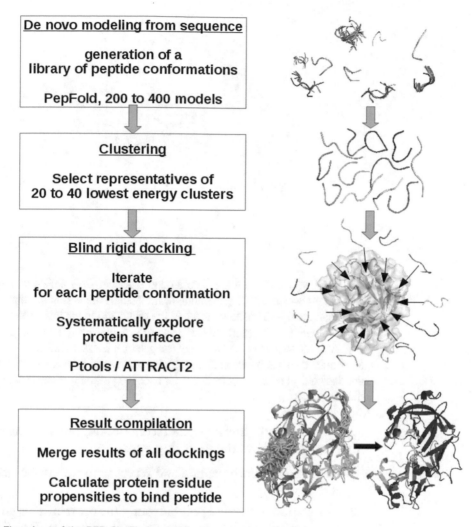

Fig. 4 Flowchart of the PEP-SiteFinder protocol

2.1.2 Selection of Peptide Representative Conformations

Performed using the python hcluster library applied to the BCscores, using as distance $(1 - BCscore)$, complete linkage and a cut-off value of 0.85. In order to limit the calculation cost, the number of cluster representatives selected for further steps is limited to 20.

2.1.3 Blind Docking Using PTools and the ATTRACT2 Force Field

For each peptide conformation, the ATTRACT docking protocol is applied, as implemented in the PTools library. Details on the ATTRACT force field can be found in Fiorucci et al. [21]. Briefly, the ATTRACT force field uses a coarse-grained representation in which all atoms from the backbone are kept while side chains are represented by up to two pseudo-atoms. The energy function is the sum of two contributions, the electrostatic energy and a pairwise soft Lennard-Jones potential. The PTools library [22] is freely

Fig. 5 Generation of starting points around a protein. Only a slice is depicted. Two parameters drive the generation of the points. The first is the distance to the protein (d1), expressed in terms of the radius of the peptide conformation. Here, starting points generated using values of 2 (*yellow*, *red*) and 3 (*gray*) are depicted. The second is the density of the points. It is expressed as the minimal separation distance between them (d2). Here values of 5 (*yellow*) or 10 Å (*red*) are depicted. In practice, PEP-SiteFinder uses the parameters corresponding to the *red points*

available on github (https://github.com/ptools/ptools). The docking protocol involves the following steps:

1. Reduction of both peptide and receptor to the coarse-grained representation.

2. Generation of starting points around the protein. This generation depends on two parameters, as illustrated in Fig. 5. The first is the distance of the starting points to the surface of the protein (d1). It is dependent on the radius of the peptide conformation. The second is the separation distance between the starting points that cannot be less than a cut-off value (d2). A sensitivity analysis of these two parameters has shown that using twice the radius of the peptide (i.e., the minimal distance to the protein ensuring no peptide protein clash of the initial complex), and a separation distance of 10 Å provides a good tradeoff between performance and calculation cost.

3. For each starting point, 260 peptide orientations are generated.

4. For each orientation, a minimization is performed, allowing the peptide to move only using its translational and rotational degrees of freedom (rigid-body docking). All minimizations are independent and can be run in parallel on a cluster. Using on the order of 200 cores, docking simulations take on the order of 1 min for most targets.

5. For each docking experiment, the poses are sorted and ranked by energy. Redundant solutions are filtered. Note that the possibly huge number of poses make an exact clustering approach unmanageable. Instead, PEP-SiteFinder relies on a fast and sequential clustering approach starting from the lowest energy pose, removing a pose if it has a RMSD of less than 1 Å with the poses of lower energy not discarded. To keep the algorithm in $O(n)$ with respect to the possibly large number of poses, only the latest 50 poses are compared to a new pose.

2.1.4 Protein Residue Propensities to Interact with the Peptide

The results of all the rigid docking performed for all the peptide representative conformations are merged, and poses are ranked by energy. The 50 best poses are used to determine protein residue propensities to bind the peptide. For each pose, protein residues at the peptide interface are defined as the residues having at least one heavy atom at a distance of less than 5 Å of any peptide heavy atom. The propensity of a residue is then calculated as the fraction of times it has been at the peptide–protein interface.

2.2 Protocol Assessment

The PEP-SiteFinder protocol relies on several approximations. A first one is the rigid docking approach. Although representative conformations can approximate in a satisfactory manner the conformation of the peptide in complex, the limited number of the conformations selected introduces some degradation in the quality of the conformations effectively in use (*see* Fig. 2d). The second one is the use of a simplified representation associated with a very simple force field. The impact of such a representation can be assessed by an auto validation where the experimental conformation is docked on the holo conformation of the protein. Results over all the 103 peptiDB complexes are illustrated in Fig. 6. Clearly, the best poses generated can, in some cases, differ largely from the experimental one—large ligand RMSD values. However, the fraction of native contacts matching these poses looks much better. Indeed for 94 cases over 103 at least one pose overlaps with the experimental one for more than 50% of the interacting residues. Thus, it is important to note that high accuracy poses are not requested to obtain correct propensities. As illustrated in Fig. 6 for the case of PDB entry 1YWO, a large RMSD value (over 16 Å) results from the flip of the peptide, but is still located in the correct binding site, resulting in correct estimation of the propensity of interaction of the residues in the binding site.

Therefore, despite these approximations, the rigid docking protocol is able to infer useful information about the interaction. Figure 7 shows the results in terms of propensities obtained for the nonredundant subset of 41 complexes of the peptiDB. A first observation is that for noninteracting residues (middle), the propensities tend to be low, as expected. However, a second observation is that the distribution of the values of the propensities of

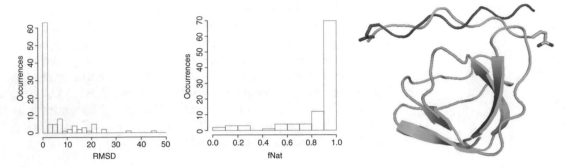

Fig. 6 Positive control over the 103 complexes of peptidb. Only the peptide holo conformation undergoes blind docking onto the receptor holo conformation. *Left*: Peptide RMS deviation without superimposition (Å). *Middle*: Fraction of the native contacts identified by the best pose. *Right*: Example of a flipped pose that leads to the correct identification of contacts. PDB id: 1YWO, *magenta*: experimental peptide pose, *cyan*: best score pose

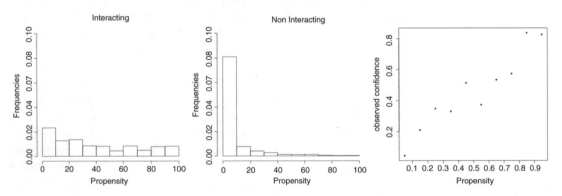

Fig. 7 Protocol assessment over nonredundant 41 complexes of the peptiDB. *Left*: distribution of the residue propensities to interact with the peptide for the protein residues actually interacting with the peptide. *Middle*: distribution of the residue propensities to interact with the peptide for the protein surface residues not interacting with the peptide. *Right*: A posteriori analysis of the probability a residue is effectively at protein peptide interface, as a function of the propensity

interacting residues (left) is also, slightly, shifted to low values, indicating that the protocol fails to identify some interacting residues. However, larger propensity values are also associated with a better confidence in the prediction of the interacting residues which makes it possible to analyze the results in a critical manner. Indeed, propensity values of more than 0.7 correspond to a correct prediction for 80 % of the cases.

2.3   Case Studies

In the following, we present two examples to illustrate the behavior of PEP-SiteFinder and to emphasize some of its limits.

2.3.1   Kelch-Like ECH-Associated Protein 1 (Keap1) Interaction with a Nonamer Peptide

The first example is the complex between mouse Kelch-like ECH-associated protein 1 (Keap1) and a 9-mer peptide derived from mouse nuclear factor erythroid 2 related factor 2 (Nrf2). The structure of this complex has been solved (PDB ID: 1X2R) to elucidate the molecular mechanism of the Nrf2/Keap1 interaction,

Fig. 8 PEP-SiteFinder results for the apo form of mouse Keap1 and a 9-mer peptide derived from mouse Nrf2. Protein residues are colored according to their predicted propensities, from *blue* (0) to *red* (100) and the peptide (positioned as in the experimental complex) is in *green*

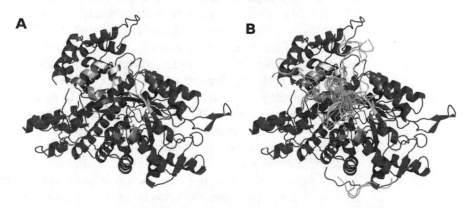

Fig. 9 PEP-SiteFinder results for the apo form of Eukaryotic ribonucleotide reductase and the FTLDADF peptide. (a) Protein residues are colored according to their predicted propensities, from *blue* (0) to *red* (100), (b) the peptide from the experimental conformation is in *magenta* and the 50 best predicted models are in *cyan*

which is implicated in a regulation pathway of the cellular response against oxidative stress [23]. The apo conformation of the protein is known (PDB ID: 1X2J). As shown in Fig. 8, application of PEP-SiteFinder to 1X2J using the peptide sequence "LDEETGEFL" identifies a candidate region that actually corresponds to that of the experimental structure of the complex (PDB ID: 1X2R). Looking more in detail at the propensity values, 25 residues have a propensity index of more than 0.9, i.e., associated with a good confidence.

2.3.2 Eukaryotic Ribonucleotide Reductase in Interaction with a Heptapeptide

The second example is the complex between Eukaryotic ribonucleotide reductase (RR), a target for cancer therapy and the FTLDADF peptide, which was designed to inhibit RR [24]. We have run PEP-SiteFinder using as apo conformation the chain A of PDB entry 2CVX. As shown in Fig. 9a, the candidate patch identified by

PEP-SiteFinder does not correspond to the region interacting with the peptide (magenta in Fig. 9b). Instead, the patch corresponds to the ADP-binding site (Note that the ADP was not present during the simulation). Looking more in detail, one observes however that only weak propensities values are returned. Only for 12 residues are the propensities higher than 0.7 and none exceeds 0.8, suggesting a medium confidence of the prediction. Actually, looking at the 50 best poses, one observes that alternate peptide positions are proposed, among which one matches the experimental interaction. This highlights the necessity of a critical analysis of the results returned by PEP-SiteFinder.

3   Discussion and Perspectives

The PEP-SiteFinder protocol relies on one core assumption and two major approximations. The assumption is that the receptor and the peptide interact. This has important consequences since PEP-SiteFinder will return results for whatever peptide sequence and receptor structure specified, not considering the affinity or the specificity of the interaction. Ways to estimate these are however presently unclear. Concerning the approximations, the first one is the use of a coarse-grained representation both to sample peptide suboptimal conformations and to score the peptide-protein interactions. The second one is the use of a rigid docking procedure. Their combination makes it possible to set up a procedure that samples exhaustively protein surface to identify candidate patches of interaction in a reasonable amount of calculation time. Doing so, an expectation is that peptide protein interactions involve specific and favorable enough interaction mechanisms, detectable even introducing fuzziness into the evaluation process. It is thus interesting that such a protocol, while simplistic, is able in most cases to infer useful information. However, it is also important to remember, as illustrated in the example cases, that a critical analysis of the results is necessary. It is also obvious that the approximations consented result in a lack of accuracy in the details of the peptide-protein interaction, and cannot lead, in general, to high accuracy complex modeling. This protocol must be considered the first step to drive further investigations. Several ways to move farther are described in other contributions of this book. Interestingly, the suboptimal sampling approach used to generate representative peptide conformations can easily be adapted to the generation of peptide-protein complexes, to fold a peptide in the vicinity of the protein receptor given a candidate binding patch, as illustrated in Fig. 10. This service, now available online at http://bioserv.rpbs. univ-paris-diderot.fr/services/PEP-FOLD3, is the subject of our present efforts [25].

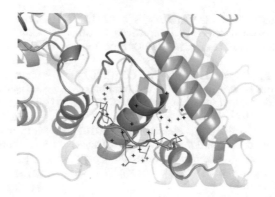

Fig. 10 Example of preliminary results obtained by folding of a peptide in the vicinity of a protein, as applied to the FTLDADF peptide binding to the Eukaryotic ribonucleotide reductase, specifying a candidate patch of interaction. Candidate starting points to fold the peptide correspond to dots. The best generated peptide conformation (ranked 5 according to lowest energy) is in blue and has a RMSD = 2.72 Å to the experimental conformation in wheat (PDB ID: 2ZLF)

Acknowledgments

This work has been supported by the French IA bioinformatics BipBip grant (ANR-10-BINF-0003), French Institute for Bioinformatics (IFB) (ANR-11-INBS-0013), and INSERM UMR-S 973 recurrent funding.

References

1. Vlieghe P, Lisowski V, Martinez J et al (2010) Synthetic therapeutic peptides: science and market. Drug Discov Today 15(1-2):40–56

2. Zambrowicz A, Timmer M, Polanowski A et al (2013) Manufacturing of peptides exhibiting biological activity. Amino Acids 44(2):315–320

3. Kaspar AA, Reichert JM (2013) Future directions for peptide therapeutics development. Drug Discov Today 18(17-18):807–817

4. Hancock RE, Sahl HG (2006) Antimicrobial and host-defense peptides as new anti-infective therapeutic strategies. Nat Biotechnol 24(12):1551–1557

5. Vetter I, Davis JL, Rash LD et al (2011) Venomics: a new paradigm for natural products-based drug discovery. Amino Acids 40(1):15–28

6. Caboche S, Pupin M, Leclère V et al (2008) NORINE: a database of nonribosomal peptides. Nucleic Acids Res 36(Database issue):D326–D331

7. Dietrich U, Dürr R, Koch J (2013) Peptides as drugs: from screening to application. Curr Pharm Biotechnol 14(5):501–512

8. Fosgerau K, Hoffmann T (2015) Peptide therapeutics: current status and future directions. Drug Discov Today 20(1):122–128

9. London N, Movshovitz-Attias D, Schueler-Furman O (2010) The structural basis of peptide-protein binding strategies. Structure 18(2):188–199

10. Berman HM, Westbrook J, Feng Z et al (2000) The Protein Data Bank. Nucleic Acids Res 28(1):235–242

11. Ma B, Kumar S, Tsai CJ et al (1999) Folding funnels and binding mechanisms. Protein Eng 12(9):713–720

12. Boehr DD, Nussinov R, Wright PE (2009) The role of dynamic conformational ensembles in biomolecular recognition. Nat Chem Biol 5(11):789–796

13. Changeux JP, Edelstein S (2011) Conformational selection or induced fit 50 years of debate resolved. F1000 Biol Rep 3:19

14. Sugase K, Dyson HJ, Wright PE (2007) Mechanism of coupled folding and binding of an intrinsically disordered protein. Nature 447(7147):1021–1025

15. Csermely P, Palotai R, Nussinov R (2010) Induced fit, conformational selection and independent dynamic segments: an extended view of binding events. Trends Biochem Sci 35(10):539–546

16. Bachmann A, Wildemann D, Praetorius F et al (2011) Mapping backbone and side-chain interactions in the transition state of a coupled protein folding and binding reaction. Proc Natl Acad Sci U S A 108(10): 3952–3957

17. Maupetit J, Derreumaux P, Tufféry P (2010) A fast method for large-scale de novo peptide and miniprotein structure prediction. J Comput Chem 31(4):726–738

18. Shen Y, Maupetit J, Derreumaux P et al (2014) Improved PEP-FOLD approach for peptide and miniprotein structure prediction. J Chem Theory Comput 10(10):4745–4758

19. Orengo CA, Michie AD, Jones S et al (1997) CATH—a hierarchic classification of protein domain structures. Structure 5(8):1093–1108

20. Guyon F, Tufféry P (2014) Fast protein fragment similarity scoring using a Binet-Cauchy kernel. Bioinformatics 30(6):784–791

21. Fiorucci S, Zacharias M (2010) Binding site prediction and improved scoring during flexible protein-protein docking with ATTRACT. Proteins 78:3131–3139

22. Saladin A, Fiorucci S, Poulain P et al (2009) PTools: an opensource molecular docking library. BMC Struct Biol 9:27

23. Padmanabhan B, Tong KI, Ohta T et al (2006) Structural basis for defects of Keap1 activity provoked by its point mutations in lung cancer. Mol Cell 21(5):689–700

24. Xu H, Fairman JW, Wijerathna SR et al (2008) The structural basis for peptidomimetic inhibition of eukaryotic ribonucleotide reductase: a conformationally flexible pharmacophore. J Med Chem 51(15):4653–4659

25. Lamiable A, Thévenet P, Rey J et al (2016) PEP-FOLD3: faster de novo structure prediction for linear peptides in solution and in complex. Nucleic Acids Res 44(W1):W449–W454. doi:10.1093/nar/gkw329

Part II

Peptide Docking

Chapter 4

Template-Based Prediction of Protein-Peptide Interactions by Using GalaxyPepDock

Hasup Lee and Chaok Seok

Abstract

We introduce a web server called GalaxyPepDock that predicts protein-peptide interactions based on templates. With the continuously increasing size of the protein structure database, the probability of finding related proteins for templates is increasing. GalaxyPepDock takes a protein structure and a peptide sequence as input and returns protein-peptide complex structures as output. Templates for protein-peptide complex structures are selected from the structure database considering similarity to the target protein structure and to putative protein-peptide interactions as estimated by protein structure alignment and peptide sequence alignment. Complex structures are then built from the template structures by template-based modeling. By further structure refinement that performs energy-based optimization, structural aspects that are missing in the template structures or that are not compatible with the given protein and peptide are refined. During the refinement, flexibilities of both protein and peptide induced by binding are considered. The atomistic protein-peptide interactions predicted by GalaxyPepDock can offer important clues for designing new peptides with desired binding properties.

Key words Protein-peptide interaction, Template-based modeling, Structure refinement, Peptide structure flexibility

1 Introduction

Protein-peptide interactions play important roles in many biological processes such as signaling pathways, protein cellular localization, immune response, and posttranslational modifications [1–3], and are related to human diseases such as cancer. Protein-peptide interactions may be modulated by small chemicals or synthetic peptides because of the relatively small interface areas. Therefore, there are a lot of interests in developing therapeutic agents by targeting protein-peptide interactions [4, 5].

The atomistic protein-peptide interactions revealed by protein-peptide complex structures can help to design therapeutic molecules. Protein-peptide complex structures can be determined by experimental methods such as X-ray crystallography and NMR, but the structures resolved by experimental methods so far cover

Ora Schueler-Furman and Nir London (eds.), *Modeling Peptide-Protein Interactions: Methods and Protocols*, Methods in Molecular Biology, vol. 1561, DOI 10.1007/978-1-4939-6798-8_4, © Springer Science+Business Media LLC 2017

only a small portion of possible protein-peptide interactions [6]. Predicting protein-peptide complex structures by computational means can be one of the alternatives for obtaining structures. Various protein-peptide docking programs have been developed, and they can be classified into local docking and global docking. Local docking methods perform docking near the predefined binding site of protein, so putative protein-peptide complex structure may be required as input. On the other hand, global docking methods search the whole protein surface to dock peptide, so predefined binding site or putative complex structure is not required and only a protein structure and a peptide sequence are needed as input. Some protein-peptide docking programs are available as web servers. For example, Rosetta FlexPepDock [7] and PepCrawler [8] are web servers for local docking and PEP-SiteFinder [9] and CABS-dock [10] are for global docking. PEP-SiteFinder generates initial peptide structures by PEP-FOLD [11] and then predicts the complex structure by rigid-body docking with ATTRACT [12]. CABS-dock generates initial peptide structures randomly and predicts the complex structure by a replica exchange Monte Carlo method. The above-mentioned programs are ab initio docking programs that do not use information on evolutionarily related protein-peptide complex structures.

It has been discussed previously that many protein-peptide interactions are stabilized through conserved binding sites of the protein and short linear motifs of the peptide [13]. It can be therefore conceived that homologous proteins may be used as templates for predicting protein-peptide complex structures. As increasing number of protein-peptide complex structures are being deposited in the Protein Data Bank (PDB), the probability of finding similar protein-peptide complexes for a given target complex is also increasing. For example, 87% of the nonredundant protein-peptide complex structures in the PeptiDB set [14] have similar proteins, with a protein TM-score > 0.6, among the experimentally determined structures released prior to the given complex. Therefore, it is timely to have a template-based protein-peptide docking method, and such a method will find more applications as the structure database grows.

In this chapter, we introduce the GalaxyPepDock server [15] that performs template-based protein-peptide docking. It is available as a part of the GalaxyWEB server [16]. GalaxyPepDock takes a protein structure and a peptide sequence as input. The server searches the protein-peptide structure database to find available experimental structures of related proteins complexed with peptides, and then generates protein-peptide complex structures based on the selected templates. The structures are further refined by minimizing GALAXY energy. The GALAXY energy is a hybrid energy that combines physicochemical energy terms derived from molecular mechanics force field, knowledge-based energy terms derived from statistics of atom pair interactions in the structure database, and restraint energy terms derived from information on the interactions found in homologous complexes. The rigid-body translational and rotational displacements of peptide relative to protein and internal structural flexibilities of protein and peptide are all allowed to change during the optimization. The method was tested successfully on the PeptiDB benchmark set and on recently released targets of critical assessment of prediction of interactions (CAPRI) ([17], http://www.ebi.ac.uk/msd-srv/capri/round28/round28.html). The server shows much superior predictions of protein-peptide complex structures compared to the existing ab initio docking methods when related complex structures are available in the database. The server also returns an estimated accuracy of the prediction to help users to decide whether ab initio docking is necessary for their problems or not.

2 Materials

1. A personal computer or device and a web browser are required to access the GalaxyWEB server through the Internet. A JavaScript-enabled web browser is highly recommended to see the results on the web browser. The server compatibility was tested on Google Chrome, Firefox, Safari, and Internet Explorer.

2. The following input material is required to use GalaxyPepDock on GalaxyWEB. To run GalaxyPepDock for predicting protein-peptide interactions, a structure file in standard PDB format for the protein of interest and a sequence file in FASTA format for the peptide of interest are required. The number of amino acids of the protein should be less than 900 and that of the peptide less than 30. The input peptide sequence file must contain 20 standard amino acids in one-letter codes. Example input files (Fig. 1, Label 1) can be obtained from the GalaxyPepDock web page.

3 Methods

3.1 The GalaxyPepDock Protocol

GalaxyPepDock runs in two stages:

1. *Template selection*: Templates for predicting protein-peptide interactions are selected from the PepBind database [18] based on protein structure similarity measured by TM-score [19] and protein-peptide interaction similarity measured by interaction similarity score (S_{inter}) [15]. Up to ten complex structures with score >90% of the maximum value are selected as templates and used for model-building.

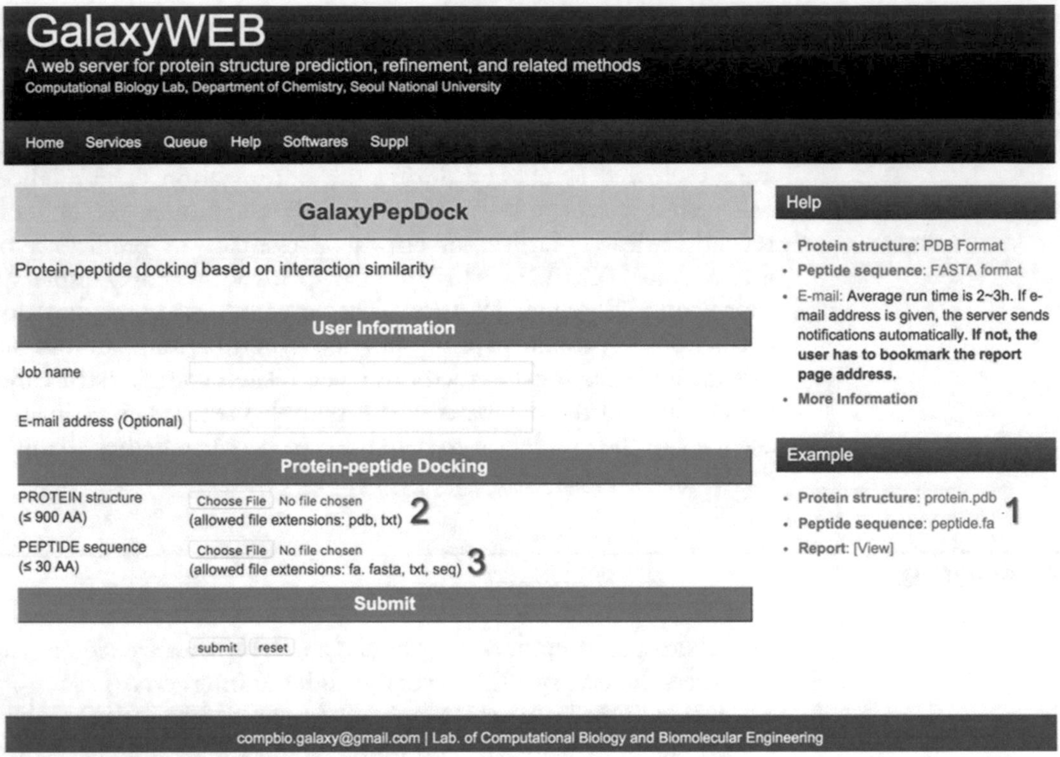

GalaxyWEB
A web server for protein structure prediction, refinement, and related methods
Computational Biology Lab, Department of Chemistry, Seoul National University

Home   Services   Queue   Help   Softwares   Suppl

GalaxyPepDock

Protein-peptide docking based on interaction similarity

User Information

Job name

E-mail address (Optional)

Protein-peptide Docking

PROTEIN structure   Choose File   No file chosen
(≤ 900 AA)   (allowed file extensions: pdb, txt)   **2**

PEPTIDE sequence   Choose File   No file chosen
(≤ 30 AA)   (allowed file extensions: fa, fasta, txt, seq)   **3**

Submit

submit   reset

Help

- Protein structure: PDB Format
- Peptide sequence: FASTA format
- E-mail: **Average run time is 2~3h. If e-mail address is given, the server sends notifications automatically. If not, the user has to bookmark the report page address.**
- More Information

Example

- Protein structure: protein.pdb   **1**
- Peptide sequence: peptide.fa
- Report: [View]

compbio.galaxy@gmail.com | Lab. of Computational Biology and Biomolecular Engineering

Fig. 1 The GalaxyPepDock input page

2. *Complex structure optimization*: 50 models are first generated by the model-building tool of the GalaxyTBM template-based modeling method [20] for each selected template. After model-building, ten structures are selected by GALAXY energy and are subject to further refinement by the GalaxyRefine model refinement method [21]. The refinement protocol refines both internal structures of protein and peptide as well as relative displacements of protein and peptide.

For further details refer to the GalaxyPepDock paper [15].

3.2 Peptide-Binding Prediction by GalaxyPepDock

1. Go to GalaxyWEB, http://galaxy.seoklab.org. Click "PepDock" in the "Services" tab at the top of the page.

2. In the "User Information" section, enter job name (defaults to "None"). If the user provides email address, the server will send progress reports of submitted job automatically. Otherwise, the user should bookmark the report page after submitting a job.

3. In the "Protein-peptide Docking" section, provide a standard PDB-formatted protein structure file (Fig. 1, Label 2) and a FASTA-formatted peptide sequence file (Fig. 1, Label 3).

4. Press the submit button to queue the job. If the submission is successful, a "Submission Information" page will appear (Fig. 2a).

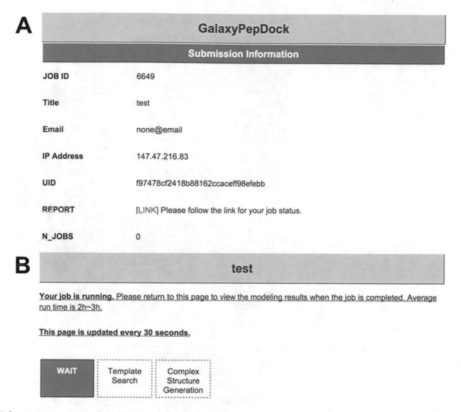

Fig. 2 (**a**) A summary page showing the submission information of a GalaxyPepDock job. (**b**) An example report page showing of the status of the GalaxyPepDock job

5. Click "LINK" of the submission information page to access the report page. The report page will be refreshed every 30 s, updating the status of the submitted job. When the job is completed, the predicted results will be presented. The average run time of GalaxyPepDock is about 2–3 h (Fig. 2b).

6. *Predicted protein-peptide complex structures*: Predicted structures of the query protein-peptide complex can be visualized on the report page using PV (http://biasmv.github.io/pv/), a JavaScript protein viewer, if the web browser supports JavaScript (Fig. 3). Users can zoom in and out by scrolling mouse wheel and change the focusing center by double clicking. Template structures selected from the database of protein-peptide complex structures to be used in the prediction are shown in light colors; protein and peptide structures are in light red and blue, respectively. Different protein-peptide complex model structures can be seen by clicking the model number in the "View in PV" line (Fig. 3, Label 1). Predicted protein-peptide complex structures can also be downloaded in PDB-formatted files for further analyses (Fig. 3, Label 2).

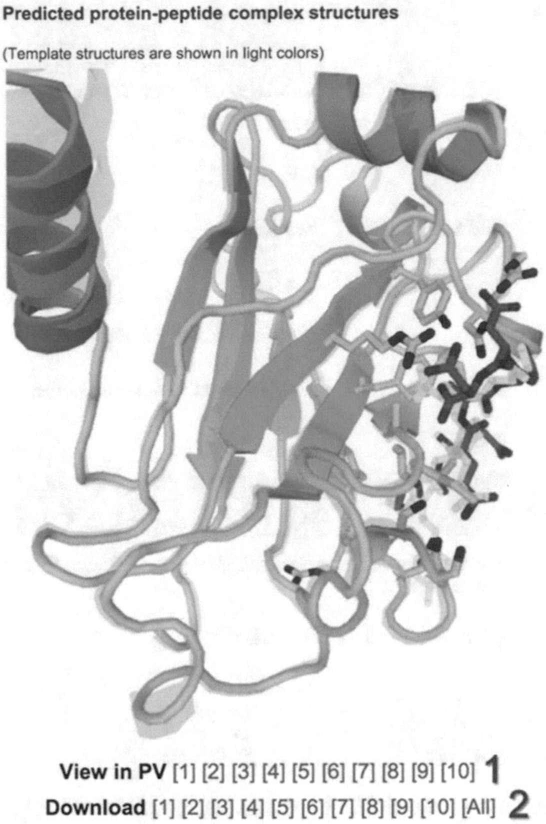

Predicted protein-peptide complex structures

(Template structures are shown in light colors)

View in PV [1] [2] [3] [4] [5] [6] [7] [8] [9] [10] **1**
Download [1] [2] [3] [4] [5] [6] [7] [8] [9] [10] [All] **2**

Fig. 3 An example of the "Predicted protein-peptide complex structures" section on the GalaxyPepDock report page

7. *Additional information*: Additional information on predicted models and intermediate results generated during the GalaxyPepDock run is provided in a table (Fig. 4a). Structures of protein template and peptide template are given as PDB IDs and can also be downloaded (Fig. 4a, Label 1 and 2, respectively). Sequences and alignments of the query and the template used for the prediction are provided (Fig. 4a, Label 3) for both protein and peptide (Fig. 4b). Structure similarity between the predicted protein structure and the protein template structure is presented in terms of TM-score [19] and RMSD (Fig. 4a, Label 4). A score called interaction similarity score (S_{inter}) [15] that was designed to describe the similarity of the amino acids of the query complex aligned to the interacting residues of the template complex is reported for each prediction. This is to give an idea on the degree of the relative differences in similarity to the selected templates among different models (Fig. 4a, Label 5).

8. *Predicted binding site residues*: Binding site residues of the protein taken from the predicted complex structure (Fig. 4a,

A Additional information

Model	Protein template	Peptide template	Sequences& alignments	Protein structure similarity (TM-score/RMSD)	Interaction similarity score	Estimated accuracy	Predicted binding site residues in protein
1	1QSC_A	1QSC_D	LINK	0.985 / 0.65	55	0.735	LINK
2	1QSC_A	1QSC_D	LINK	0.985 / 0.66	55	0.735	LINK
3	1QSC_A	1QSC_D	LINK	0.985 / 0.66	55	0.735	LINK
4	1QSC_A	1QSC_D	LINK	0.985 / 0.69	55	0.735	LINK
5	1CA9_A	1CA9_G	LINK	0.988 / 0.57	42	0.709	LINK
6	1CA9_A	1CA9_G	LINK	0.988 / 0.64	42	0.709	LINK
7	1CA9_A	1CA9_G	LINK	0.988 / 0.59	42	0.709	LINK
8	1CA9_E	1CA9_G	LINK	0.989 / 0.53	16	0.652	LINK
9	1CA9_E	1CA9_G	LINK	0.989 / 0.59	16	0.652	LINK
10	1CA9_E	1CA9_G	LINK	0.989 / 0.63	16	0.652	LINK

B
```
Query protein:  -----------AMADLEQKVLEMEASTYDGVFIWKISDFPR
Templ protein:  QLERSIGLKDLAMADLEQKVLEMEASTYDGVFIWKISDFAR
Query peptide:  -PQQATDD
Templ peptide:  YPIQET--
```

C
```
41  PRO
44  PHE
60  ARG
62  TYR
66  ASP
67  GLY
70  ARG
77  PHE
114 PHE
115 ARG
116 PRO
117 ASP
120 SER
121 SER
122 SER
123 PHE
131 ASN
132 ILE
133 ALA
134 SER
135 GLY
136 CYS
137 PRO
```

Fig. 4 An example of the "Additional information" section on the GalaxyPepDock report page. (**a**) A summary table showing the results of the protein-peptide complex structure predictions. (**b**) An example of structure/sequence alignments between the query protein/peptide and the template protein/peptide. (**c**) An example of the list of predicted binding residues of protein

Label 7 and Fig. 4c) and the estimated prediction accuracy of the binding site (Fig. 4a, Label 6) are provided (*see* **Note 1**). Those residues with any heavy atom within 5 Å from any peptide heavy atom in the predicted structure are reported as binding residues.

9. A GalaxyPepDock help page is also available; click the "Help" tab at the top of the page, and click "GalaxyPepDock" on the right of the help page. A more detailed description of the prediction method of GalaxyPepDock can be found in the original paper [15].

3.3 Case Study

We describe here the GalaxyPepDock results when tested on the CAPRI target 67, a complex of the third WW domain of human Nedd4 and the first PPXY motif of ARRDC3 (PDB ID: 4N7H) [22]. The test was of course performed without including the known crystal structure. In the template selection stage, the server selected a template of high structural similarity to the input protein structure (TM-score = 0.733) and of high interaction similarity to the target complex as estimated by the structure alignment to the input protein structure and sequence alignment to the input peptide sequence ($S_{inter} = 182$). The initial model generated from the template had the ligand RMSD and interface RMSD from the

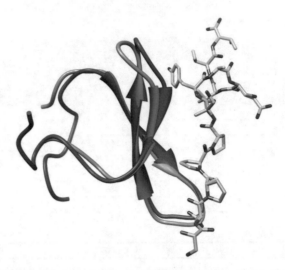

Fig. 5 An input protein structure and peptide sequence and selected template. Input protein and peptide are shown in *green* and *light green*, and selected template is shown in *magenta* and *pink*

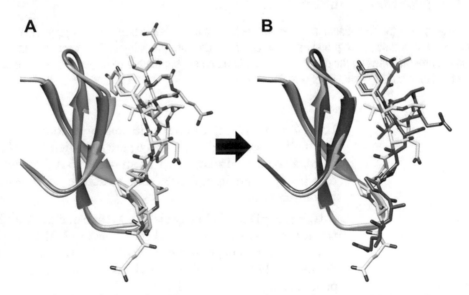

Fig. 6 Native structure (*yellow*) and (**a**) initial model (*green*) and (**b**) refined model (*dark green*)

crystal structure of 2.9 and 1.5 Å, respectively, and the fraction of native contacts of 0.5. This prediction corresponds to acceptable accuracy by the CAPRI criterion (Fig. 5). The initial model was further refined in the complex optimization stage, and the quality of the model was improved from acceptable accuracy to medium accuracy by the CAPRI criterion, with improvements in ligand RMSD/interface RMSD/(fraction of native contacts) to 1.8 Å/1.0 Å/0.688 (Fig. 6). GalaxyPepDock successfully predicted the following key interactions found in this target: hydrophobic interactions among tryptophan, phenylalanine, isoleucine, proline, and valine residues and polar interaction between

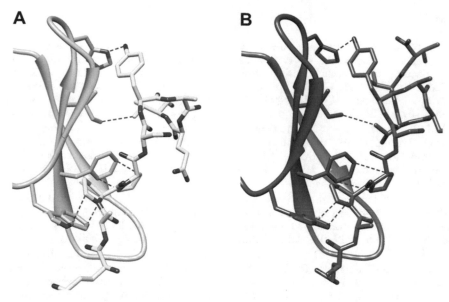

Fig. 7 Key interactions of (**a**) native structure and (**b**) GalaxyPepDock model

histidine and tyrosine residues (Fig. 7). The fraction of correctly predicted binding site residues is 0.929. This result shows that when reasonably close templates can be found with protein TM-score > 0.7, GalaxyPepDock can predict the binding site and interactions quite reliably.

4 Note

1. The estimated accuracy returned by GalaxyPepDock means an estimated fraction of correctly predicted binding site residues. This value was calculated by plugging in the intermediate data generated during the prediction to the linear regression relationship between the intermediate data and the prediction accuracy for the targets of the PeptiDB test set [14]. A low value of the estimated prediction accuracy implies that proper templates could not be found, and the current similarity-based method may not be accurate. If a very low value of estimated accuracy is returned, the user is recommended to try other ab initio protein-peptide docking methods such as PEP-SiteFinder [9] or CABS-dock [10] that do not rely on template information.

Acknowledgment

This work was supported by National Research Foundation of Korea grants funded by the Ministry of Science, ICT & Future Planning (No. 2013R1A2A1A09012229).

References

1. London N, Raveh B, Schueler-Furman O (2013) Peptide docking and structure-based characterization of peptide binding: from knowledge to know-how. Current opinion in structural biology 23(6):894–902. doi:10.1016/j.sbi.2013.07.006

2. Petsalaki E, Russell RB (2008) Peptide-mediated interactions in biological systems: new discoveries and applications. Current opinion in biotechnology 19(4):344–350. doi:10.1016/j.copbio.2008.06.004

3. Yan C, Zou X (2015) Predicting peptide binding sites on protein surfaces by clustering chemical interactions. Journal of computational chemistry 36(1):49–61. doi:10.1002/jcc.23771

4. Yang Y, Ludwig RL, Jensen JP, Pierre SA, Medaglia MV, Davydov IV, Safiran YJ, Oberoi P, Kenten JH, Phillips AC, Weissman AM, Vousden KH (2005) Small molecule inhibitors of HDM2 ubiquitin ligase activity stabilize and activate p53 in cells. Cancer cell 7(6):547–559. doi:10.1016/j.ccr.2005.04.029

5. Soni V, Cahir-McFarland E, Kieff E (2007) LMP1 TRAFficking activates growth and survival pathways. Advances in experimental medicine and biology 597:173–187. doi:10.1007/978-0-387-70630-6_14

6. Raveh B, London N, Zimmerman L, Schueler-Furman O (2011) Rosetta FlexPepDock ab-initio: simultaneous folding, docking and refinement of peptides onto their receptors. PloS one 6(4), e18934. doi:10.1371/journal.pone.0018934

7. London N, Raveh B, Cohen E, Fathi G, Schueler-Furman O (2011) Rosetta FlexPepDock web server—high resolution modeling of peptide-protein interactions. Nucleic acids research 39(Web Server issue):W249–W253. doi:10.1093/nar/gkr431

8. Donsky E, Wolfson HJ (2011) PepCrawler: a fast RRT-based algorithm for high-resolution refinement and binding affinity estimation of peptide inhibitors. Bioinformatics 27(20):2836–2842. doi:10.1093/bioinformatics/btr498

9. Saladin A, Rey J, Thevenet P, Zacharias M, Moroy G, Tuffery P (2014) PEP-SiteFinder: a tool for the blind identification of peptide binding sites on protein surfaces. Nucleic acids research 42(Web Server issue):W221–W226. doi:10.1093/nar/gku404

10. Kurcinski M, Jamroz M, Blaszczyk M, Kolinski A, Kmiecik S (2015) CABS-dock web server for the flexible docking of peptides to proteins without prior knowledge of the binding site. Nucleic acids research 43(Web Server issue):W419–W424. doi:10.1093/nar/gkv456

11. Thevenet P, Shen Y, Maupetit J, Guyon F, Derreumaux P, Tuffery P (2012) PEP-FOLD: an updated de novo structure prediction server for both linear and disulfide bonded cyclic peptides. Nucleic acids research 40(Web Server issue):W288–W293. doi:10.1093/nar/gks419

12. Zacharias M (2005) ATTRACT: protein-protein docking in CAPRI using a reduced protein model. Proteins 60(2):252–256. doi:10.1002/prot.20566

13. Trabuco LG, Lise S, Petsalaki E, Russell RB (2012) PepSite: prediction of peptide-binding sites from protein surfaces. Nucleic acids research 40(Web Server issue):W423–W427. doi:10.1093/nar/gks398

14. London N, Movshovitz-Attias D, Schueler-Furman O (2010) The structural basis of peptide-protein binding strategies. Structure 18(2):188–199. doi:10.1016/j.str.2009.11.012

15. Lee H, Heo L, Lee MS, Seok C (2015) GalaxyPepDock: a protein-peptide docking tool based on interaction similarity and energy optimization. Nucleic acids research 43(W1):W431–W435. doi:10.1093/nar/gkv495

16. Ko J, Park H, Heo L, Seok C (2012) GalaxyWEB server for protein structure prediction and refinement. Nucleic acids research 40(Web Server issue):W294–W297. doi:10.1093/nar/gks493

17. Lensink MF, Wodak SJ (2013) Docking, scoring, and affinity prediction in CAPRI. Proteins 81(12):2082–2095. doi:10.1002/prot.24428

18. Das AA, Sharma OP, Kumar MS, Krishna R, Mathur PP (2013) PepBind: a comprehensive database and computational tool for analysis of protein-peptide interactions. Genomics, proteomics & bioinformatics 11(4):241–246. doi:10.1016/j.gpb.2013.03.002

19. Zhang Y, Skolnick J (2004) Scoring function for automated assessment of protein structure template quality. Proteins 57(4):702–710. doi:10.1002/prot.20264

20. Ko J, Park H, Seok C (2012) GalaxyTBM: template-based modeling by building a reliable core and refining unreliable local regions. BMC bioinformatics 13:198. doi:10.1186/1471-2105-13-198

21. Heo L, Park H, Seok C (2013) GalaxyRefine: protein structure refinement driven by side-chain repacking. Nucleic acids research 41(Web Server issue):W384–W388. doi:10.1093/nar/gkt458

22. Qi S, O'Hayre M, Gutkind JS, Hurley JH (2014) Structural and biochemical basis for ubiquitin ligase recruitment by arrestin-related domain-containing protein-3 (ARRDC3). The Journal of biological chemistry 289(8):4743–4752. doi:10.1074/jbc.M113.527473

Application of the ATTRACT Coarse-Grained Docking and Atomistic Refinement for Predicting Peptide-Protein Interactions

Christina Schindler and Martin Zacharias

Abstract

Peptide-protein interactions are abundant in the cell and form an important part of the interactome. Large-scale modeling of peptide-protein complexes requires a fully blind approach; i.e., simultaneously predicting the peptide-binding site and the peptide conformation to high accuracy. Here, we present one of the first fully blind peptide-protein docking protocols, pepATTRACT. It combines a coarse-grained ensemble docking search of the entire protein surface with two stages of atomistic flexible refinement. pepATTRACT yields high-quality predictions for 70% of the cases when tested on a large benchmark of peptide-protein complexes. This performance in fully blind mode is similar to state-of-the-art local docking approaches that use information on the location of the binding site. Limiting the search to the peptide-binding region, the resulting pepATTRACT-local approach further improves the performance. Docking scripts for pepATTRACT and pepATTRACT-local can be generated via a web interface at www.attract. ph.tum.de/peptide.html. Here, we explain how to set up a docking run with the pepATTRACT web interface and demonstrate its usage by an application on binding of disordered regions from tumor suppressor p53 to a partner protein.

Key words Fully blind peptide-protein docking, Peptide flexibility, Peptide-protein complex formation, Proteome-wide modeling, Coarse-graining, Ensemble docking, Flexible interface refinement, Docking minimization, Molecular dynamics refinement

1   Introduction

Proteins are involved in almost all biological processes within and between cells. In the last decades, it has become clear that most proteins do not carry out their function individually but rather as parts of larger, often transiently formed assemblies. Increasing efforts have gone into characterizing these protein-protein interaction networks. Interactomes for several model organisms have already been drafted [1–5]. Protein-protein interactions can be divided into two main categories: domain-domain and peptide-domain interactions. The former refers to the binding of two

Ora Schueler-Furman and Nir London (eds.), *Modeling Peptide-Protein Interactions: Methods and Protocols*, Methods in Molecular Biology, vol. 1561, DOI 10.1007/978-1-4939-6798-8_5, © Springer Science+Business Media LLC 2017

globular domains from different proteins, whereas in the latter, a peptidic motif interacts with a globular domain from another protein. Peptidic-binding motifs can be short, isolated peptides, but are more often located as part of intrinsically disordered protein (IDP) regions [6, 7]. Such regions can be found in approximately 20–50% of all eukaryotic proteins [8]. In recent years, the discovery of the various functions of IDPs and IDP regions demonstrated that a lack of structural constraints facilitates several biological processes. Similarly, peptide-protein complexes benefit from the intrinsic flexibility and plasticity of the interaction and are therefore key players in many cellular pathways [9]. Peptide-protein interactions are found in signaling, apoptosis, immune response, and other (regulatory) pathways and possibly account for up to 40% of the interactome [10]. The vital role in the cell and the smaller interface compared to domain-domain interactions potentially make peptide-protein complexes promising targets for modulation by small molecules [11–15]. However, rational peptidic drug design requires a detailed understanding of the interaction and hence atomistic structural knowledge of the complex. The transient nature and the inherent flexibility of peptide-mediated interactions make it challenging to resolve the three-dimensional structure by experimental methods such as X-ray crystallography or NMR. In addition, due to the large number of known and possible peptide-protein complexes, it is unlikely that high-resolution structures for all of them will become available in the near future. To overcome these difficulties and to complement experimental efforts, a range of computational approaches have been developed for modeling of peptide-protein interactions. If the structure of the isolated protein is available, computational peptide-protein docking methods can be used to generate models of the bound complex.

Peptide-protein docking aims to predict the three-dimensional structure of the complex based on the unbound (apo) structure of the protein and the peptide sequence. Due to the intrinsic flexibility of peptides, unbound structures for them are not available in the general case. Hence, predicting the structure of a peptide-protein complex requires both predicting the peptide-binding site on the protein and the conformation of the bound peptide to high accuracy. A large variety of binding site prediction tools have been developed to date [16–24]. However, typical binding site predictors do not model the bound peptide structure (at high precision) [25]. In contrast, local docking approaches, including several target-specific approaches, sample possible peptide conformations at a previously identified binding site only [26–39]. Local docking approaches often yield high-quality predictions when tested on peptide-protein docking benchmarks [25] and these approaches are also very useful for refinement. Still, for proteome-wide, high-throughput modeling, it is desirable for peptide-protein docking methods to be fully blind meaning that they should predict the peptide-binding site and the peptide conformation simultaneously without prior information. Recently, several fully blind protocols have been developed and tested on known peptide-protein complexes [40–44]. Our peptide-protein docking approach pepAT-TRACT [43] combines coarse-grained and atomistic docking with different flexibility mechanisms in the ATTRACT docking engine.

1.1 The ATTRACT Docking Engine

The ATTRACT docking engine [45–47] (Fig. 1) can perform structural modeling for a variety of biomolecular interactions and has been applied to protein-protein, protein-DNA [48], protein-RNA [49], and protein-small molecule complexes [50]. It has been used successfully to predict targets in various rounds of the blind protein-protein docking challenge CAPRI [51–53]. ATTRACT distinguishes itself from other docking programs by its coarse-grained force field, the possible use of protein flexibility throughout the docking search, and the simultaneous docking of any number of (protein) bodies. It was recently expanded to fitting protein molecules in low-resolution cryo-EM maps [54] and also supports incorporating experimental information on interface residues and contacts. Flexible interface refinement of ATTRACT-generated models can be performed by the iATTRACT protocol [55]. A part of the functionality in ATTRACT has been made easily accessible to non-expert users through a web-interface that facilitates setting up protein-protein docking scripts (www.attract.ph.tum.de) [47].

The empirical, coarse-grained protein model in ATTRACT is intermediate between a residue-level and full atomistic description. It represents each amino acid of a protein by up to four pseudo atoms. The protein main chain is represented by two pseudo atoms per residue (located at the backbone nitrogen and backbone oxygen atoms, respectively). Small amino acid side chains (Ala, Asp, Asn, Cys, Ile, Leu, Pro, Ser, Thr, Val) are represented by one pseudo atom (geometric mean of side chain heavy atoms). Larger and more flexible side chains are represented by two pseudo atoms to better account for the shape and dual chemical character of some side chains. Effective interactions between pseudoatoms A and B are described by soft distance-dependent Lennard-Jones (LJ) type potentials of the following form

$$
V_{AB}\left(r_{ij}\right) = \begin{cases} \in_{AB}\left[\left(\dfrac{\sigma_{AB}}{r_{ij}}\right)^{8} - \left(\dfrac{\sigma_{AB}}{r_{ij}}\right)^{6}\right] + \dfrac{q_i q_j}{\varepsilon r_{ij}} & \text{for attractive pairs} \\[3em] -\in_{AB}\left[\left(\dfrac{\sigma_{AB}}{r_{ij}}\right)^{8} - \left(\dfrac{\sigma_{AB}}{r_{ij}}\right)^{6}\right] + \dfrac{q_i q_j}{\varepsilon r_{ij}} & \text{for repulsive pairs if } r_{ij} > r_{min} \\[3em] 2e_{min} + \in_{AB}\left[\left(\dfrac{\sigma_{AB}}{r_{ij}}\right)^{8} - \left(\dfrac{\sigma_{AB}}{r_{ij}}\right)^{6}\right] + \dfrac{q_i q_j}{\varepsilon r_{ij}} & \text{otherwise.} \end{cases}
$$

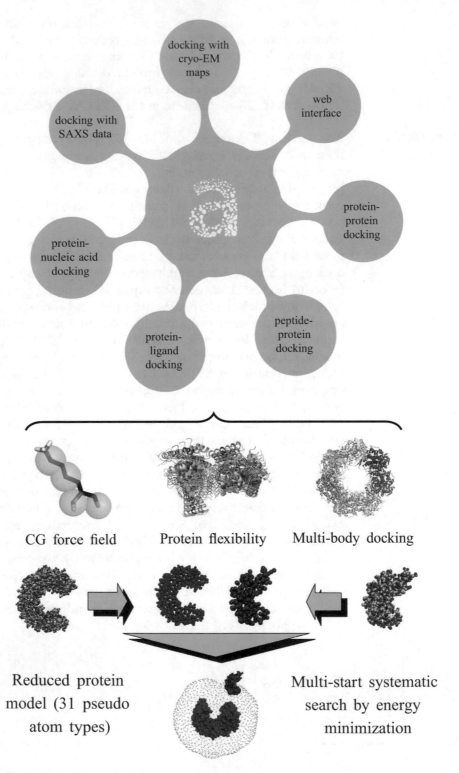

Fig. 1 The ATTRACT docking engine

where σ_{AB} and ϵ_{AB} are effective pairwise radii and attractive or repulsive Lennard-Jones parameters. At the distance r_{min} between two pseudo atoms, the standard LJ-potential has the energy e_{min}. A Coulomb type term accounts for electrostatic interactions between real charges (Lys, Arg, Glu, Asp) damped by a distance-dependent dielectric constant ($\varepsilon = 15r$). In contrast to the original force field [4], this form allows for purely repulsive interacting pseudo-atom pairs. The attractive and repulsive parameters for each pseudo-atom pair were iteratively optimized by minimizing the root-mean-square-deviation of near-native docking minima and by comparing the scoring of near-native minima with many high-scoring decoy complexes [45, 56]. The LJ interaction potentials were parameterized to consider both surface complementarity and physicochemical properties of protein-protein interfaces. The coarse-grained representation is coupled to a simplified, smoothened energy landscape that contains fewer docking energy minima and allows for much more rapid and fully converged energy minimization compared with an atomic resolution representation. Calculations can be further accelerated by precalculating the potential energy around one of the protein partners on a grid (de Vries, unpublished data). This fast coarse-grained docking approach makes it possible to scan hundreds of thousands configurations in the initial docking stage. ATTRACT also provides coarse-grained models for nucleic acids and their interactions with their protein partners [48, 49].

Several different possibilities are available to model flexibility throughout the different docking stages in ATTRACT. Global backbone flexibility (e.g., domain-domain motion) can be included by energy minimization along the directions of precalculated soft normal modes for each partner structure [46, 57]. The normal modes are derived from an elastic network model. This approach mimics an induced fit upon binding of the partners. For modeling a conformational selection-type binding process, a set of multiple rigid conformations can be supplied for each partner allowing selection of the most likely conformation during docking (ensemble docking). Initial rigid body docking solutions can be optimized with the flexible interface refinement method iATTRACT [55]. For efficient refinement, iATTRACT combines simultaneous local optimization of the interface with large-scale whole body translation and rotation of the ligand protein relative to the receptor. A structure-based force field for the unbound protein structures is generated on the fly and applied to the flexible interface atoms throughout the refinement process. The structure-based intramolecular force field contains harmonic potentials for controlling bond lengths and bond angles as well as a double-quadratic potential to represent steric repulsion between nonbonded atoms. The force field representation allows motions that result in changes of dihedral torsion angles and does not include any attractive nonbonded interactions within the protein. The structure-based potential is specific for the particular protein and optimal in the vicinity of the unbound structure that serves as reference. The form of the soft repulsive potential allows for partial atom-atom overlap (of atoms within one protein partner but not between atoms of separate partners) during the refinement. For the intermolecular interaction between partners, an atomistic force field based on the OPLS parameters is used. Applied to a large benchmark set of protein-protein complexes, the iATTRACT refinement method resulted in significant improvements of the quality of many docking solutions with respect to the native complexes. A major effect on increasing the fraction of native contacts up to 70% was observed [55]. Even though in many cases only small conformational changes at the interface were detected, these changes were decisive to removing steric barriers for larger scale whole body movements of the ligand protein. Hence in iATTRACT, increased local flexibility lowers barriers for triggering simultaneous global large-scale motions, which can result in improved surface complementarity and a larger fraction of native contacts.

1.2 Fully Blind Modeling of Peptide-Protein Complexes with pepATTRACT

Peptide-protein interactions constitute a large fraction of the interactome but due to their abundance and the inherent flexibility, only a limited number of complexes have been characterized experimentally. The high level of flexibility and the small size of the interface have proven to be obstacles for peptide-protein docking and until recently, many methods carried out local docking, relying on prior information about the peptide-binding site. Our peptide-protein docking method pepATTRACT is one of the first fully blind flexible protocols. The protocol allows for global searches of the entire protein surface given the protein structure and the peptide sequence. It identifies the binding site and simultaneously predicts the bound peptide conformation for a large variety of complexes. Our protocol is intermediate between binding site prediction tools and high-resolution refinement methods and can be easily combined with either method for improved accuracy. We believe that this fast method could be useful for large-scale studies and the design of peptide-based inhibitors. Especially, interactions of globular proteins with disordered segments could also be modeled with this approach.

1.2.1 Protocol Description

pepATTRACT [43] combines different flexibility mechanisms in the ATTRACT engine in a very efficient manner. The protocol uses a coarse-grained force field, ensemble docking, flexible interface refinement, and a final molecular dynamics refinement to model protein and peptide flexibility. A schematic overview of the protocol is shown in Fig. 2. First, three idealized peptide conformations (α-helical, extended, and polyproline-II) are generated from sequence and used as an ensemble during an initial global search of the protein surface. This ensemble is composed of the

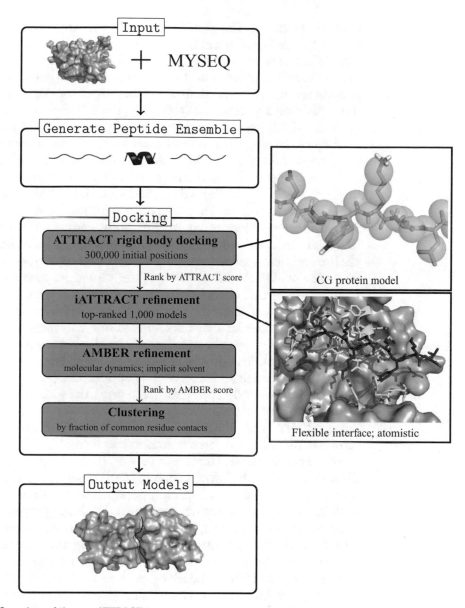

Fig. 2 Overview of the pepATTRACT protocol

three most common conformations found in peptide-protein complexes [58]. For each of the conformations, 100,000 different starting positions are explored in a potential energy minimization (ensemble docking). During this stage, we employ the coarse-grained ATTRACT model to represent the protein and the peptide. We also precalculate the potential energy around the protein on a grid to speed up the calculations. The internal structure of the protein and the peptide is kept rigid at this stage, but the coarse-grained representation implicitly takes side-chain flexibility into account and partly compensates for inaccuracies in the initial peptide structures. The rigid-body docking models are ranked by their ATTRACT score and the top-ranked models are selected for flexible refinement (typically 1000 models). During iATTRACT flexible interface refinement, the placement of the protein and the peptide is optimized and clashes resulting from the coarse-grained representation are relaxed. At this stage, the interface region of the protein and the peptide are treated as fully flexible, while simultaneously optimizing the center of mass position and orientation of the peptide. Finally, we run molecular dynamics refinement on each docking solution with Amber 14 [59]. A Generalized-Born implicit solvent model (igb = 8) is used with the ff14SB version of the AMBER force field. First, we perform a short minimization *in vacuo* with the sander program (500 steps) using a short cutoff (9 Å) and a small step size (dx = 0.0000001) to relax possible clashes and deformations resulting from the structure-based force field used in iATTRACT. Then two short molecular dynamics simulations are carried out with the pmemd.cuda program for 1000 and 2500 steps at temperatures $T = 400$ and 350 K, respectively. During the molecular dynamics simulations, Cα intra-molecular distances for the protein and intermolecular distances between protein and peptide backbone atoms are restrained to prevent large deformations and dissociation of the peptide. The intra-molecular distances are controlled by a harmonic potential with equilibrium value set to the initial distance in the structure with force constant 2 kcal/mol/Å^2 and for deviations of larger than 3.5 Å with a linear response function and force constant 2 kcal/mol/Å^2. The intermolecular distances are allowed to vary by 10 Å with respect to the distance found in the initial structure. Deviations between 10 and 13.5 Å are restrained by a harmonic potential with a force constant 0.25 kcal/mol/Å^2 and further deviations by a linear potential with a force constant 0.25 kcal/mol/Å. Then the structures are minimized for 5000 steps with a large cutoff using the pmemd.cuda program without restraints. Finally, the energy is evaluated for the complex and the individual docking partners by the sander program ("one-step minimization"). The binding interaction energy score is calculated by subtracting the energy of the free monomers from the energy of the complex $\Delta E = E_{complex} - (E_{protein} + E_{peptide})$. We cluster the refined models by fraction of common contacts [60] and rank the clusters by the average AMBER interaction energy score of their top-ranked four members.

1.2.2 Benchmarking pepATTRACT's Performance

We tested the pepATTRACT approach on 80 previously characterized peptide-protein complexes from the peptiDB benchmark [58]. Overall, pepATTRACT yielded near-native models of the complex for 70% of the docking cases in a fully blind prediction manner and the top-ranked 10 clusters contained a near-native cluster in 68% of the successful cases [43]. During the initial coarse-grained docking stage, the surface of the protein was scanned and in many cases the top-ranked solutions identified the binding site correctly. The coarse-grained rigid body docking stage achieved a success rate of 58 % among the top-ranked 1000 models. Subsequent iATTRACT refinement increased this total success rate to 68 % and also succeeded in refining near-native structures to sub-angstrom precision. Moreover, iATTRACT helped to resolve minor clashes resulting from transitioning between a coarse-grained to an atomistic force field. MD refinement with Amber optimized the structures further and decreased the interface-RMSD of the docking solutions on average by 0.44 Å. More importantly, it significantly improved the scoring of near-native solutions compared to the ranking based on the ATTRACT coarse-grained force field. We also developed a local docking approach, pepATTRACT-local by combining pepATTRACT with ambiguous interaction restraints as often used in the HADDOCK program [61, 62]. In this way, information about the binding site from either experiments or bioinformatic predictions can be incorporated in the docking protocol and enhance the accuracy of the prediction. pepATTRACT's performance in fully blind mode was comparable to two previously published local docking methods [35, 36] and clearly surpassed their performance in local mode [43].

The success of the pepATTRACT protocol is based on a versatile combination of different docking stages and flexibility mechanisms. The protocol allows a high level of detail and accuracy in the final stages but at the same time is computationally efficient enough to screen 300,000 initial positions in a matter of minutes in the initial search stage. The large sampling in the rigid body phase provides placements at the native binding site even of nonoptimal peptide structures that can then be relaxed to near-native models in the subsequent flexible refinement stages. We believe that identifying many good initial global placements of the peptide and refining these is possibly more efficient than trying to sample all degrees of freedoms of the peptides right from the start. Our results on the peptiDB benchmark showed that the coarse-grained ATTRACT protein model is also applicable to modeling peptide-protein interactions. This is probably due to common interface design principles [14, 63, 64] and hence the intrinsic similarity between protein-protein and peptide-protein complexes. Yet, it might be possible to further improve the performance by designing specific parameters for peptide-protein interactions employing a similar optimization procedure as used for the design of the original form [45, 56]. Interestingly, we found that using an ensemble of only three peptide conformations already yields high-quality predictions. This again underlines the fundamental interface design principles found in nature [14, 58]. Still, for predictions of sub-angstrom precision for backbone and side chains, in many cases, more extensive peptide modeling and refinement has to be performed (e.g., [35]).

To make the protocol available to the scientific community, we created a web interface for pepATTRACT (www.attract.ph.tum.de/peptide.html). The web interface generates a docking script that performs the rigid body sampling stage and the flexible interface refinement starting from the structure of the unbound protein and the peptide sequence. The peptide-protein docking can then run on the user's machine with a local installation of the ATTRACT program (ATTRACT is available for download at www.t38.ph.tum.de). Hence, the docking script generated by the web interface provides an easy entry point for non-expert users into fast peptide-protein docking with pepATTRACT. The web-interface also offers the possibility of including experimental information in the docking run and restricting the search for the peptide binding site to a portion of the protein's surface (pepATTRACT-local). This is achieved by specifying important residues on the protein as active residues for ambiguous interaction restraints [61, 62]. Furthermore, conformational change on the protein side can be included by providing multiple protein structures for an ensemble docking approach. Note that the molecular dynamics refinement stage is not included in the docking script. Scripts for the AMBER refinement can be requested from the authors. The whole protocol, including molecular dynamics refinement, typically runs overnight on a standard Desktop PC.

2 Materials

1. Atomic 3D structure of protein of interest in PDB file format (www.pdb.org).

2. Sequence of peptide of interest in one-letter code.

3. Optional: information on protein residues involved in binding (literature research).

4. PC with Unix-based OS (Linux/Mac) and at least 2–3 GB RAM.

5. ATTRACT software (available at www.t38.ph.tum.de).

6. Molecular viewer (PyMOL, VMD, Rasmol etc.).

3 Methods

For modeling the structure of a peptide-protein complex with the fully blind docking protocol pepATTRACT, the following steps have to be performed:

1. Installation of the ATTRACT software (www.t38.ph.tum.de) and a molecular viewer.

2. Providing the protein structure and the sequence of the peptide of interest.

3. Generating a docking script and input files with the pepAT-TRACT web interface (www.attract.ph.tum.de/peptide.html).

4. Running the docking script on the user's local machine.

5. Analyzing the docking results.

Subsequently, the individual steps are described in more detail. In an example application, we demonstrate how pepATTRACT can be used to model peptidic interactions in the field of cancer research.

3.1 Installation of the ATTRACT Software

There are two possibilities of using the ATTRACT software: installing ATTRACT directly from the source code or downloading an ATTRACT VirtualBox. The virtual box has ATTRACT and all its dependencies installed. Note that ATTRACT can only be compiled and installed directly on Unix-based OS (Linux/Mac). In contrast, the ATTRACT VirtualBox can be used on a large number of operating systems (including Windows, Linux, Macintosh, and Solaris).

3.1.1 ATTRACT VirtualBox

1. Download and install VirtualBox (Oracle, www.virtualbox.org).

2. Download the ATTRACT VM from www.attract.ph.tum.de/services/ATTRACT/ATTRACT.vdi.gz and unpack the file.

3. Open the VirtualBox program. Add the ATTRACT.vdi file to VirtualBox (click the "New" button, pick the option "Use an existing virtual hard disk file," and select the ATTRACT.vdi file, follow the instructions). Start the virtual machine to use ATTRACT.

3.1.2 Building and Installing ATTRACT from Source Code

Example commands are given for Ubuntu OS (Canonical Ltd., www.ubuntu.com).

1. Download the ATTRACT source code from www.attract.ph.tum.de/services/ATTRACT/attract.tgz.

2. Open a terminal and unpack the source code (tar xzf attract.tgz).

3. Install g++ and gfortran (sudo apt-get install g++ gfortran).

4. Install numpy and scipy (sudo apt-get install python-numpy python-scipy).

5. Install pdb2pqr (sudo apt-get install pdb2pqr).

6. Go into attract/bin, type "make clean" and then "make all" or "make all -j 4" (if you have four cores on the local computer).

7. Edit your ".bashrc" file (typically found in the home directory) and add the following lines:

   • export ATTRACTDIR=/home/yourname/attract/bin *(i.e., wherever you installed ATTRACT)*.

   • export ATTRACTTOOLS = $ATTRACTDIR/../tools.

- export PYTHONPATH = $PYTHONPATH:/usr/share/pdb2pqr.

8. Type "source ~/.bashrc."

3.1.3 Installing a Molecular Viewer

The molecular viewer PyMOL (Schroedinger LLC, www.pymol.org) can be installed on Ubuntu by entering "sudo apt-get install pymol" in the terminal. Open a PDB file with it by typing "pymol myprotein.pdb."

3.2 Input Data

The user needs an atomic 3D structure of the protein of interest in PDB file format. Such a file can be obtained, e.g., from the Protein Data Bank (www.pdb.org). Alternatively, good homology models with sufficient sequence similarity can be used (e.g., from I-TASSER [65] or MODELLER [66]). At the moment, unnatural and modified amino acids and cofactors are not supported (*see* **Note 1**).

3.3 Obtaining a pepATTRACT Docking Script and Running It

1. Go to www.attract.ph.tum.de/peptide.html (Fig. 3).
2. Upload the PDB file of the protein.
3. Enter the sequence of the peptide in one-letter code (standard amino acids only).
4. Specify the number of cores to be used for the calculation (default = 1, Section "Computation").
5. Specify optional parameters (*see* **Notes 2–6**).
6. Hit the "Get configuration" button.
7. Download the archive mydocking.tgz and unpack it.
8. Run the protocol by double-click (typical run time 1–4 h).

For changing the standard settings for building ATTRACT, *see* **Note 7, 8**.

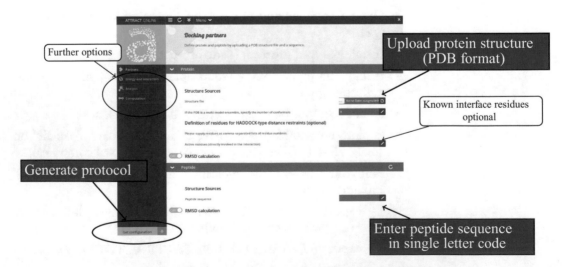

Fig. 3 Instructions to the pepATTRACT web interface (www.attract.ph.tum.de/peptide.html)

3.4 Analysis of Docking Results

If the docking script runs without problems, the final docking models can be found in the file result.pdb. This file can be opened with a molecular viewer to visualize and analyze predicted contacts between peptide and protein.

3.5 Application to Tumor Suppressor p53

We demonstrate the usage of pepATTRACT by docking the disordered regions of tumor suppressor p53 to three of its interaction partners. p53 plays a crucial role in preventing cancer formation. It is engaged in DNA repair, apoptosis, and cell cycle arrest and is subject to tight regulation. In the absence of cellular stress, MDM2 binds to an N-terminal disordered region of p53 to inhibit its function and mark it for proteosomal degradation. The crystal structure of this interaction was characterized experimentally (PDB ID: 1YCR) [67]. A structure of apo MDM2 is also available in the Protein Data Bank (PDB ID: 4HBM) [68]. We also predicted two other complexes that involve the C-terminal disordered region of p53: the interactions with the Set9 methyl-tranferase and the Sir2 deacetylase. Both complexes have been resolved by X-ray crystallography (PDB ID: 1XQH, 1MA3) [69, 70]. For Set9, we also identified an unbound structure (PDB ID: 1NC6) [71], for Sir2, we repacked the side chains of the protein with the program Scwrl [72] and used this structure during docking. In all cases, we used the pepATTRACT web-interface to generate the docking scripts as described above and refined the top 1000 models in a short molecular dynamics simulation with AMBER [43]. Near-native docking results in comparison with the experimental crystal structures are shown in Fig. 4. The binding site was identified correctly by a large number of solutions for all cases and near-native structures were among the top-ranked ten clusters for MDM2 and Sir2 and among the top ranked 50 clusters for Set9 (the clusters were ranked by interaction energy; an alternative to looking at the clusters ranked by energy would be to examine the largest clusters). The interface backbone atoms of these near-native docking models only deviated by ≈2 Å from the native complex. The docking models also correctly predicted important contacts between the protein and the peptide (e.g., the crucial role for tryptophan in the interaction with MDM2, Fig. 4) and the structural class of the bound peptide (extended or helical). Even though not all residues of the peptide were inferred at high precision, the key interactions (e.g., a lysine residue in docking to Sir2, Fig. 4) were reproduced by the model. The docking models could be used to extract the predicted interface residues on the protein and to validate/investigate these in mutational studies or truncation experiments. Once a model has been validated experimentally, the predicted structure can be used for further protein/peptide/peptidomimetic design studies.

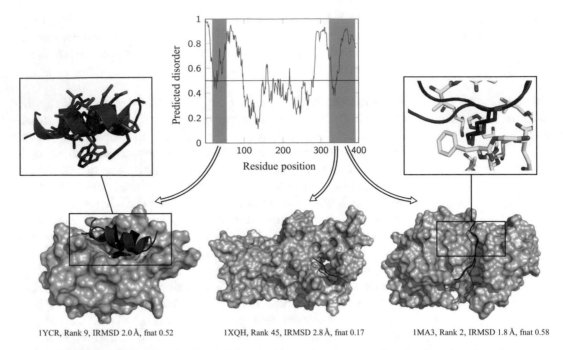

Fig. 4 pepATTRACT docking results for modeling the interactions between disordered regions from tumor suppressor p53 and MDM2, Set9 or Sir2. The protein is shown in surface representation, the docked peptide in *red*. For comparison, the crystal structure is shown in *black*. For each case, the PDB ID of the bound complex, the cluster rank, the interface-RMSD (IRMSD), and the fraction of native contacts (fnat) of the docking model are listed. The disorder in the p53 sequence was predicted with the IUPred web server [73, 74]

4   Notes

1. *Structure preparation.*

    ATTRACT does not support unnatural and modified amino acids and cofactors to date. All HETATM entries in the PDB file will be ignored. The user has to change modified amino acids like phospho-threonine manually to threonine (and HETATM to ATOM); otherwise, these atoms will not be considered during docking. In the future, we plan to offer support for modified amino acid and protein-bound ions and cofactors into the docking process. If multiple conformations are resolved for certain residues, the alternative conformations have to be deleted and the residue names should be changed accordingly (e.g., ASER to SER). Furthermore, if the asymmetric unit contains more than one copy of the protein of interest, these should be deleted as well (unless it is the biological assembly). If the PDB file contains multiple models of the protein, the number of models has to be specified in the web interface ("number of conformers," *see* also **Note 3**).

2. *Using experimental information.*

The web interface allows the user to restrict the docking search to a proportion of the protein surface/known peptide binding site (pepATTRACT-local). If certain protein residues are known to be important in peptide binding, e.g., from mutational experiments, they can be specified as active residues ("Partners" section). This will ensure that only solutions in which these residues are in contact with the peptide are generated.

3. *Protein flexibility.*

Many proteins undergo conformational changes upon binding to a partner molecule. Although the effect is smaller in peptide-protein interactions than in protein-protein complexes, it affects the prediction quality of our semirigid peptide docking approach. The pepATTRACT protocol allows accounting for conformational change on the protein side by providing an ensemble of possible conformations (ensemble docking). This option is also suitable, if the protein structure has been homology modeled or derived from NMR experiments. Snapshots from MD simulations could also be used as an ensemble in docking. Instead of uploading a PDB file containing a single protein structure, the user can upload a multi-model PDB file to the web interface and specify the number of conformers (number of models in the PDB file). It is necessary to superimpose all the models in the file; otherwise, the docking will give incorrect results. Unfortunately, predicting when conformational change happens is very difficult. Protein flexibility can be tested to a certain degree by performing MD simulations or normal mode calculations prior to docking.

4. *Enhanced peptide modeling.*

During the initial search, pepATTRACT uses just three types of peptide conformations to identify an initial binding position and binding orientation. Further conformational changes and adaptation to the binding pocket are possible during the refinement steps. However, it is likely that the performance can be enhanced by using more types of basic peptide conformations in the initial rigid body docking. Such an extension is unlikely to degrade the speed of the approach, since it only affects the rapid initial search. First, the user needs to set up a docking run with the pepATTRACT web interface as described before and to download the embedded parameter file from the web interface instead of the docking archive (e.g., attract pep-embedded.web). Then this parameter file should be uploaded at www.attract.ph.tum.de/services/ATTRACT/upload.html, choose Standard interface for "Data model." Now modify the number of conformers for the "Ligand" protein (i.e., the peptide) from three to the number of conformations used and upload the multi-model PDB file with the peptide conformations as the structure for the "Ligand" protein. Generate the docking script ("Get configuration"), download the docking archive from the web interface, unpack it, and execute the docking script by double-click. Peptide conformations can be obtained from fragment libraries [75], by ab-initio structure prediction [76] or as snapshots from molecular dynamics simulations. The user needs to make sure that all peptide conformation used contain the same number of residues and that all models are aligned to the first model.

5. *Refinement.*

The iATTRACT refinement can be disabled in the "Energy and Interaction" section of the web-interface (enabled by default). As a default in the web interface, 50 structures are selected for refinement. To increase this number, the user has to change the field "Number of structures to collect as PDB file" in the "Analysis" section. We recommend using 1000 structures for iATTRACT and if possible for a final molecular dynamics based refinement. It is likely that the final MD refinement step can be considerably improved with respect to the procedure described above. Both more extensive MD simulations and conformational optimization in implicit solvent but also possible extension to explicit solvent (e.g., limited to the immediate vicinity of the docking placement) are routes that could be explored. With the availability of graphical processing units (GPUs), it is possible to perform such simulations on hundreds or thousands of putative docking geometries.

6. *Analysis and benchmarking.*

The final models are converted to PDB file format for visual inspection by the user (type pymol results.pdb to look at the structures). The structures could then be used as starting structures in molecular dynamics simulations (like described above and in [43]). The web interface offers the opportunity to benchmark pepATTRACT's performance by docking known peptide-protein complexes and evaluating whether the experimental structure can be reproduced. For this, reference PDB files containing the protein structure and the peptide structure can be provided (RMSD calculation in "Partners" section of the web interface). Standard CAPRI evaluation criteria like ligand-RMSD, interface-RMSD, and fraction of native contacts can be selected for automatic computation in the "Analysis" section (the results can be found in the files results.lrmsd, results.irmsd, results.fnat). We highly recommend testing the pepATTRACT protocol for specific biological systems whenever possible using similar experimentally resolved complex structures as benchmark cases.

7. *Memory requirements.*

During docking, a precalculated potential energy grid is loaded into shared memory. This requires at least 2 GB of

RAM, for larger proteins, the demand might be higher. Failures of the protocol are often a result of insufficient memory.

8. *Large protein structures.*

The ATTRACT software is compiled with default settings limiting the number of atoms in a protein to 10,000. This limit can be expanded by changing the file $ATTRACTDIR/max.h and increasing the variable MAXATOM (maximum number of atoms in protein) and if necessary MAXRES (maximum number of residues), TOTMAXATOM, TOTMAXRES etc. to the desired number. Then the user has to recompile by going to $ATTRACTDIR and typing make clean and make all. We recommend checking the OPLS converted file myprotein-aa.pdb to find out how many atoms the protein has during docking with ATTRACT. Users should keep in mind that docking larger proteins requires more memory and longer run times. For large proteins, it might also be necessary to increase the number of starting conformations (default 100,000 per peptide conformation) by modifying the line with randsearch.py in the docking script.

Acknowledgments

The authors are grateful for financial support from the Center for Integrated Protein Science Munich (CIPSM).

References

1. Rual JF et al (2005) Towards a proteome-scale map of the human protein–protein interaction network. Nature 437:1173–1178

2. Giot L et al (2003) A protein interaction map of Drosophila melanogaster. Science 302:1727–1736

3. Uetz P et al (2000) A comprehensive analysis of protein–protein interactions in Saccharomyces cerevisiae. Nature 403:623–627

4. Ito T et al (2001) A comprehensive two-hybrid analysis to explore the yeast protein interactome. Proc Natl Acad Sci U S A 98:4569–4574

5. Li S et al (2004) A map of the interactome network of the metazoan C. elegans. Science 303:540–543

6. Fuxreiter M, Tompa P, Simon I (2007) Local structural disorder imparts plasticity on linear motifs. Bioinformatics 23:950–956

7. Dyson HJ, Wright PE (2005) Intrinsically unstructured proteins and their functions. Nat Rev Mol Cell Biol 6:197–208

8. Dunker AK et al (2000) Intrinsic protein disorder in complete genomes. Genome Inf 11:161–171

9. Pawson T, Nash P (2003) Assembly of cell regulatory systems through protein interaction domains. Science 300:445–452

10. Petsalaki E, Russell R (2008) Peptide-mediated interactions in biological systems: new discoveries and applications. Curr Opin Biotechnol 19:344–350

11. Vagner J, Qu H, Hruby VJ (2008) Peptidomimetics, a synthetic tool of drug discovery. Curr Opin Chem Biol 12:292–296

12. Neduva V, Russell RB (2006) Peptides mediating interaction networks: new leads at last. Curr Opin Biotechnol 17:465–471

13. Vanhee P et al (2011) Computational design of peptide ligands. Trends Biotechnol 29:231–239

14. London N, Raveh B, Schueler-Furman O (2013) Druggable protein-protein interactions – from hot spots to hot segments. Curr Opin Chem Biol 17:952–959

15. Rubinstein M, Niv MY (2009) Peptidic modulators of protein-protein interactions: progress and challenges in computational design. Biopolymers 91:505–513

16. Dundas J et al (2006) CASTp: computed atlas of surface topography of proteins with structural and topographical mapping of functionally annotated residues. Nucleic Acids Res 34:W116–W118

17. Petsalaki E et al (2009) Accurate prediction of peptide binding sites on protein surfaces. PLoS Comput Biol 5, e1000335

18. Ben-Shimon A, Eisenstein M (2010) Computational mapping of anchoring spots on protein surfaces. J Mol Biol 402:259–277

19. Dagliyan O et al (2011) Structural and dynamic determinants of protein–peptide recognition. Structure 19:1837–1845

20. Trabuco LG et al (2012) PepSite: prediction of peptide-binding sites from protein surfaces. Nucleic Acids Res 40:W423–W427

21. Lavi A et al (2013) Detection of peptide-binding sites on protein surfaces: the first step toward the modeling and targeting of peptide-mediated interactions. Proteins Struct Funct Bioinf 81:2096–2105

22. Verschueren E et al (2013) Protein–peptide complex prediction through fragment interaction patterns. Structure 21:789–797

23. Saladin A et al (2014) PEP-SiteFinder: a tool for the blind identification of peptide binding sites on protein surfaces. Nucleic Acids Res 42:W221–W226

24. Yan C, Zou X (2015) Predicting peptide binding sites on protein surfaces by clustering chemical interactions. J Comp Chem 36:49–61

25. London N, Raveh B, Schueler-Furman O (2013) Peptide docking and structure-based characterization of peptide binding: from knowledge to know-how. Curr Opin Struct Biol 23:894–902

26. Tubert-Brohman I et al (2013) Improved docking of polypeptides with Glide. J Chem Inf Model 53:1689–1699

27. Rosenfeld R et al (1995) Flexible docking of peptides to class I major histocompatibility-complex receptors. Genet Anal Biomol Eng 12:1–21

28. Bordner AJ, Abagyan R (2006) Ab initio prediction of peptide-MHC binding geometry for diverse class I MHC allotypes. Proteins Struct Funct Bioinf 63:512–526

29. Yanover C, Bradley P (2011) Large-scale characterization of peptide-MHC binding landscapes with structural simulations. Proc Natl Acad Sci U S A 108:6981–6986

30. Antes I, Siu SWI, Lengauer T (2006) DynaPred: a structure and sequence based method for the prediction of MHC class I binding peptide sequences and conformations. Bioinformatics 22:e16–e24

31. Niv MY, Weinstein H (2005) A flexible docking procedure for the exploration of peptide binding selectivity to known structures and homology models of PDZ domains. J Am Chem Soc 127:14072–14079

32. Staneva I, Wallin S (2009) All-atom Monte Carlo approach to protein–peptide binding. J Mol Biol 393:1118–1128

33. London N et al (2011) Identification of a novel class of farnesylation targets by structure-based modeling of binding specificity. PLoS Comput Biol 7, e1002170

34. Raveh B, London N, Schueler-Furman O (2010) Sub-angstrom modeling of complexes between flexible peptides and globular proteins. Proteins Struct Funct Bioinf 78:2029–2040

35. Raveh B et al (2011) Rosetta FlexPepDock ab-initio: simultaneous folding, docking and refinement of peptides onto their receptors. PLoS One 6, e18934

36. Trellet M, Melquiond AS, Bonvin AM (2013) A unified conformational selection and induced fit approach to protein–peptide docking. PLoS One 8, e58769

37. Antes I (2010) DynaDock: a new molecular dynamics-based algorithm for protein-peptide docking including receptor flexibility. Proteins Struct Funct Bioinf 78:1084–1104

38. Donsky E, Wolfson HJ (2011) PepCrawler: a fast RRT-based algorithm for high-resolution refinement and binding affinity estimation of peptide inhibitors. Bioinformatics 27:2836–2842

39. Luitz MP, Zacharias M (2014) Protein-ligand docking using Hamiltonian replica exchange simulations with soft core potentials. J Chem Inf Model 54:1669–1675

40. Kurcinski M et al (2015) CABS-dock web server for the flexible docking of peptides to proteins without prior knowledge of the binding site. Nucleic Acids Res 43:W419–W424

41. Ben-Shimon A, Niv MY (2015) AnchorDock: blind and flexible anchor-driven peptide docking. Structure 23:929–940

42. Lee H et al (2015) GalaxyPepDock: a protein–peptide docking tool based on interaction similarity and energy optimization. Nucleic Acids Res 43:W431–W435

43. Schindler CEM, de Vries SJ, Zacharias M (2015) Fully blind peptide-protein docking with pepATTRACT. Structure 23:1507–1515

44. Rentzsch R, Renard BY (2015) Docking small peptides remains a great challenge: an assessment using AutoDock Vina. Brief Bioinf 16(6):1045–1056

45. Zacharias M (2003) Protein–protein docking with a reduced protein model accounting for side-chain flexibility. Protein Sci 12:1271–1282

46. May A, Zacharias M (2005) Accounting for global protein deformability during protein–protein and protein–ligand docking. Biochim Biophys Acta Proteins Proteomics 1754:225–231

47. de Vries SJ et al (2015) A web interface for easy flexible protein–protein docking with ATTRACT. Biophys J 108:462–465

48. Setny P, Bahadur R, Zacharias M (2012) Protein-DNA docking with a coarse-grained force field. BMC Bioinf 13:228

49. Setny P, Zacharias M (2011) A coarse-grained force field for protein-RNA docking. Nucleic Acids Res 39:9118–9129

50. May A, Zacharias M (2008) Protein-ligand docking accounting for receptor side chain and global flexibility in normal modes: evaluation on kinase inhibitor cross docking. J Med Chem 51:3499–3506

51. May A, Zacharias M (2007) Protein–protein docking in CAPRI using ATTRACT to account for global and local flexibility. Proteins Struct Funct Bioinf 69:774–780

52. de Vries S, Zacharias M (2013) Flexible docking and refinement with a coarse-grained protein model using ATTRACT. Proteins Struct Funct Bioinf 81:2167–2174

53. Lensink MF, Wodak SJ (2013) Docking, scoring, and affinity prediction in CAPRI. Proteins Struct Funct Bioinf 81:2082–2095

54. de Vries SJ, Zacharias M (2012) Attract-EM: a new method for the computational assembly of large molecular machines using cryo-EM maps. PLoS One 7, e49733

55. Schindler CEM, de Vries SJ, Zacharias M (2015) iATTRACT: simultaneous global and local interface optimization for protein-protein docking refinement. Proteins Struct Funct Bioinf 83:248–258

56. Fiorucci S, Zacharias M (2010) Binding site prediction and improved scoring during flexible protein–protein docking with ATTRACT. Proteins Struct Funct Bioinf 78:3131–3139

57. May A, Zacharias M (2008) Energy minimization in low-frequency normal modes to efficiently allow for global flexibility during systematic protein–protein docking. Proteins Struct Funct Bioinf 70:794–809

58. London N, Movshovitz-Attias D, Schueler-Furman O (2010) The structural basis of peptide-protein binding strategies. Structure 18:188–199

59. Case D et al (2014) AMBER 14. University of California, San Francisco, CA

60. Rodrigues JPGLM et al (2012) Clustering biomolecular complexes by residue contacts similarity. Proteins Struct Funct Bioinf 80:1810–1817

61. Nilges M (1993) A calculation strategy for the structure determination of symmetric demers by 1H NMR. Proteins Struct Funct Bioinf 17:297–309

62. Dominguez C, Boelens R, Bonvin AMJJ (2003) HADDOCK: a protein-protein docking approach based on biochemical or biophysical information. J Am Chem Soc 125:1731–1737

63. Vanhee P et al (2009) Protein-peptide interactions adopt the same structural motifs as monomeric protein folds. Structure 17:1128–1136

64. Watkins AM, Wuo MG, Arora PS (2015) Protein-protein interactions mediated by helical tertiary structure motifs. J Am Chem Soc 137:11622–11630

65. Roy A, Kucukural A, Zhang Y (2010) I-TASSER: a unified platform for automated protein structure and function prediction. Nat Protoc 5:725–738

66. Eswar N et al (2007) Comparative protein structure modeling using Modeller. Curr Protoc Protein Sci Chapter 2:Unit 2.9

67. Kussie PH et al (1996) Structure of the MDM2 oncoprotein bound to the p53 tumor suppressor transactivation domain. Science 274:948–953

68. Michelsen K et al (2012) Ordering of the N-terminus of human MDM2 by small molecule inhibitors. J Am Chem Soc 134:17059–17067

69. Chuikov S et al (2004) Regulation of p53 activity through lysine methylation. Nature 432:353–360

70. Avalos JL et al (2002) Structure of a Sir2 enzyme bound to an acetylated p53 peptide. Mol Cell 10:523–535

71. Kwon T et al (2003) Mechanism of histone lysine methyl transfer revealed by the structure of SET7/9AdoMet. EMBO J 22:292–303

72. Krivov GG, Shapovalov MV, Dunbrack RL (2009) Improved prediction of protein side-chain conformations with SCWRL4. Proteins Struct Funct Bioinf 77:778–795

73. Dosztanyi Z et al (2005) The pairwise energy content estimated from amino acid composition discriminates between folded and intrinsically unstructured proteins. J Mol Biol 347:827–839

74. Dosztanyi Z et al (2005) IUPred: web server for the prediction of intrinsically unstructured regions of proteins based on estimated energy content. Bioinformatics 21:3433–3434

75. Vanhee P et al (2011) BriX: a database of protein building blocks for structural analysis, modeling and design. Nucleic Acids Res 39:D435–D442

76. Thevenet P et al (2012) PEP-FOLD: an updated de novo structure prediction server for both linear and disulfide bonded cyclic peptides. Nucleic Acids Res 40:W288–W293

Chapter 6

Highly Flexible Protein-Peptide Docking Using CABS-Dock

Maciej Paweł Ciemny*, Mateusz Kurcinski*, Konrad Jakub Kozak, Andrzej Kolinski, and Sebastian Kmiecik

Abstract

Protein-peptide molecular docking is a difficult modeling problem. It is even more challenging when significant conformational changes that may occur during the binding process need to be predicted. In this chapter, we demonstrate the capabilities and features of the CABS-dock server for flexible protein-peptide docking. CABS-dock allows highly efficient modeling of full peptide flexibility and significant flexibility of a protein receptor. During CABS-dock docking, the peptide folding and binding process is explicitly simulated and no information about the peptide binding site or its structure is used. This chapter presents a successful CABS-dock use for docking a potentially therapeutic peptide to a protein target. Moreover, simulation contact maps, a new CABS-dock feature, are described and applied to the docking test case. Finally, a tutorial for running CABS-dock from the command line or command line scripts is provided. The CABS-dock web server is available from http://biocomp.chem.uw.edu.pl/CABSdock/.

Key words Protein-peptide interactions, Molecular docking, CABS, Peptide binding, Peptide design, Computational modeling

1 Introduction

Protein-peptide interactions play a predominant role in cell function and they can be found in a variety of signaling pathways involved in cellular localization, immune response or protein expression, and degradation. Because of their association with cellular regulatory mechanisms, erroneous protein-peptide interactions are speculated to be pathogenic in a number of diseases (e.g., cancer, autoimmune diseases). The possible applications in biomedical research (targeted drug design) make the understanding of protein-peptide interactions a critical issue for further advances in the field [1, 2]. Characterization of protein-peptide interactions is difficult due to their large complexity and transient and dynamic

---

*Authors contributed equally with all other contributors.

Ora Schueler-Furman and Nir London (eds.), *Modeling Peptide-Protein Interactions: Methods and Protocols*, Methods in Molecular Biology, vol. 1561, DOI 10.1007/978-1-4939-6798-8_6, © Springer Science+Business Media LLC 2017

nature. Despite extensive computational and experimental studies in this area, peptide-mediated cellular regulation mechanisms have not been fully described or understood.

Among computational approaches, molecular docking is commonly used to predict the structure of protein-peptide complexes. Handling large conformational changes during docking is one of the most challenging and important issues in the field [3, 4]. Modeling of protein-peptide interactions usually follows two steps realized by separate protocols: (1) prediction of binding site location on the protein surface [5–8], and (2) local protein-peptide docking (i.e., modeling of the peptide backbone in the binding site) [9–14]. The CABS-dock method [15, 16] unifies these two steps into one efficient docking simulation. In the CABS-dock single simulation run, a fully flexible peptide explores the entire surface of a flexible protein receptor in search for a binding site (no information about the binding site is used). Such high modeling efficiency is achieved thanks to the simulation engine based on the CABS prediction platform [17–19]. Alongside with the Rosetta platform [20], CABS currently offers perhaps the most efficient means for modeling significant conformational changes, successfully tested in protein-peptide on-the-fly docking [21].

This chapter provides a tutorial for the CABS-dock server and for its possible applications. Subheading 2 gives a short description of the CABS-dock methodology, together with the information and instructions required to successfully perform a CABS-dock run. It is followed by Subheading 3 which serves as a step-by-step guide with example docking results and analysis. A description of simulation contact maps, the new CABS-dock feature, is also provided along with the examples of use. Subsequently, possible schemes for incorporating CABS-dock in the multi-stage modeling of protein-peptide interaction are given. Finally, a tutorial how to use the CABS-dock server from the command line or command line scripts is provided. Additional comments on the procedure or method itself are provided in Subheading 4.

2 Materials

2.1 CABS-Dock Server Methodology

The CABS-dock web server (freely available at http://biocomp. chem.uw.edu.pl/CABSdock/) provides an interface for the CABS-dock method for protein-peptide docking together with up-to-date documentation and benchmark examples [15]. Several illustrative examples of CABS-dock applications have also been described in [16, 21]. Here only the basic CABS-dock features are outlined. The CABS-dock server protocol is based on the CABS model [17–19, 22] for coarse-grained simulations of protein dynamics and protein structure prediction. The model employs a reduced representation of the protein chain (*see* Fig. 1). The protein is represented with a set of pseudo-atoms: each residue is described by beads corresponding to the alpha carbon (**CA**), beta carbon (**B**), and side chain (**S**) (*see* Fig. 1).

Fig. 1 Representation of protein and peptide chains in CABS-dock. CABS-dock uses all-atom and coarse-grained modeling tools merged with procedures enabling transition between both resolutions. The figure shows comparison between all-atom (*left*) and CABS coarse-grained representation (*right*) for an example 4-residue peptide. In the CABS model, a single residue is represented by two atoms (alpha and beta carbon, colored in *black*) and two pseudo-atoms (side chain, colored in *orange*, and center of the peptide bond, colored in *green*)

To define the hydrogen bonds properly, an additional pseudo-atom representing the geometric center of the virtual CA-CA bond is also included. The knowledge-based force-field used for calculations was derived from statistical potentials based on known protein structures. Sampling of the conformation space is executed with a Replica Exchange Monte Carlo protocol. The CABS-dock docking procedure may be divided into four stages: (1) flexible docking based on the CABS model resulting in 10,000 models, (2) initial filtering resulting in 1000 models, (3) selection of 10 representative (top-ranked) models using structural clustering, (4) all-atom model reconstruction of ten top-ranked models combined with local optimization of their structure. All those sets of models can be downloaded from the server web site for their visualization or analysis.

2.2 Running the CABS-Dock Server

A fully functional, up-to-date version of the CABS dock method is available as an automated server accessible via standard internet browsers [15, 16]. No registration is required to use CABS-dock. To run the automated docking procedure on the CABS-dock server it is sufficient to provide:

1. A 3D model of the protein receptor in the PDB format (the protein model should be provided in the standard PDB format; if the protein receptor structure is stored in the PDB databank, it is sufficient to provide its code only); for additional protein input hints *see* **Note 1**.

2. A peptide sequence and, optionally, peptide secondary structure in the one letter code; for additional peptide input hints *see* **Note 2**.

The screenshots of the CABS-dock web server interface are shown in Fig. 2. Docking results may be further improved by providing additional information about the protein complex (assigning regions of increased flexibility or excluded from docking, *see* **Note 3**).

Fig. 2 Screenshots of the CABS-dock server. The figure shows the main page input panel (**a**) and example output panels (**b**–**d**). The buttons to be selected to *see* these panels are marked by *red rectangles* and *arrows*

3 Methods

3.1 A Case Study of Docking a Peptide Containing the LXXLL Motif to PPARγ

This case study presents an example of CABS-dock docking performed with default server settings. The docking uses the protein receptor structure: peroxisome proliferator-activated receptor gamma (PPARγ) (PDB code of the unbound receptor form: 2HWQ) and the sequence of the peptide that contains the LXXLL motif of a cofactor protein crucial for the biological action of

PPARγs [23]. Such a complex has been hypothesized to be responsible for the decoupling of insulin sensitization from adipogenesis in type-2 diabetes patients. The hypothesis was positively validated in vitro. A candidate for a partial PPARγ agonist was synthesized and crystallized by Burgermeister et al. [24] (PDB code of the complex: 2FVJ). The complex structure has been explored experimentally because of its potential for developing new therapies with fewer adverse effects on diabetes patients.

3.1.1 Input and Job Submission

The "submit new job" form was completed in the following manner to attempt docking the peptide to the protein receptor:

1. Protein tab: "2HWQ:A" (this instructs the server to access the "A" chain of the 2HWQ structure). For additional hints regarding the input of a protein receptor structure, *see* **Note 1**.

2. Peptide tab: HKLVQLLTTT (this is the one-letter code sequence of the peptide containing the LXXLL docking motif of the protein cofactor). For additional hints regarding the input of a peptide sequence, *see* **Note 2**.

3. Optional tab:

   - Project name: "2HWQ:A tutorial" (used to identify the project in the server queue).

   - Peptide secondary structure: "CHHHHHHHCC"; this is the experimentally derived preferred secondary structure of the peptide. For additional hints *see* **Note 2**.

   - Additionally, an e-mail address may be provided. It will be used to notify the user on project status.

The run is started with the "Submit" button. The server will redirect the user to an auto-refreshing site with details on project status. Alternatively, it is possible to run the docking from the terminal command line using the following command (for further details on command-line job submission, *see* Subheading 3.4):

```
curl -H "Content-Type: application/json" -X POST -d '
{"receptor_pdb_code":"2HWQ:A",
"ligand_seq":"HKLVQLLTTT","ligand_ss":"CHHHHHHHCC",    "project_name":"2HWQ:A
  tutorial", "email":"mail@host.com"}'
http://biocomp.chem.uw.edu.pl/CABSdock/REST/add_job/
```

3.1.2 Analysis of Results

The results of docking may be either interactively viewed on the CABS-dock server or downloaded from the project site as a zipped folder with all the resulting files, *see* **Note 4**. The basic output provided by the CABS-dock server interface consists of ten top-ranked models (CABS-dock ranking is largely based on the outcome of structural clustering, for details *see* [15, 16]). The ten top-ranked models are also stored in the zipped folder (in the form of PDB files named "model_(number).pdb"). The structures of models resulting from the docking performed for this case study are presented in Fig. 3 together with the crystallographic structure of the protein (extracted from the 2FVJ PDB entry).

To analyze the quality of the resulting structures, calculation of RMSD values can be performed using for example VMD software [25]. A detailed tutorial for the VMD analysis of CABS-dock results is provided in the supplementary data in [16]. Our analysis below was performed using this tutorial to calculate RMSD values: first for the ten top-ranked models, and second for the 10,000 models obtained in the CABS-dock simulation.

The RMSD values for the ten top-ranked models to the crystal structure of the peptide (from the 2FVJ complex) are presented in Table 1. The lowest RMSD value of 1.29 Å was obtained for the model ranked as the sixth out of ten models (*see* Fig. 3). Obviously, in the best case scenario the model with the lowest RMSD is ranked first. However, this is rarely the case as ranking the models is a very

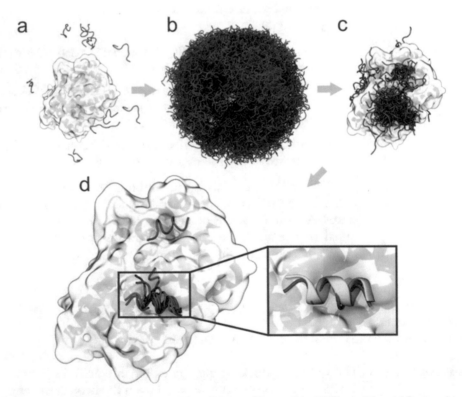

Fig. 3 Steps of the CABS-dock docking procedure illustrated by peptide-PPARγ docking. (**a**) Ten random peptide structures placed in random positions around the PPARγ structure. (**b**) Ten thousand peptide structures generated in the CABS-dock docking simulation. (**c**) Thousand models filtered from the previous set. (**d**) Ten top-ranked models (according to the structural clustering analysis) resulting from the docking. The close-up frame shows the best fitting model (RMSD value of 1.29 Å) out of the ten top-ranked models. The peptide models resulting from docking are shown in *orange*, the crystallographic peptide structure is shown in *yellow*, and the protein receptor is represented by its surface with elements of the secondary structure visible

Table 1
The RMSD values of ten top-ranked models to the crystallographic structure

Index of top-ranked models	RMSD value
1	9.705
2	3.444
3	3.801
4	9.987
5	9.242
6	**1.290**
7	3.610
8	1.778
9	31.353
10	26.036

The entry for the best fitting structure is marked in bold. The provided RMSD values are root mean square deviations calculated on the peptides after superposition of the receptor molecules

complex and yet unsolved problem (the scoring problem has been discussed in ref. 16).

As briefly described in Subheading 3, CABS-dock flexible docking produces a total of 10,000 models. For all these models RMSD values can also be easily calculated and plotted, for example against their CABS-energy values. Such analysis (showing for example whether the top-ranked models are also the lowest-RMSD models) is presented in Fig. 4. The lowest RMSD model from all the 10,000 models has the accuracy of 1.00 Å and belongs to the set of near-native low-energy models. As shown in Fig. 4, apart from the low-energy and low-RMSD set of structures, there is also another low-energy set with RMSD around 9 Å. These structures also have their representatives in the set of ten top-ranked models (i.e., models number 1, 4, and 5, *see* Table 1). The analysis of those cases proves that they fit into the appropriate binding site of the receptor. However, the peptide conformation differs from that of the crystallographic structure. With models 1 and 4 the C and N termini of the peptide are flipped, and in model 5 the peptide is bent and does not form a helix.

Please note that in this test case: (1) in several top-ranked models the actual binding site of the receptor protein was not found, and (2) the CABS-dock ranking procedure works relatively well (the lowest RMSD out of ten top scored models is only slightly higher than out of 10,000 models). Obviously, these two points

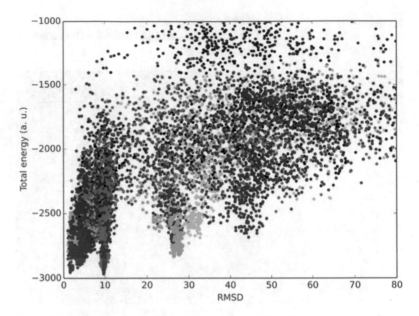

Fig. 4 CABS-energy and RMSD values for all (10,000) models obtained in peptide-PPARγ docking. The *colors* of the *dots* represent ten trajectories of a single docking simulation

may be not satisfied in other docking cases and the detailed statistics of CABS-dock performance on a large benchmark data set is presented in detail in [15, 16]. Finally, it is important to note that the CABS-dock procedure is a Monte Carlo-based algorithm, which may lead to different results in different runs.

3.2 Simulation Contact Maps

3.2.1 Maps: An Overview

The CABS-dock server provides an additional tool for the analysis of docking simulations in the form of contact maps. These maps depict the frequencies at which a pair of receptor/peptide residues interacts during simulation. Such information may be utilized to investigate the binding mechanism and three-dimensional structures of intermediates that occur on complex formation (as presented in our study of the folding and binding of a disordered peptide [26]). It can also provide clues about potential mutation sites to alter the binding affinity of the peptide.

An archive with CABS-dock simulation contact maps (maps. tar) can be downloaded as part of the ZIP file with the results (*see* **Note 4**). The contact maps are both given in the MAP file format (txt files, *see* **Note 5**) and PNG images. The file names correspond to maps presenting contact frequencies of the following sets of models:

1. cluster_(number)—models classified to a particular cluster in structural clustering. Cluster numbering corresponds to model numbering (i.e., model_6.pdb is a representative model of the models grouped into the sixth cluster. The clusters are ranked according to their CABS-score).

2. trajectory_(number)—trajectory models (each of the trajectories contains 1000 models). Each CABS-dock job contains ten trajectories.

3. top1000—top 1000 models selected after initial filtering.

4. trajectory_all—all models from the ten trajectories (10,000 models in total).

In the PNG images, contact frequencies are denoted by colors (example maps are presented in Subheading 3.2.2 below). Residue numbers and chain identifiers are marked on map borders. All the maps were derived from the distances of gravity centers of the side chains (in the CABS CG representation) and the contact cutoff was set to 4.5 Å.

3.2.2 Example Maps for Docking a Peptide Containing the LXXLL Motif to PPARγ

Because of their importance in further studies of the complex as well as potential significance in drug design, contact maps are one of the most informative results of the CABS-dock docking procedure. Most importantly, they may be used to predict residue-residue contacts that are crucial for the interaction, which for example can be subsequently used in peptide design.

According to experimental studies of the PPARγ-SRC-1 (a coactivator protein with the LXXLL motif) complex [27], the interaction site on the receptor protein is formed by the following residues: L468, L318, T297, Q314, L311, V315, K301, and E471. The docking LXXLL motif, which was experimentally determined to interact with PPARγs [23], is represented by residues 3–7 of the peptide used in the docking.

The contact maps for all the models (10,000 models in total, from the ten trajectories) and cluster number 6 (the representative of this cluster is the lowest RMSD model from the top scored models) are presented in Fig. 5. The maps show that the peptide residues comprising the motif, and the receptor residues creating native contacts in the crystallographic structure form the most persistent contacts during the CABS-dock docking simulation. The map prepared for all the simulation models (Fig. 5a) shows that most of the (final) contacts are in the expected contact area.

Another informative way to visualize the engagement of particular residues in protein-peptide interaction during docking simulation is to prepare a histogram of residue contacts. The histogram can be prepared by summing up contact frequencies from the maps (available in MAP txt files) over the peptide residues. Two histograms for PPARγ receptor residues for all the models and cluster number 6 models are presented in Fig. 6. The peaks found on both histograms correspond to residues crucial for the modeled interaction which form the interaction site of the receptor. The histogram for all the structures (Fig. 6a) contains "background" noise resulting from peptide sampling of the receptor surface in search for the best binding position. Some of those interactions are more persistent (e.g., residue 259) and may take part in

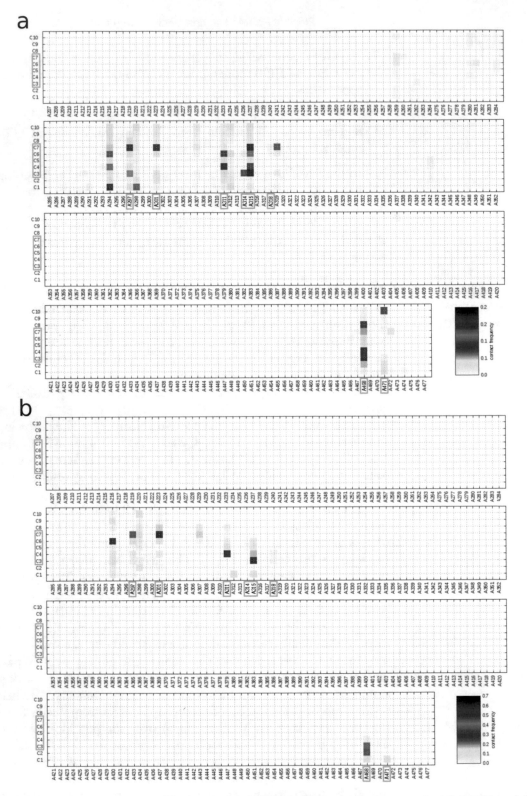

Fig. 5 Protein-peptide contact maps from peptide-PPARγ docking. (**a**) Contact map for all 10,000 models. (**b**) Contact map for the models from cluster number 6 whose representative was the model best fitting the experimental structure. The *columns* represent amino acids of the receptor, and the *rows* are amino acids of the peptide. The residues reported in the literature to form the interaction site of the complex are marked in *green*. Contact frequencies are marked according to the color maps below each of the maps. The maps were divided into four elements for clarity of presentation

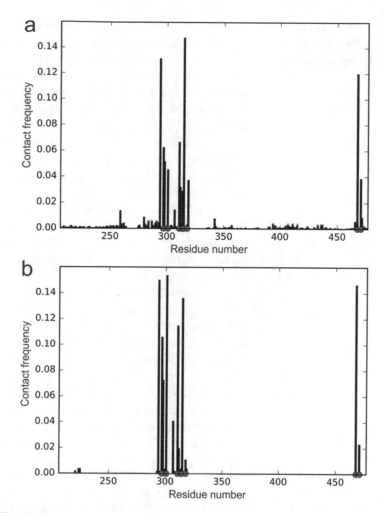

Fig. 6 Histograms of protein-peptide contacts from peptide-PPARγ docking. The normalized histograms show frequencies of contact for each of the receptor residues with the peptide: (**a**) for all 10,000 models, (**b**) for the models from cluster 6 whose representative was the best fitting model. The *green* markers represent residues that were reported in the literature to form a pocket for the LXXLL-peptide motif on the surface of PPARγ

intermediate complex formation while most of them are likely to be accidental. Although not all of the expected contacts were present in the resulting structures, it is clear that the most important interactions are well preserved and visible in the models. It is also possible that further all-atom refinement of the complex structure may lead to enhancement of the interaction site details that are not clear in the CG representation (*see* Subheading 3.3 below).

Finally, note that the contact map analysis of the folding and binding of a disordered peptide (simulated using CABS-dock methodology) has also been presented in [26].

3.3 CABS-Dock: Possible Applications and Future Advances

It is expected that computational techniques will play an important role in the rational design of peptide therapeutics [3]. Peptides make very promising candidates for drugs as they can adopt multiple shapes and various chemical features through careful design. Moreover, the design and synthesis of peptide drugs is relatively simple, so large libraries of peptides may be easily scanned to look for optimal peptide design.

The CABS-dock server may be used as an initial docking tool in a multistage docking procedure. Perhaps the most straightforward CABS-dock application is to use it as a tool for determining the initial structure(s) of a protein-peptide complex that may be used as an input for further refinement by local docking methods [9–14]. As shown before on a large protein-peptide benchmark dataset [15, 16], for the majority of cases CABS-dock produced models with high or medium accuracy (for example sufficient for structure refinement by Rosetta FlexPepDock [10, 12]). Another conclusion from the benchmark analysis was that CABS-dock accuracy can be significantly improved by its combination with exact scoring methods. By default, top-ranked models produced in the CABS-dock procedure are reconstructed to all-atom representation and refined using MODELLER [28] procedures and ranked by the DOPE score [29]. Since the reconstruction and the final all-atom refinement may significantly alter the quality of models, other techniques (better suited for the reconstruction and optimization of CABS-dock coarse-grained models) may be highly useful.

Future CABS-dock improvements also include its integration with methods for the prediction of peptide binding-sites [5–8] or extending the CABS-dock functionality to user-guided docking (by providing a possibility of pointing residues that belong to the binding site). Narrowing the conformation space to the selected neighborhood should result in the better sampling of near-native states, and thus in increasing the chances for building high accuracy models. Virtually any structural information may be utilized by CABS-dock as distance restraints or filters. Therefore, CABS-dock is well suited to be integrated as an efficient sampling tool with computational pipelines for modeling protein-peptide interactions, including methods for de novo peptide design [30, 31] or template-based docking [32].

Finally, CABS-dock could be used in hierarchical protein-protein docking protocols composed of three modeling steps:

1. Reduction of the protein-protein docking problem to protein-peptide docking. This starts from the arbitrary selection of the receptor protein and bound protein, followed by the identification of "hot segment(s)" of the bound protein, i.e., a short epitope that contributes the most to the protein-protein interaction [33, 34].

2. CABS-dock docking of "hot segment(s)" [33, 34], i.e., peptide(s).

3. Reconstruction and adjustment of the remaining receptor structure to the docked peptide-like fragment.

Peptide-like "hot segment(s)" can be of various length and can represent more than one fragment of the original structure, provided that they can be realistically replaced by a continuous peptide chain. In the context of the potential application of CABS-dock in protein-protein docking described above, one can also easily design a simple sequential procedure for the efficient modeling of amyloid aggregation.

3.4 Running CABS-Dock from the Command Line

Except for using the web interface (available at http://biocomp. chem.uw.edu.pl/CABSdock/), the CABS-dock server can also be operated from the command line or scripts using REST-full service. This option is recommended for handling multiple jobs by users experienced in Bash and python scripting.

3.4.1 Submitting a Job with the PDB Code of a Protein Receptor

To submit a job for a chosen protein, e.g., 2GB1, and a peptide sequence, e.g., SFDG, with default parameters, the following command or python script should be run:

- **command line:**

```
curl -H "Content-Type: application/json" -X POST -d
'{"receptor_pdb_code":"2GB1", "ligand_seq":"SFGD"}'
http://biocomp.chem.uw.edu.pl/CABSdock/REST/add_job/
```

- **python script:**

```
import requests
import json
url = ' http://biocomp.chem.uw.edu.pl/CABSdock/
REST/add_job/'
data = {
   "receptor_pdb_code": "2GB1",
   "ligand_seq": "SFGD",
}
response = requests.post(url, data=data)
```

The PDB file corresponding to "receptor_pdb_code" will be automatically downloaded from the PDB database. On success, a job identifier assigned to the submitted job "jid" will be returned. Jid will be used as a query for the job status and results later on. Otherwise, for example if the pdb code does not exist or input data do not fulfill requirements, error will be signaled.

3.4.2 Submitting a Job with a User-Provided PDB File

Instead of the PDB code, a PDB file can be attached to the query in the following ways:

- **command line:**

```
curl -X POST -F data='{"ligand_seq":"SFGD"}' -F
file=@path_to_pdb_file.pdb
http://biocomp.chem.uw.edu.pl/CABSdock/REST/add_job/
```

- **python script:**

```
import requests
import json
url = ' http://biocomp.chem.uw.edu.pl/CABSdock/
REST/add_job/'
files = {'file': open('path_to_pdb_file.pdb')}
data = {
    "ligand_seq": "SFGD",
}
response = requests.post(url, files=files, data=data)
```

3.4.3 Overriding Default Parameters

To override default parameters, additional options may be posted, i.e.,

- **command line:**

```
curl -H "Content-Type: application/json" -X POST -d
'{"receptor_pdb_code":"2IV9",
"ligand_seq":"SFGD","project_name":"my_project1",
"email":"mail@host.com",
"ligand_ss":"CCHHC",
"simulation_cycles":"100", "show_job":True,
"excluded_regions":[{"start":"100","end":"340","chain
":"A"}],
"flexible_regions":[{"start":"101","end":"202","chain":
"B","flexibility":"full"}]}'
http://biocomp.chem.uw.edu.pl/CABSdock/REST/add_job/
```

- **python script:**

```
import requests
import json
url = ' http://biocomp.chem.uw.edu.pl/CABSdock/
REST/add_job/'
files = {'file': open('your_PDB_file.pdb')}
#or use PDB code in var data
data = {
    "receptor_pdb_code": "2IV9",
    #or use PDB file in var files
    "ligand_seq": "SFGD",
    "email": "mail@host.com",
    "show": True,
    "project_name":"my_project1",
    "excluded_regions":[
        {
            "start": "1000",
            "end": "2000",
            "chain": "A"
        }
    ],
    "flexible_regions":[
        {
            "start": "101",
            "end": "202",
            "chain": "A",
            "flexibility": "full"
        },
```

```
            {
                "start": "300",
                "end": "370",
                "chain": "B",
                "flexibility": "moderate"
            },
        ]
    }
#request with file
response = requests.post(url, files=files, data=data)
#request without file
#response = requests.post(url, data=data)
```

Additional Parameters Include

1. project_name—name of the project used for job identification, i.e., in the queue.

2. email—email used to inform the user about job progress.

3. ligand_ss—ligand secondary structure.

4. simulation_cycles—number of simulation cycles: the default is 100 and the maximum is 200.

5. show_job—boolean value (True or False) indicating whether to show a job on the queue page.

6. excluded_regions—array of excluded regions. Each excluded region represents a selected receptor residue that is unlikely to interact with the peptide and should contain the following fields: start position, end position, and chain

7. flexible_regions—array of flexible regions. The flexibility of the region is changed by removing distance restraints that keep the receptor structure in a near native conformation. Each element of the array contains start position, end position, chain, and flexibility. Flexibility can be either full or moderate.

3.4.4 Getting Job Status

To check the status of a job, a job identifier ("jid") should be provided:

- **command line:**

```
curl -I
"http://biocomp.chem.uw.edu.pl/CABSdock/REST/
status/somejobidentifier"
```

- **python script:**

```
import requests
import json
url = 'http://biocomp.chem.uw.edu.pl/CABSdock/REST/
status/somejobidentifier'
response = requests.post(url)
```

As a result, one of the following statuses will be returned:

- done—job is finished and the results are ready.

- pending/running/pre_quere—job is in progress.

- error—the job identifier does not exist.

More detailed information about the job can be obtained by running:

- **command line:**

```
curl -I
"http://biocomp.chem.uw.edu.pl/CABSdock/REST/job_
info/somejobidentifier"
```

- **python script:**

```
import requests
import json
url = 'http://biocomp.chem.uw.edu.pl/CABSdock/REST/
job_info/somejobidentifier'
response = requests.post(url)
```

Additional information includes job configuration that was provided on submission and more details about the status. The following fields will be listed in the result:

1. del—job results will be kept on the server until this date.
2. excluded—list of excluded regions sent on job submission.
3. flexible—list of flexible regions sent on job submission.
4. ligand_sequence—ligand sequence sent on job submission.
5. ligand_ss—ligand secondary structure sent on job submission.
6. project_name—name assigned to the project on job submission.
7. receptor_sequence—receptor sequence sent on job submission.
8. ss_psipred—secondary structure predicted by psipred.
9. status—one of the possible job statuses as described in the section Getting job status.
10. status_change—time of last status change.

3.4.5 Getting Job Results: Essential Information

Essential information for each model includes:

1. Average RMSD.
2. Max RMSD.
3. Cluster density.
4. Number of elements.
5. Model data.
6. Information about submitted data.

See the next chapter for more information.

To obtain essential information, the job identifier ("jid") must be provided:

- **command line:**

```
curl -i
"http://biocomp.chem.uw.edu.pl/CABSdock/REST/get_
job/somejobidentifier"
```

  We strongly recommend that curl with compression should be sent:

```
curl -i -H 'Accept-Encoding: gzip,deflate'
"http://biocomp.chem.uw.edu.pl/CABSdock/REST/get_
job/somejobidentifier"
```

- **python script:**

```
import requests
import json
url = 'http://biocomp.chem.uw.edu.pl/CABSdock/REST/
get_job/somejobidentifier'
response = requests.post(url)
```

Optional parameters for filtering the results can be attached to the query. The parameters must specify the attribute used for filtering ("value") and the allowed range of values for the attribute ("min" and "max"). The following attributes can be used for filtering:

1. density—cluster density.
2. rmsd—average RMSD.
3. maxrmsd—maximum RMSD.
4. counts—number of elements in a cluster.

Exemplary use of filtering:

- **command line:**

```
curl -i -X POST -d '{"filter":"density","min":
"10","max":"20"}'
"http://biocomp.chem.uw.edu.pl/CABSdock/REST/
get_job/somejobidentifier"
```

- **python script:**

```
import requests
import json
url = 'http://biocomp.chem.uw.edu.pl/CABSdock/REST/
get_job/somejobidentifier'
data = {
    "value":"rmsd",
    "min":"5",
    "max":"12"
    }
response = requests.post(url, data=data)
```

All information for each model includes:

1. Average RMSD.
2. Max RMSD.
3. Cluster density.
4. Number of elements.
5. Model data.
6. Information about submitted data.

   and additionally:

1. Cluster data.
   To get cluster data or trajectory data, *see* the next sections.
   To obtain all information, the job identifier ("jid") must be provided:

- **command line:**

```
curl -i
"http://biocomp.chem.uw.edu.pl/CABSdock/
REST/get_job_all/somejobidentifier"
```

We strongly recommend that curl with compression should be sent:

```
curl -i -H 'Accept-Encoding: gzip,deflate'
"http://biocomp.chem.uw.edu.pl/CABSdock/REST/
get_job_all/somejobidentifier"
```

- **python script:**

```
import requests
import json
url = 'http://biocomp.chem.uw.edu.pl/CABSdock/REST/
get_job_all/somejobidentifier'
response = requests.post(url)
```

Additional filtering can be applied to the query as described in the previous section.

To get information about a chosen cluster, the job identifier together with the cluster number corresponding to the model number should be submitted:

- **command line:**

```
curl -i
"http://biocomp.chem.uw.edu.pl/CABSdock/REST/get_
cluster/somejobidentifier/clusterNumber"
```

We strongly recommend that curl with compression should be sent:

```
curl -i -H 'Accept-Encoding: gzip,deflate'
"http://biocomp.chem.uw.edu.pl/CABSdock/REST/get_
cluster/somejobidentifier/clusterNumber"
```

- **python script:**

```
import requests
import json
url = 'http://biocomp.chem.uw.edu.pl/CABSdock/REST/
get_cluster/somejobidentifier/clusterNumber'
response = requests.post(url)
```

The cluster number must be in the range [1, 10]. As a result, cluster data and additional information about the cluster (average and maximum RMSD, cluster density, and number of elements) will be returned.

3.4.0 Getting Trajcotory Information

Trajectory data can be obtained by sending a query with the attached job identifier and model number in the range [1, 10]:

- **command line:**

```
curl -i
"http://biocomp.chem.uw.edu.pl/CABSdock/REST/get_
trajectory/somejobidentifier/modelNumber"
```

We strongly recommend that curl with compression should be sent:

```
curl -i -H 'Accept-Encoding: gzip,deflate'
"http://biocomp.chem.uw.edu.pl/CABSdock/REST/get_
trajectory/somejobidentifier/modelNumber"
```

- **python script:**

```
import requests
import json
url = 'http://biocomp.chem.uw.edu.pl/CABSdock/REST/get_
trajectory/somejobidentifier/modelNumber'
response = requests.post(url)
```

Additionally, a section of the trajectory model can be selected by:

- **command line:**

```
curl -i
"http://biocomp.chem.uw.edu.pl/CABSdock/REST/get_
selected_trajectory/somejobidentifier/modelNumber/
start/end"
```

- **python script:**

```
import requests
import json
url = 'http://biocomp.chem.uw.edu.pl/CABSdock/REST/
get_selected_trajectory/somejobidentifier/modelNum-
ber/start/end'
response = requests.post(url)
```

3.4.9 Examples

Example 1 (default settings)

The first example shows how to submit a job with the following data:

- Peptide sequence: SSRFESLFAG.

- Peptide secondary structure: CHHHHHHHHC.

- Receptor input structure: PDB ID, 2 AM9, crystal structure of the human androgen receptor in the unbound form.

and the default CABS-dock server settings.

- **command line:**

```
curl -H "Content-Type: application/json" -X POST -d
'{"receptor_pdb_code":"2 AM9", "ligand_seq":"SSRFESLFAG",
"ligand_ss":"CHHHHHHHHC"}'
"http://biocomp.chem.uw.edu.pl/CABSdock/REST/add_
job/"
```

- **python script:**

```
import requests
import json
url = 'http://biocomp.chem.uw.edu.pl/CABSdock/REST/
add_job/'
data = {
    "receptor_pdb_code": "2 AM9"
    "ligand_seq": "SSRFESLFAG",
    "ligand_ss": "CHHHHHHHHC"
}
response = requests.post(url, data=data)
```

Example 2 (increasing the flexibility of selected receptor fragments)

The second example shows how to increase the flexibility of selected receptor fragments.

For each selected residue, one of two settings of flexibility (moderate or full) can be set. Technically, this is achieved by changing the default distance restraints used to keep the receptor structure near to the input conformation. The assignment of moderate flexibility decreases the strength of restrains, while full flexibility assignment removes all the restraints imposed on the selected residue.

Data used in the example:

1. Peptide sequence: HPQFEK.

2. Peptide secondary structure: CHHHCC.

3. Receptor input structure: PDB ID: 2RTM, crystal structure of biotin binding protein in the unbound form.

Additional options:

1. Using the CABS-dock "Mark flexible regions" option, ten residues (45–54) forming the flexible loop are selected and the fully flexible setting is assigned to those residues.

Important: Numbering in the PDB format must be used.

- **command line:**

```
curl -H "Content-Type: application/json" -X POST -d
'{"receptor_pdb_code":"2RTM", "ligand_seq":"HPQFEK",
"ligand_ss":"CHHHCC", "flexible_regions":[{"start"
:"45","end":"54","chain": "A","flexibility":"full"}]}'
http://biocomp.chem.uw.edu.pl/CABSdock/REST/add_job/
```

- **python script:**

```
import requests
import json
url = 'http://biocomp.chem.uw.edu.pl/CABSdock/REST/
add_job/'
data = {
    "receptor_pdb_code": "2RTM"
    "ligand_seq": "HPQFEK",
    "ligand_ss": "CHHHCC"
    "flexible_regions":[
        {
            "start": "45",
            "end": "54",
            "chain": "A",
            "flexibility": "full"
        }
    ]
}
response = requests.post(url, data=data)
print response.text
```

Example 3 (excluding binding modes from docking search)

The third example focuses on excluding binding modes form docking search. In the default mode, CABS-dock allows peptides to explore the entire receptor surface. However, in certain modeling cases it is known that some parts of the protein are not accessible (for example due to binding to other proteins) and therefore it could be useful to exclude these regions from the search.

Data used in the example:

1. Peptide sequence: PQQATDD.

2. Peptide secondary structure: CEECCCC.

3. Receptor input structure: PDB ID: 1CZY:C, tumor necrosis factor receptor associated protein 2 in the unbound form.

Additional options:

1. 1CZY protein is a trimer and 1CZY:C forms contacts with 1CZY:A (according to the http://ligin.weizmann.ac.il/cma/ server for the analysis of protein-protein interfaces). Therefore, the residues in the C chain (the input protein) listed above which are responsible for contacts with A and B chains can be

excluded from the docking search using the CABS-dock "Exclude regions" option.

- **command line:**

```
curl -H "Content-Type: application/json" -X POST -d
'{"receptor_pdb_code":"1CZY:C", "ligand_seq":"PQQATDD",
"ligand_ss":"CEECCCC", "excluded_regions":[
{"start":"334","end":"335","chain":"C"},
{"start":"338","end":"338","chain":"C"},
{"start":"341","end":"342","chain":"C"},
{"start":"345","end":"345","chain":"C"},
{"start":"350","end":"350","chain":"C"},
{"start":"385","end":"386","chain":"C"},
{"start":"416","end":"418","chain":"C"},
{"start":"420","end":"421","chain":"C"},
{"start":"458","end":"458","chain":"C"} ]}'
http://biocomp.chem.uw.edu.pl/CABSdock/REST/add_job/
```

- **python script:**

```
import requests
import json
url = 'http://biocomp.chem.uw.edu.pl/CABSdock/REST/
add_job/'
files = {'file': open('your_PDB_file.pdb')}
#or use PDB code in var data
data = {
    "receptor_pdb_code": "1CZY:C"
    "ligand_seq": "PQQATDD",
    "ligand_ss": "CEECCCC"
    "excluded_regions":[
        {
            "start": "334",
            "end": "335",
            "chain": "C",
        },
        {
            "start": "338",
            "end": "338",
            "chain": "C",
        },
        {
            "start": "341",
            "end": "342",
            "chain": "C",
        },
        {
            "start": "345",
            "end": "345",
            "chain": "C",
        },
        {
            "start": "350",
            "end": "350",
            "chain": "C",
        },
```

```
        {
            "start": "385",
            "end": "386",
            "chain": "C",
        },
        {
            "start": "416",
            "end": "418",
            "chain": "C",
        },
        {
            "start": "420",
            "end": "421",
            "chain": "C",
        },
        {
            "start": "458",
            "end": "458",
            "chain": "C",
        }
    ]
}
response = requests.post(url, data=data)
```

4   Notes

1. The CABS-dock server requires a user-provided protein receptor structure in the PDB format or the PDB code of the receptor (the file will be automatically downloaded to the server from the PDB database). The chain of the protein receptor must be shorter than 500 amino acids. The backbone must be complete; however side chain atoms may be missing. Any nonstandard amino acids in the protein receptor will be changed to their standard counterparts.

2. The peptide sequence input must be 4–30 amino acids in length and consist of standard amino acids only. It is also possible to provide the secondary structure of the peptide in the standard one-letter code (C—coil, H—helix, E—extended) using the "Optional" tab (if not, the secondary structure will be predicted with PsiPred). The structure may be experimentally derived or based on any sequence-based prediction method. Please note that "overprediction" of regular structures (H, E) was shown to be more likely to give incorrect results of docking than their underprediction. If the secondary structure is not known, it is better to supply it as a list of "C" (coil assignments). More information on how the secondary structure information is used in the simulations is provided in reference [35].

3. On top of standard input settings the CABS-dock server provides an advanced input panel that enables additional features to tailor simulation conditions to the user's needs. These features

include: (a) Custom adjusted run time: the user is allowed to lengthen the simulation run time, which may save time in case of small complexes or lead to better results for large complexes, where the standard setting may be insufficient to cover the whole conformational space. (b) Selection of flexible regions of the receptor: the user may mark some of the residues of the receptor to be granted more conformational flexibility than in the standard settings. By default receptor residues are flexible, but limited to only near-native conformations, which is suitable for most docking applications. Additional flexibility may be adjusted to semi- or full flexible to model more accurately regions believed to change their conformation on peptide binding. (c) Exclusion from sampling the receptor regions unlikely to be involved in peptide binding: the user may select some of the receptor residues believed not to take part in peptide binding. This feature is useful when the receptor molecule contains more than one binding spot and only one needs to be investigated (i.e., in receptors containing dimerization sites) or when part of the receptor is inaccessible to the peptide in vivo (i.e., receptors embedded in the membrane). Illustrative examples of using these advanced features are provided in [16].

4. All CABS-dock results can be downloaded in a single ZIP archive file available from the "Docking predictions results" tab. The ZIP archive file contains the simulation trajectories, clusters of models, and the top-ranked models (representatives of the clusters). All the provided structures are in PDB format files and the top-ranked models are provided in all-atom resolution. The trajectories and cluster model coordinates are provided in C-alpha representation only. The ZIP archive also contains simulation contact maps (discussed in Subheading 3.2).

5. The contact maps are stored as PNG figures and MAP files. The MAP file is a text file (txt) that consists of three columns: the first two list the residues of the protein receptor and the peptide, respectively. In each row, the third column gives the frequency of the contact between the residues in the first two columns. An example fragment of a MAP file format is presented below:

...

```
A224   C7    0.0117647
A224   C8    0.0117647
A224   C9    0
A225   C1    0
A225   C10   0.0117647
A225   C2    0
A225   C3    0
A225   C4    0
```

A225   C5   0

A225   C6   0

A225   C7   0.0117647

...

Each of the residues in the receptor protein is paired with each residue of the peptide, so the number of rows in the file is (number of protein residues)×(number of peptide residues).

Acknowledgments

The authors acknowledge support from the National Science Center grant [MAESTRO 2014/14/A/ST6/00088].

References

1. Tsomaia N (2015) Peptide therapeutics: targeting the undruggable space. Eur J Med Chem 94:459–470

2. Fosgerau K, Hoffmann T (2015) Peptide therapeutics: current status and future directions. Drug Discov Today 20:122–128

3. Diller DJ, Swanson J, Bayden AS, Jarosinski M, Audie J (2015) Rational, computer-enabled peptide drug design: principles, methods, applications and future directions. Future Med Chem 7:2173–2193

4. London N, Raveh B, Schueler-Furman O (2013) Peptide docking and structure-based characterization of peptide binding: from knowledge to know-how. Curr Opin Struct Biol 23:894–902

5. Yan C, Zou X (2015) Predicting peptide binding sites on protein surfaces by clustering chemical interactions. J Comput Chem 36:49–61

6. Verschueren E, Vanhee P, Rousseau F, Schymkowitz J, Serrano L (2013) Protein-peptide complex prediction through fragment interaction patterns. Structure 21:789–797

7. Saladin A, Rey J, Thevenet P, Zacharias M, Moroy G, Tuffery P (2014) PEP-SiteFinder: a tool for the blind identification of peptide binding sites on protein surfaces. Nucleic Acids Res 42:W221–W226

8. Lavi A, Ngan CH, Movshovitz-Attias D, Bohnuud T, Yueh C, Beglov D, Schueler-Furman O, Kozakov D (2013) Detection of peptide-binding sites on protein surfaces: the first step toward the modeling and targeting of peptide-mediated interactions. Proteins 81:2096–2105

9. Antes I (2010) DynaDock: a new molecular dynamics-based algorithm for protein-peptide docking including receptor flexibility. Proteins 78:1084–1104

10. London N, Raveh B, Cohen E, Fathi G, Schueler-Furman O (2011) Rosetta FlexPepDock web server—high resolution modeling of peptide-protein interactions. Nucleic Acids Res 39:W249–W253

11. Trellet M, Melquiond AS, Bonvin AM (2013) A unified conformational selection and induced fit approach to protein-peptide docking. PLoS One 8, e58769

12. Raveh B, London N, Zimmerman L, Schueler-Furman O (2011) Rosetta FlexPepDock ab-initio: simultaneous folding, docking and refinement of peptides onto their receptors. PLoS One 6, e18934

13. Trellet M, Melquiond AS, Bonvin AM (2015) Information-driven modeling of protein-peptide complexes. Methods Mol Biol 1268:221–239

14. Donsky E, Wolfson HJ (2011) PepCrawler: a fast RRT-based algorithm for high-resolution refinement and binding affinity estimation of peptide inhibitors. Bioinformatics 27:2836–2842

15. Kurcinski M, Jamroz M, Blaszczyk M, Kolinski A, Kmiecik S (2015) CABS-dock web server for the flexible docking of peptides to proteins without prior knowledge of the binding site. Nucleic Acids Res 43:W419–W424

16. Blaszczyk M, Kurcinski M, Kouza M, Wieteska L, Debinski A, Kolinski A, Kmiecik S (2015) Modeling of protein-peptide interactions using the CABS-dock web server for binding site search and flexible docking. Methods 93:72–83

17. Jamroz M, Kolinski A, Kmiecik S (2013) CABS-flex: server for fast simulation of protein

structure fluctuations. Nucleic Acids Res 41:W427–W431

18. Blaszczyk M, Jamroz M, Kmiecik S, Kolinski A (2013) CABS-fold: server for the de novo and consensus-based prediction of protein structure. Nucleic Acids Res 41:W406–W411

19. Jamroz M, Kolinski A, Kmiecik S (2014) Protocols for efficient simulations of long-time protein dynamics using coarse-grained CABS model. Methods Mol Biol 1137:235–250

20. Das R, Baker D (2008) Macromolecular modeling with rosetta. Annu Rev Biochem 77:363–382

21. Ciemny MP, Debinski A, Paczkowska M, Kolinski A, Kurcinski M, Kmiecik S (2016) Protein-peptide molecular docking with large-scale conformational changes: the p53-MDM2 interaction. Sci Rep 6:37532

22. Kmiecik S, Gront D, Kolinski M, Wieteska L, Dawid AE, Kolinski A (2016) Coarse-grained protein models and their applications. Chem Rev 116:7898–7936

23. Heery DM, Kalkhoven E, Hoare S, Parker MG (1997) A signature motif in transcriptional co-activators mediates binding to nuclear receptors. Nature 387:733–736

24. Burgermeister E, Schnoebelen A, Flament A, Benz J, Stihle M, Gsell B, Rufer A, Ruf A, Kuhn B, Marki HP, Mizrahi J, Sebokova E, Niesor E, Meyer M (2006) A novel partial agonist of peroxisome proliferator-activated receptor-gamma (PPARgamma) recruits PPARgamma-coactivator-1alpha, prevents triglyceride accumulation, and potentiates insulin signaling in vitro. Mol Endocrinol 20:809–830

25. Hsin J, Arkhipov A, Yin Y, Stone JE, Schulten K (2008) Using VMD: an introductory tutorial. Curr Protoc Bioinformatics 5:57

26. Kurcinski M, Kolinski A, Kmiecik S (2014) Mechanism of folding and binding of an intrinsically disordered protein as revealed by ab ini-

tio simulations. J Chem Theory Comput 10:2224–2231

27. Nolte RT, Wisely GB, Westin S, Cobb JE, Lambert MH, Kurokawa R, Rosenfeld MG, Willson TM, Glass CK, Milburn MV (1998) Ligand binding and co-activator assembly of the peroxisome proliferator-activated receptor-gamma. Nature 395:137–143

28. Eswar N, Webb B, Marti-Renom MA, Madhusudhan MS, Eramian D, Shen M-Y, Pieper U, Sali A (2007) Comparative protein structure modeling using MODELLER. Curr Protoc Protein Sci 2:1–31

29. Shen MY, Sali A (2006) Statistical potential for assessment and prediction of protein structures. Protein Sci 15:2507–2524

30. Unal EB, Gursoy A, Erman B (2010) VitAL: Viterbi algorithm for de novo peptide design. PLoS One 5, e10926

31. Bhattacherjee A, Wallin S (2013) Exploring protein-peptide binding specificity through computational peptide screening. PLoS Comput Biol 9, e1003277

32. Lee H, Heo L, Lee MS, Seok C (2015) GalaxyPepDock: a protein-peptide docking tool based on interaction similarity and energy optimization. Nucleic Acids Res 43:W431–W435

33. London N, Raveh B, Schueler-Furman O (2013) Druggable protein-protein interactions—from hot spots to hot segments. Curr Opin Chem Biol 17:952–959

34. London N, Raveh B, Movshovitz-Attias D, Schueler-Furman O (2010) Can self-inhibitory peptides be derived from the interfaces of globular protein-protein interactions? Proteins 78:3140–3149

35. Kmiecik S, Kolinski A (2017) One-dimensional structural properties of proteins in the coarse-grained CABS model. In: Prediction of protein secondary structure. Springer, New York, pp 83–113

Chapter 7

AnchorDock for Blind Flexible Docking of Peptides to Proteins

Michal Slutzki, Avraham Ben-Shimon, and Masha Y. Niv

Abstract

Due to increasing interest in peptides as signaling modulators and drug candidates, several methods for peptide docking to their target proteins are under active development. The "blind" docking problem, where the peptide-binding site on the protein surface is unknown, presents one of the current challenges in the field. AnchorDock protocol was developed by Ben-Shimon and Niv to address this challenge.

This protocol narrows the docking search to the most relevant parts of the conformational space. This is achieved by pre-folding the free peptide and by computationally detecting anchoring spots on the surface of the unbound protein. Multiple flexible simulated annealing molecular dynamics (SAMD) simulations are subsequently carried out, starting from pre-folded peptide conformations, constrained to the various precomputed anchoring spots.

Here, AnchorDock is demonstrated using two known protein-peptide complexes. A PDZ-peptide complex provides a relatively easy case due to the relatively small size of the protein, and a typical peptide conformation and binding region; a more challenging example is a complex between $USP7_{N-term}$ and a p53-derived peptide, where the protein is larger, and the peptide conformation and a binding site are generally assumed to be unknown. AnchorDock returned native-like solutions ranked first and third for the PDZ and USP7 complexes, respectively. We describe the procedure step by step and discuss possible modifications where applicable.

Key words Peptide folding, Peptide docking, Binding site prediction, Anchors, Surface mapping, Simulated annealing molecular dynamics

1 Introduction

Protein-peptide interactions occur widely in nature: peptides bind to proteins and exert different biological effects. Peptides act as hormones [1, 2], neurotransmitters [2, 3], various bioactives from food [4], antibacterial agents [5] and more. Furthermore, peptides can be also rationally designed with the goal of modulating protein-protein interactions [6, 7]. In recent years, peptides are increasingly used for therapeutic purposes [8–10]. In 2015, there were about 140 FDA-approved peptidic drugs on the market, with

Ora Schueler-Furman and Nir London (eds.), *Modeling Peptide-Protein Interactions: Methods and Protocols*, Methods in Molecular Biology, vol. 1561, DOI 10.1007/978-1-4939-6798-8_7, © Springer Science+Business Media LLC 2017

hundreds more under development [8]. These peptides vary in length (from a few to above 40 amino acids) and in indications, which include short bowel syndrome, Cushing's disease, and some types of cancer [10].

Low oral bioavailability and poor metabolic stability are some of the peptides' disadvantages, but these are counterbalanced by the straightforward and relatively inexpensive synthesis, and peptides' ability to bind to interfaces not easily targeted by small molecules, potentially achieving higher specificity [9]. Thus, peptides provide novel opportunities for alternative or complementary drug design strategies to those pursued by the design of small molecule drugs. Specifically, protein-like "building blocks" make naturally occurring and synthetic peptides particularly suitable to modulate protein-protein interactions. Lack of structural data on protein-peptide complexes calls for advanced modeling approaches. However, state-of-the-art docking tools designed for small molecules are not well suited even for short peptides [11]. Due to a large number of rotatable bonds in peptides and the flatter binding sites compared to binding pockets for small molecules, specialized peptide docking tools are required.

Indeed, multiple peptide docking methods have been recently developed. Some of these methods are mainly geared toward refinement of an initial pose (such as Rosetta FlexPepDock [12–15], PepCrawler [16], and HADDOCK [17, 18]). Other methods are knowledge-based, namely, use structurally analogous peptide-protein complexes (i.e., GalaxyPepDock [19]) or experimentally known interactions between part of the peptide and the protein (i.e., PDZ domain-binding proteins [20] and MHC peptides [21, 22]). The most difficult methodological challenge is that of a "blind" docking approach, when no prior knowledge of the binding site is used. This was recently tackled by CABS-dock [23], pepATTRACT [24], and AnchorDock [25], which will be described below.

Previously, SAMD simulations starting from an extended peptide conformation and using a single knowledge-based constraint regarding C-terminus location of the peptide were successfully applied to PDZ domains [20]. To generalize the method, namely—to overcome the need for knowledge-based input, the AnchorDock protocol was developed [25]. AnchorDock allows protein and peptide flexibility and uses the ANCHORSmap algorithm [26] to locate possible binding positions of individual peptide residues on the protein. These positions are used to reduce the SAMD conformational search, as the pre-folded peptides conformations are tethered to these anchors via constraints.

AnchorDock was previously used to dock peptides into 13 unbound proteins (also available in their bound state) for validation purposes [25]. In ten cases, the method yielded solutions with backbone (bb) RMSD below 2.2 Å, and in nine of those, the best solution was ranked among the top 10. In round 29 of CAPRI

Table 1
Protein-peptide complexes used for demonstration of the protocol

Case	Name	PDB complex	PDB apo	Peptide sequence	Protein size in AAs
1	Tax-Interacting Protein-1 (TIP-1) PDZ domain and Y-iCAL36 peptide	4NNM	2VZ5	YPTSII	102
2	USP7 and p53-derived peptide [28]	2FOJ	2F1W	GARAHSS	138

(A Critical Assessment of Predicted Interaction [27]), AnchorDock was the only method that provided an acceptable model for target T65 [25].

Our goal here is to demonstrate step-by-step usage of the AnchorDock protocol. We do so by presenting two illustrative cases of protein—peptide docking, starting from the unbound protein conformation, and the amino acid sequence of the peptide (Table 1). The first case is an interaction between the PDZ domain of Tax-Interacting Protein-1 (TIP-1) and its peptidic inhibitor (PDB_ID 4NNM). TIP-1 is a human protein, composed of a single PDZ domain. It is found in the brain in asterocytes and neurons and is involved in cystic fibrosis disease. In the complex we model in this chapter, TIP-1 interacts with its inhibitory peptide, which was designed based on the C-terminus of its naturally binding partner, CAL (CFTR—Cystic Fibrosis Transmembrane conductance Regulator Associated Ligand) [24, 29]. We chose this example for demonstration purposes because it belongs to the large and important family of PDZ domains which is known to interact with many different peptides through a carboxylic termini-mediated interaction [30, 31]. We have studied PDZ domains before [20, 30, 32] and have successfully docked peptides to other members of the PDZ domain family using the specialized PDZ-DocScheme [20] and the general AnchorDock protocol [25]. The second example we chose for illustrating the AnchorDock protocol is somewhat more challenging—the complex between the N-terminal domain of USP7 and a p53-derived peptide [28], a larger protein target with the main direct contact to the peptide's sidechain made by a C-terminal serine.

2   Materials

2.1   Software

1. GROMACS version 4.5.5 [33]—a molecular dynamics package for simulation of macromolecules.

2. PyMol version 1.3—a molecular visualization software that is used in this protocol to remove hydrogens and to calculate RMSD between a docking solution and a native structure.

3. **DeepView—Swiss-PdbViewer** v4.1.0 http://www.expasy. org/spdbv/ [34]—a molecular visualization software that is used in the protocol to fill the missing sidechains atoms. Freely available.

4. **Maestro 10.2** (Schrodinger LLC.)—a molecular modeling environment that was used for data analysis, visualizations, and figures preparation.

2.2 Hardware

A computer cluster that contains 128 processors, in a Sun Greed Engene (SGE), was used for the calculations [8]. Short calculations such as file preparations, peptide rotations, anchors calculations, and clustering were performed on single processors (8 CPUs), while folding and docking were calculated in parallel in the SGE queue.

2.3 Peptide Sequences

The amino acid sequence of the peptide is used. AnchorDock was tested on peptides smaller than 14 amino acids [25]. Peptide sequences used in this protocol appear in Table 1.

2.4 ANCHORSmap

We use the ANCHORSmap protocol [26], developed in the lab of Dr. Miriam Eisenstein at the Weizmann Institute of Science, Israel, to find putative anchoring spots, representing possible interaction sites between a peptide (as an anchor) and a protein (as anchor-surrounding residues on the protein surface). In this protocol, a protein can be introduced by several conformations according to a multiple protein conformation approach (MPCA). The method first detects local surface minima (LSM) and distributes probes for amino acids in the vicinity of each LSM, then applies simultaneous minimization (SM) and clusters spatially adjacent probes. Free energy of the probes is calculated with correction for dielectric shielding (to represent the effect created by the rest of the peptide behind the probe) in the electrostatic energy term. Finally, probes for all protein conformations are averaged into one set of anchors, which are then ranked according to their energy-based score. Twelve percent of the highest scored anchors are used as input for AnchorDock.

3 Methods

The steps of the AnchorDock protocol are illustrated for two protein-peptide complexes that are listed in Table 1.

1. Peptide preparation for docking:

    (a) The extended form of the peptide is generated using Maestro or PyMol (*see* **Note 1**). Specifically, we created the extended structure of YPTSII and of GARAHSS.

    (b) Simulated Annealing Molecular Dynamics (SAMD) is performed in GROMACS to produce peptide conformations in the unbound state. See parameters of the molecular dynamics simulation in Table 2. The annealing protocol is repeated 25 times the number of residues in the peptide.

(c) The snapshots, collected every 1 ps from all trajectories at the temperature range between 285 and 295 K, are clustered according to backbone RMSD using the gromos clustering algorithm [36], and ranked based on the average energy of the 15 lowest energy structures in the cluster.

(d) The second best scoring conformation, shown to be a preferable starting conformation [25], is used as input for

Table 2
Parameters for molecular dynamics for different stages of the protocol

	Peptide folding	Receptor conformations	First cycle of docking	Second cycle of docking
Force field		AMBER96		
Implicit solvent		GBSA, OBC		
Thermostat		V-rescale		
PBS		No		
Distance restraint potential	Low = 0, low1 = 0.5 up2 = 1.5 fac = 10	Low = 0, low1 = 0.15, up2 = 1.5, fac = 250		
Minimization	Conjugate gradient			
Maximal force	1.0 kJ/mol			
Max. steps	20,000	5000	5000	5000
Write every		10 steps		
SAMD Integration time step		5 fs		
Write every		200 steps		
VDW and electrostatic cutoff		0 nm	1.5 nm	1.8 nm
Annealing times, ps	Peptide size dependent, calculated as in Figure S4 in [25]	0, 70, 215, 225, 250	0, 100, 250, 350, 500, 580, 680, 690, 700, 750	0, 5, 80, 205, 250, 320, 455, 500, 570, 705, 750, 820, 955, 1000
Annealing temperatures, K		290, 450, 275, 290, 290	275, 600, 275, 550, 275, 500, 275, 275, 290, 290	290, 209, 450, 290, 290, 600, 290, 290, 550, 290, 290, 500, 290, 290
Snapshot frequency	1 ps	0.5 ps	0.5 ps	0.5 ps

PBC periodic boundary conditions, *OBC* Onufriev-Bashford-Case method for Born radii [35], *SAMD* simulated annealing molecular dynamics

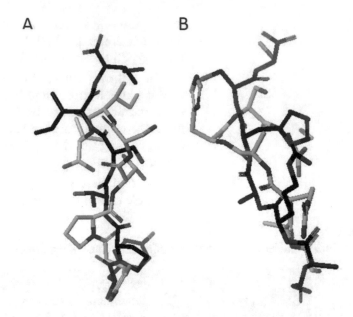

Fig. 1 The second conformation of unbound peptide as modeled using AnchorDock peptide structure prediction protocol (end of stage 1). (**a**) Inhibitory peptide based on CAL (case 1); (**b**) p53 fragment (case 2). These conformations were used as input for the peptide docking protocol. The bound peptide conformations are shown in black

docking (*see* **Note 2**). The conformations used for our example calculations are shown in Fig. 1.

2. Protein preparation for docking:

   (a) The unbound structure of the protein is downloaded from the Protein Data Bank (PDB_ID 2VZ5 and 2F1W for cases 1 and 2, respectively). It is possible to use a homology model (as was successfully done by the related PDZ-DockScheme [20], but has not been evaluated for AnchorDock). No hydrogens should be included since they are introduced by GROMACS through virtual hydrogen representation.

   (b) Heteroatoms and water molecules are removed using PyMol, Maestro, or other structure visualization software. The 2VZ5.pdb file contains also anisotropic temperature factors for some atoms, which should be removed as well.

   (c) Missing sidechains atoms are added (with Swiss-PdbViewer, for example).

   (d) N- and C- termini of the protein can be defined as frozen (not allowed to move during the simulation). If there is a missing segment in the protein, the atoms before and after the missing segment should be frozen as well (*see* **Note 3**).

3. The files are converted into GROMACS formats .gro and .itp, for structure and topology representations, respectively. Protonation state of peptide histidines can be defined at this stage [37]. For case 2 peptide, the δ-nitrogen protonation (HID) was used.

4. Protein/peptide flexibility is defined as follows: all atoms in the protein (defined by shell radius of 250–400 Å to cover entire mid-size proteins) are allowed to move freely, except backbone nitrogens that can move in a radius of 1.5 Å and are constrained by a distance restraint potential at further distances (Table 2).

5. Multiple conformations of the protein:

   (a) The unbound protein is simulated for 1 ns, using SAMD parameters detailed in Table 2.

   (b) The 0.5 ps snapshot conformations obtained in 5a are clustered. The goal is to obtain 3–4 clusters, each containing at least 15 structures. To this end, the RMSD cutoff can usually be set to 0.6–0.8 Å. The variability between different conformations in the test case examples is illustrated in Fig. 2.

6. Protein conformations from **step 5** are subjected to the ANCHORSmap protocol [26] to obtain an anchors list (*see* **Note 4**): The appropriate probe for each residue in the peptide sequence is selected from 14 anchor probes available in ANCHORSmap: Arg, Ile, His, Asp, Tyr, Trp, Lys, Gln, Phe, Leu, Asn, Met, Val, and Glu (*see* **Note 5**). In addition to the

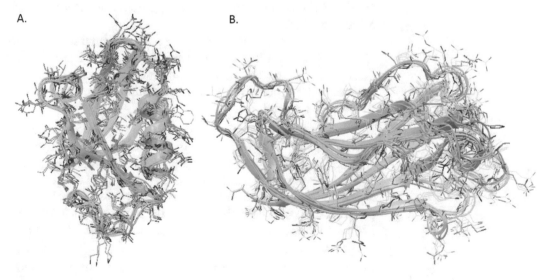

A.   B.

Fig. 2 Receptor conformations. For each receptor, the unbound conformation is in gray and four additional fluctuating conformations are presented as cartoons and as sticks (**a**) TIP-1 (case 1) in *green*; (**b**) N-term domain of USP7 (case 2) in *blue*

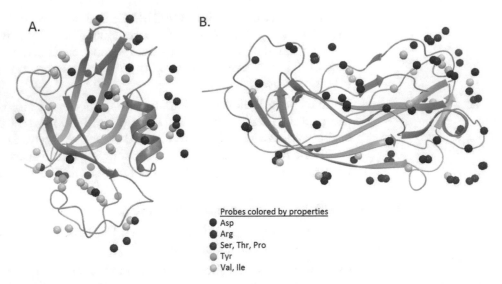

Fig. 3 Anchors on (**a**) TIP-1 (case 1); (**b**) N-term domain of USP7 (case 2). The anchors are colored according to their properties as detailed in the figure

peptide side chains, peptide termini should be represented. The carboxylic group at the C-terminus of the peptide is known to play an important role in many interactions and can be modeled by an Asp probe (e.g., in PDZ domains, as in our case 1). In our calculations, we did not use any probe for the N-terminus, but it can be approximately modeled using Asn probe. For case 1 we used Tyr, Val, Asn, Ile, and Asp probes and for case 2 we used Val, Arg, His, Asn, and Asp probes. In Fig. 3 the anchors for each case protein are demonstrated. *See* also **Note 6**.

At this point, the following are available: protein structure(s), peptide conformation(s) (*see* **Note 2**), and a list of anchors. When the goal is evaluation of protocol performance, peptide pose from a known complex structure should be available as well.

7. Rotations of the peptide around each anchor: for each anchor spot, each conformation of the peptide is placed 5 Å from the anchor. AnchorDock creates copies of the peptide rotated around the axes by increments of 36°; one axis is determined as the normal from the anchoring spot toward a centroid of the protein and the other is the peptide normal from the atom that fits the anchor spot to the centroid of the peptide (*see* **Note 7**).

8. To obtain a reduced number of productive peptide conformations and orientations from those generated in **step 7**, a short and crude anchor-driven cycle of docking is performed:

   (a) The above peptide rotations for each anchor are used as initial positions for 0.75 ns SAMD with parameters summarized in Table 2. The protein is flexible as described in **step 4**. The anchor residue of the peptide is constrained to the anchoring spot by energy penalty applied at distances larger than 5 Å.

(b) The snapshots are collected from the last 50 ps of all trajectories in temperature range of 285–295 K and are clustered using the gromos clustering algorithm [36]. Only clusters containing at least 15 structures are used for further analysis. An RMSD matrix is built with a cutoff of 2.5 Å. An energy-based score is calculated to rank the clusters: the central solution of each cluster is selected as a representative structure and is assigned the energy average of the 15 lowest energy solutions of the cluster.

(c) The number of peptide conformations ($N_{\text{peptide confs}}$) used in the simulation, the length of the proteins (N_{aa}), and the number of anchors (N_{anchors}) determine the number of conformations (N_{sm}) that enter the refinement stage.

$$N_{\text{conf}} = N_{\text{anchors}} \times N_{\text{peptide confs}} \times 10 \text{ rotations}$$

$$N_{\text{sm}} = \sqrt{N_{\text{conf}}} \times \frac{N_{\text{aa}}}{110}$$

In our illustrative examples, we used 26 initial structures of 2VZ5/YPTSII and 32 of 2F1W/GARAHSS for the refinement stage of SAMD (stage 9).

9. Refinement SAMD simulation:

(a) A 5 ns long SAMD simulation is applied with parameters as presented in Table 2, repeating the simulation annealing protocol 5 times. Flexibility and constraints are defined as in **step 8a**.

(b) All snapshots are collected and clustered as described in **step 8b**, but omitting the first 70 ps for each trajectory and using an RMSD matrix cutoff of 2.0 Å. *See* **Note 8**.

10. Results are analyzed and can be compared to the experimental structure (when available, as in our examples):

(a) The reference complex and the docking solutions are represented in Maestro (or another molecular visualization environment).

(b) The proteins in the reference structure and the model are aligned using structural alignment option in Maestro.

(c) Heavy atom RMSD between the reference and the docked peptide conformations are calculated.

Figure 4 presents the high-scored native-like solutions for both test cases. The top-ranked solution for case 1 (PDZ domain/YPTSII) has low RMSD. In the more challenging case 2 (USP7/GARASS), a native-like solution was ranked third (Table 3).

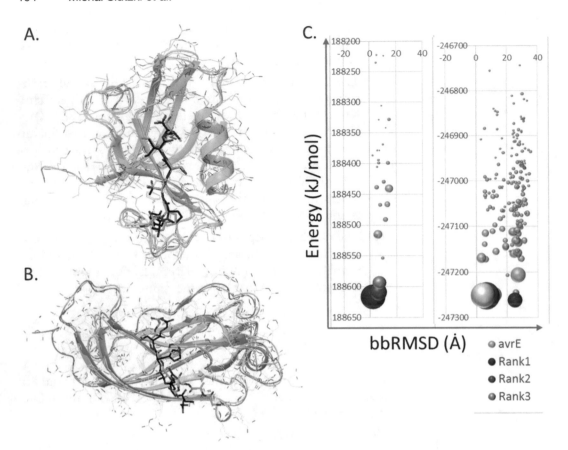

Fig. 4 Docking solution. (**a**) TIP-1 with its peptide inhibitor (case 1); (**b**) N-terminal domain of USP7 with p53 fragment (case 2); (**c**) bb-RMSD vs. energy plots (the *right* for case 1 and the *left* for case 2). Three best-scored solutions are colored *red*

Table 3
Quality of the native-like solution of the AnchorDock protocol

Case	Name	PDB complex	PDB apo	Peptide sequence	Protein Size	Native-like solution	
						Rank	Heavy atom RMSD[a]
1	Tax-Interacting Protein-1 (TIP-1) PDZ domain and Y-iCAL36 peptide	4NNM	2VZ5	YPTSII	102	1	1.3 (45 atoms)
2	USP7 and p53-derived peptide [28]	2FOJ	2F1W	GARAHSS	138	3	2.1 (46 atoms)

[a]Calculated in PyMol

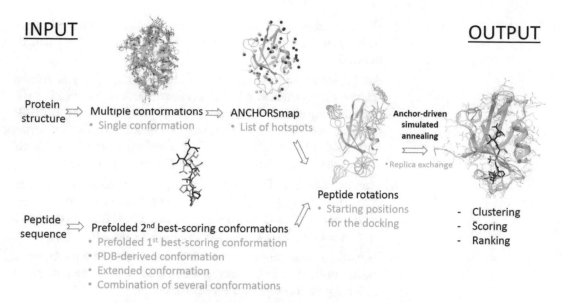

Fig. 5 AnchorDock protocol for anchor-driven simulated annealing demonstrated for the PDZ-domain case. Alternative options for different parts of the protocol are written in *gray*

In summary, prediction of peptide-protein interaction with the AnchorDock protocol was illustrated step by step, resulting in excellent solutions for blind docking to unbound proteins. The protocol and possible variations (discussed in **Notes 2, 4** and **9** below) are summarized in Fig. 5.

4   Notes

1. It is important to verify that the termini of the peptide are represented correctly. Specifically, the carboxylic terminus of the peptide mediates many important interactions (e.g., in peptide binding to PDZ domains).

2. At least one input conformation of the peptide is needed. It can be either a pre-folded conformation as discussed here, or an extended one. Alternatively, a structure of the sequence corresponding to the sequence of the peptide can be extracted from the PDB (Fig. 5). From our experience, the conformation that was scored second in the peptide folding leads to better prediction results than the top-scoring conformation [25]. In the examples used in this chapter (Table 1), using the first conformation produced very similar results for the PDZ domain case, but higher RMSD (2.4) and lower rank (4) than the second conformation (RMSD 2.1 ranked 3) for USP7/ p53 case.

3. If the protein contains unstructured regions that are known to be irrelevant to interactions with the peptide, they can be deleted.

4. An important advantage of ANCHORSmap [26] is that it searches sites on the protein surface that are specific for the peptide sequence. However, other predictions of possible binding sites (i.e., [38–40]) may be considered. For example, PeptiMap [41] predicts binding sites using small molecule probes. Knowledge-based anchors can be used as well.

5. For Ala, Cys, Ser, Thr, Pro, and Gly, no probes are available in ANCHORSmap [26]. Val probe may be used for Pro and Ala; Met probe may be used for Cys. Ser and Thr can be represented by Asn probe [26] (as was done here for case 2).

6. The initial position of anchors does not have to be exact, since during both stages of docking the peptide can move toward its optimal position within a radius of 5–6 Å.

7. If the binding pocket is deep and peptide rotation results in clashes with the protein, optimize the protocol for producing initial rotations:

   (a) Increase the number of starting peptide conformations.

   (b) Use a finer rotation grid.

   (c) Reduce the radius of the sphere for mass center calculation (e.g., 6 Å instead of 10 Å).

   (d) Decrease the distance from the anchor of the peptide starting position (e.g., 3 Å instead of 5 Å).

   (e) Of course, in some of these cases (such as (a) and (b)) the new calculation will be computationally more costly.

8. It is possible to use another cycle of SAMD for further refinement with parameters similar to those of second cycle of docking in Table 2.

9. Alternatively to SAMD, a replica exchange molecular dynamics (REMD) protocol can be used to improve sampling further [42, 43].

Acknowledgments

This study was supported by grant no. 3-9543 from the Chief Scientist Office of the Ministry of Health, Israel via the ERA-net network to M.Y.N. We thank Dr. Miriam Eisenstein for sharing the source code of ANCHORSmap, Dr. Ester Livshits for technical support, and Dr. Tali Yarnitzky and Yaron Ben Shoshan-Galeczki for critical reading of the manuscript and stimulating discussions.

References

1. Gallagher PE, Arter AL, Deng G, Tallant EA (2014) Angiotensin-(1–7): a peptide hormone with anti-cancer activity. Curr Med Chem 21:2417–2423

2. Larsen PJ, Holst JJ (2005) Glucagon-related peptide 1 (GLP-1): hormone and neurotransmitter. Regul Pept 128:97–107

3. Assas BM, Pennock JI, Miyan JA (2014) Calcitonin gene-related peptide is a key neurotransmitter in the neuro-immune axis. Front Neurosci 8:23

4. Yoshikawa M (2015) Bioactive peptides derived from natural proteins with respect to diversity of their receptors and physiological effects. Peptides 72:208–225

5. Padhi A, Sengupta M, Sengupta S, Roehm KH, Sonawane A (2014) Antimicrobial peptides and proteins in mycobacterial therapy: current status and future prospects. Tuberculosis (Edinb) 94:363–373

6. Rubinstein M, Niv MY (2009) Peptidic modulators of protein-protein interactions: progress and challenges in computational design. Biopolymers 91:505–513

7. London N, Raveh B, Schueler-Furman O (2013) Druggable protein-protein interactions—from hot spots to hot segments. Curr Opin Chem Biol 17:952–959

8. Fosgerau K, Hoffmann T (2015) Peptide therapeutics: current status and future directions. Drug Discov Today 20:122–128

9. Craik DJ, Fairlie DP, Liras S, Price D (2013) The future of peptide-based drugs. Chem Biol Drug Des 81:136–147

10. Kaspar AA, Reichert JM (2013) Future directions for peptide therapeutics development. Drug Discov Today 18:807–817

11. Rentzsch R, Renard BY (2015) Docking small peptides remains a great challenge: an assessment using AutoDock Vina. Brief Bioinform 16:1045–1056

12. London N, Raveh B, Cohen E, Fathi G, Schueler-Furman O (2011) Rosetta FlexPepDock web server—high resolution modeling of peptide-protein interactions. Nucleic Acids Res 39:W249–W253

13. Raveh B, London N, Schueler-Furman O (2010) Sub-angstrom modeling of complexes between flexible peptides and globular proteins. Proteins 78:2029–2040

14. Raveh B, London N, Zimmerman L, Schueler-Furman O (2011) Rosetta FlexPepDock ab-initio: simultaneous folding, docking and refinement of peptides onto their receptors. PLoS One 6, e18934

15. London N, Raveh B, Schueler-Furman O (2013) Peptide docking and structure-based characterization of peptide binding: from knowledge to know-how. Curr Opin Struct Biol 23:894–902

16. Donsky E, Wolfson HJ (2011) PepCrawler: a fast RRT-based algorithm for high-resolution refinement and binding affinity estimation of peptide inhibitors. Bioinformatics 27:2836–2842

17. de Vries SJ, van Dijk M, Bonvin AM (2010) The HADDOCK web server for data-driven biomolecular docking. Nat Protoc 5:883–897

18. van Zundert GC, Rodrigues JP, Trellet M, Schmitz C, Kastritis PL, Karaca E, Melquiond AS, van Dijk M, de Vries SJ, Bonvin AM (2015) The HADDOCK2.2 Web Server: user-friendly integrative modeling of biomolecular complexes. J Mol Biol 428(4):720–725

19. Lee H, Heo L, Lee MS, Seok C (2015) GalaxyPepDock: a protein-peptide docking tool based on interaction similarity and energy optimization. Nucleic Acids Res 43:W431–W435

20. Niv MY, Weinstein H (2005) A flexible docking procedure for the exploration of peptide binding selectivity to known structures and homology models of PDZ domains. J Am Chem Soc 127:14072–14079

21. Desmet J, Wilson IA, Joniau M, De Maeyer M, Lasters I (1997) Computation of the binding of fully flexible peptides to proteins with flexible side chains. FASEB J 11:164–172

22. Tong JC, Tan TW, Ranganathan S (2004) Modeling the structure of bound peptide ligands to major histocompatibility complex. Protein Sci 13:2523–2532

23. Kurcinski M, Jamroz M, Blaszczyk M, Kolinski A, Kmiecik S (2015) CABS-dock web server for the flexible docking of peptides to proteins without prior knowledge of the binding site. Nucleic Acids Res 43(W1):W419–W424

24. Amacher JF, Cushing PR, Brooks L 3rd, Boisguerin P, Madden DR (2014) Stereochemical preferences modulate affinity and selectivity among five PDZ domains that bind CFTR: comparative structural and sequence analyses. Structure 22:82–93

25. Ben-Shimon A, Niv MY (2015) AnchorDock: blind and flexible anchor-driven peptide docking. Structure 23:929–940

26. Ben-Shimon A, Eisenstein M (2010) Computational mapping of anchoring spots on protein surfaces. J Mol Biol 402:259–277

27. Janin J (2002) Welcome to CAPRI: a critical assessment of PRedicted interactions. Proteins 47:257

28. Sheng Y, Saridakis V, Sarkari F, Duan S, Wu T, Arrowsmith CH, Frappier L (2006) Molecular recognition of p53 and MDM2 by USP7/HAUSP. Nat Struct Mol Biol 13:285–291

29. Amacher JF, Zhao R, Spaller MR, Madden DR (2014) Chemically modified peptide scaffolds target the CFTR-associated ligand PDZ domain. PLoS One 9, e103650

30. Beuming T, Skrabanek L, Niv MY, Mukherjee P, Weinstein H (2005) PDZBase: a protein-protein interaction database for PDZ-domains. Bioinformatics 21:827–828

31. Songyang Z, Fanning AS, Fu C, Xu J, Marfatia SM, Chishti AH, Crompton A, Chan AC, Anderson JM, Cantley LC (1997) Recognition of unique carboxyl-terminal motifs by distinct PDZ domains. Science 275:73–77

32. Madsen KL, Beuming T, Niv MY, Chang CW, Dev KK, Weinstein H, Gether U (2005) Molecular determinants for the complex binding specificity of the PDZ domain in PICK1. J Biol Chem 280:20539–20548

33. Pronk S, Pall S, Schulz R, Larsson P, Bjelkmar P, Apostolov R, Shirts MR, Smith JC, Kasson PM, van der Spoel D, Hess B, Lindahl E (2013) GROMACS 4.5: a high-throughput and highly parallel open source molecular simulation toolkit. Bioinformatics 29:845–854

34. Guex N, Peitsch MC (1997) SWISS-MODEL and the Swiss-PdbViewer: an environment for comparative protein modeling. Electrophoresis 18:2714–2723

35. Onufriev A, Bashford D, Case DA (2004) Exploring protein native states and large-scale conformational changes with a modified generalized born model. Proteins 55:383–394

36. Daura X, Gademann K, Jaun B, Seebach D, van Gunsteren WF, Mark AE (1999) Peptide folding: when simulation meets experiment. Angew Chem Int Ed 38:236–240

37. Ben-Shimon A, Shalev DE, Niv MY (2013) Protonation states in molecular dynamics simulations of peptide folding and binding. Curr Pharm Des 19:4173–4181

38. Barillari C, Marcou G, Rognan D (2008) Hot-spots-guided receptor-based pharmacophores (HS-Pharm): a knowledge-based approach to identify ligand-anchoring atoms in protein cavities and prioritize structure-based pharmacophores. J Chem Inf Model 48:1396–1410

39. Miranker A, Karplus M (1991) Functionality maps of binding sites: a multiple copy simultaneous search method. Proteins 11:29–34

40. Neuvirth H, Raz R, Schreiber G (2004) ProMate: a structure based prediction program to identify the location of protein-protein binding sites. J Mol Biol 338:181–199

41. Lavi A, Ngan CH, Movshovitz-Attias D, Bohnuud T, Yueh C, Beglov D, Schueler-Furman O, Kozakov D (2013) Detection of peptide-binding sites on protein surfaces: the first step toward the modeling and targeting of peptide-mediated interactions. Proteins 81:2096–2105

42. Anselmi M, Pisabarro MT (2015) Exploring multiple binding modes using confined replica exchange molecular dynamics. J Chem Theory Comput 11:3906–3918

43. Ostermeir K, Zacharias M (2013) Advanced replica-exchange sampling to study the flexibility and plasticity of peptides and proteins. Biochim Biophys Acta 1834:847–853

Chapter 8

Information-Driven, Ensemble Flexible Peptide Docking Using HADDOCK

Cunliang Geng*, Siddarth Narasimhan*, João P.G.L.M. Rodrigues, and Alexandre M.J.J. Bonvin

Abstract

Modeling protein-peptide interactions remains a significant challenge for docking programs due to the inherent highly flexible nature of peptides, which often adopt different conformations whether in their free or bound forms. We present here a protocol consisting of a hybrid approach, combining the most frequently found peptide conformations in complexes with representative conformations taken from molecular dynamics simulations of the free peptide. This approach intends to broaden the range of conformations sampled during docking. The resulting ensemble of conformations is used as a starting point for information-driven flexible docking with HADDOCK. We demonstrate the performance of this protocol on six cases of increasing difficulty, taken from a protein-peptide benchmark set. In each case, we use knowledge of the binding site on the receptor to drive the docking process. In the majority of cases where MD conformations are added to the starting ensemble for docking, we observe an improvement in the quality of the resulting models.

Key words Protein-peptide docking, Flexibility, Information-driven docking, Ensemble docking, HADDOCK, Molecular dynamics simulations

1 Introduction

Peptides are receiving an increasing level of attention from the wider biological and pharmaceutical communities owing to an increase in the number of peptide-based drugs and therapeutics entering the market [1, 2]. Despite their importance, there is much to be learned about the structural and dynamical properties of peptides, in particular in the context of their interactions with other biomolecules. The binding partner that peptides associate with often plays an important role in restricting/defining their

*These authors contributed equally to this work.

Electronic supplementary material: The online version of this chapter (doi:10.1007/978-1-4939-6798-8_8) contains supplementary material, which is available to authorized users.

Ora Schueler-Furman and Nir London (eds.), *Modeling Peptide-Protein Interactions: Methods and Protocols*, Methods in Molecular Biology, vol. 1561, DOI 10.1007/978-1-4939-6798-8_8, © Springer Science+Business Media LLC 2017

conformational space. Many peptides are known to exist in an intrinsically disordered state, meaning they lack a well-defined and stable folded form on the time scales that are available to experimental methods in structural biology. As a consequence, the structures of peptides are mostly known in the context of their protein receptors, which adds to the challenge of predicting in silico, the structure of these interactions.

The two most common thermodynamic models often used to describe biomolecular recognition and binding processes are induced fit [3, 4] and conformational selection [5–7]. These models were formulated based on observations from classical titration-like experiments, aimed at studying the manner in which an estimated equilibrium is achieved upon addition of a binding partner [8]. From a structural perspective, the induced fit model can be explained as a binding mechanism where the partners induce conformational changes on each other during complex formation. The conformational selection mechanism, on the other hand, predicts bound conformations are sampled naturally by the free molecules, i.e., without induction by the partner, and that partners merely select the most favorable conformation for binding (minor structural changes, such as side-chain re-orientation, may still occur upon binding).

A combination of the aforementioned mechanisms has been exploited previously to design a protocol for information-driven protein-peptide docking using HADDOCK [9]. Briefly, this protocol uses an ensemble of starting conformations for the peptide (alpha-helix, polyproline-II, and extended) that represent "ideal" conformational states of a given peptide and have been shown to feature in a large fraction of the protein-peptide interactions deposited in the Protein Data Bank (PDB) [10]. This ensemble is then docked onto the receptor structure through restraints-guided rigid-body energy minimization, and then a fraction of the models is further optimized in successive flexible refinement stages. The flexibility of the peptide is also increased compared to default HADDOCK settings. The scoring function of HADDOCK selects, at each stage, the most favored conformations, i.e., those showing the most favorable interaction with the receptor based on a set of energy criteria. This protocol thus computationally approximates a combination of the induced fit and conformational selection models.

In the protocol presented here, we suggest a way to further improve our published protein-peptide docking protocol of HADDOCK by focusing more on the fact that peptides are inherently flexible. In addition to the three most common bound conformations, we use short MD simulations of the peptides, starting from the three conformations mentioned above, to obtain more detailed information on the conformational landscape of the free peptide. Structures selected from the MD simulations that correspond to different preferred conformational states of the free peptide supplement the three ideal structures to perform ensemble docking. Thereby, we aim at improving the conformational selection scheme in the rigid body docking stage by providing more plausible conformations of the peptide and subsequently improving the odds of success of the refinement stages. We illustrate this extended ensemble approach with cases from the benchmark of protein-peptide docking benchmark [11] and compare its performance with the standard three-conformation protocol we previously proposed [9].

2 Materials

2.1 Software Requirements

This protocol was designed and tested on a Linux cluster. Given the computational cost of molecular dynamics (MD) simulations, we recommend the use of multiple CPUs and/or GPUs. Local installation/compilation of the following programs is necessary, most of which are available for GNU/Linux and OS X operating systems:

1. *PyMOL:* PyMOL [12] is a 3D molecular structure visualization program, which can be obtained from http://pymol.org/. We use it here to generate the ideal peptide conformations (alpha-helix, polyproline-II, and extended).

2. *GROMACS*: GROMACS [13] is a molecular dynamics simulation program that includes a number of useful tools for analysis. The current protocol was run using version 5.0.4. Note that commands for versions 4.x and earlier might differ from those used here. The software is available free of charge at http://www.gromacs.org/

3. *Grace*: Grace (xmgrace) is a 2D plotting software, which provides a quick way to visualize plots generated during the execution of this protocol. It is available free of charge at http://plasma-gate.weizmann.ac.il/Grace

4. *MolProbity*: MolProbity [14] is a structure validation service that we use to assign the protonation states of Histidine residues. It can be downloaded from its GitHub repository: https://github.com/rlabduke/MolProbity. Note that other software/approaches can be used to define the charge state of Histidine residues.

5. *HADDOCK v2.2*: HADDOCK [15, 16] can be obtained free of charge for noncommercial users by filling and returning the license form available from http://www.bonvinlab.org/software/haddock2.2/download.html. Installation instructions can be found at http://www.bonvinlab.org/software/haddock2.2/installation.html. Moreover, the software can be used via a user-friendly web server [17, 18]. This protocol, however, makes use of a locally installed version of HADDOCK.

6. *Crystallography and NMR System (CNS) v1.3*: CNS [19, 20] is the engine used for energy minimization and molecular dynamics simulations in HADDOCK. Therefore, it is a main requirement for running HADDOCK. Note that HADDOCK v2.2 is designed to work with CNS v1.3, but recompiled using additional source code provided together with HADDOCK (see the cns1.3 directory in the HADDOCK distribution). The program is freely available for nonprofit users from http://cns-online.org/v1.3/

7. *NACCESS:* NACCESS is a useful tool that can be used to calculate the solvent accessible surface area of a molecule from a PDB structure file for both proteins and nucleic acids. It is free for academic users and can be obtained from http://www.bioinf.manchester.ac.uk/naccess/. A free alternative can be obtained from http://freesasa.github.io/ [21].

8. *ProFit*: ProFit is a protein least squares fitting program with many powerful features including flexible selection of fitting zones and atoms, calculation of RMS over different zones or atoms, etc. It can be obtained free of charge for academic users at http://www.bioinf.org.uk/software

2.2 Data Requirements

The structure of the receptor, preferably in the bound conformation, should be available (e.g., from the PDB, or via homology modelling) and the peptide sequence should be known. Additionally, for information-driven docking, experimental data pertaining to the interaction between the protein and peptide should be available to define the binding site on the receptor. The more information is available, the higher the chances for correct resulting models of the protein-peptide complex. Such information can be obtained from a variety of experimental techniques such as mutagenesis, chemical cross-linking, NMR chemical shift perturbations, etc. [22–24], or bioinformatics predictions (e.g., CPORT [25, 26]), all of which can be used to drive the docking in HADDOCK.

3 Methods

This protocol is divided into five major stages (summarized in Fig. 1):

1. Building the peptide in three extreme conformations.

2. Running MD simulations (50 ns) in explicit water for the peptide conformations built in **step 1**.

3. Analysis of the MD trajectories by Dihedral Principal Component Analysis (dPCA) and selection of the 30 most populated conformational states of the peptides.

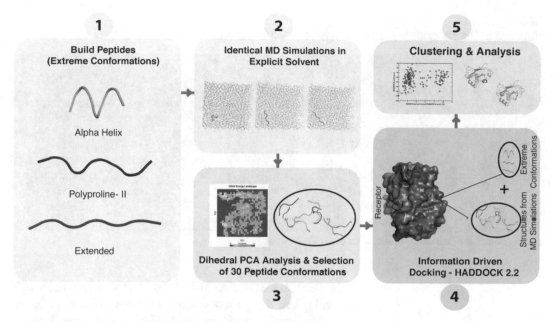

Fig. 1 Schematic overview of the workflow described in our protocol

4. Protein-peptide docking from an ensemble of 30 structures obtained from MD plus the three extreme conformations built in **step 1**, including available information on the binding sites, using HADDOCK 2.2.

5. Analysis of the docking solutions to select the best models.

To execute this protocol, the user is required to have working knowledge of a command-line interface and, preferably, experience with running MD simulations using GROMACS. In the following sections, commands are indicated in Courier font, and start with a ">" sign (note that here a command that should be given as a single line—i.e., indicated by a single ">," could span multiple lines). Text between < > in a command should be replaced by the proper selection/value.

For the sake of demonstration, unless otherwise specified, we will illustrate all the following steps using the complex of the TRAF domain of TRAF2 with the LMP1 binding peptide (PDB ID: 1CZY, *see* Table 1). All the necessary information to run this example case is provided in the supplementary material.

3.1 Generating Peptide Conformations for MD Simulations

As described in our original protein-peptide docking protocol [9] using the HADDOCK webserver, we use the *build_seq.py* script written by Robert L. Campbell to generate the starting structures of the peptides for MD simulations in PyMOL. The following steps describe this procedure, with **steps 3** and **4** describing the procedure to cap residues at the N- and C-termini. GROMACS can perform terminal capping during topology generation,

Table 1
Statistics of the six protein-peptide complexes used in the case study

Case difficulty	PDB ID complex	PDB ID free protein	Number of protein residues	Number of peptide residues	RMSD$_{bound/extended}$ (Å)
Easy (RMSD$_{bound/extended}$ ≤4 Å)	1DDV	1I2H	104	6	2.58
	1LVM	1LVB	214	6	1.54
Medium (4 Å < RMSD$_{bound/extended}$ ≤ 8 Å)	1CZY	1CZZ	168	7	1.94
	1D4T	1D1Z	101	11	3.27
Hard (RMSD$_{bound/extended}$ >8 Å)	1HC9	2ABX	74	13	11.04
	1NX1	1ALV	173	11	6.11

The classification of the case difficulty is based on Trellet et al. [9]

however, if one uses the AMBER99SB-ILDN force field [27] (which we will use in this protocol), it is necessary to manually add standard capping residues Acetyl and *N*-Methyl (abbreviated as ACE and NME). These steps can be ignored if the peptide has charged termini.

1. Start PyMOL and load the build_seq.py script from the PyMOL command line interface by typing:

```
> run build_seq.py
```

2. Build the structure:

```
> build_seq <peptide_name>, <peptide_seq>,
ss=<secondary_structure: helix, beta, or
polypro>
```

For example, to create an alpha-helical conformation of the peptide of the case 1CZY (peptide sequence: PQQATDD), type:

```
> build_seq alpha-peptide, PQQATDD, ss=helix
```

3. To add the capping residue to the N-terminus, first select the Nitrogen atom of the first residue (numbered as 2 by default in PyMOL) by typing the following command:

```
> select pk1, name n and resi 2
```

Alternatively, simply select the proper atom by clicking on it with the mouse in "editing mode," for which using a stick representation can be useful. Then select from the PyMOL menu:
"Build > Residue > Acetyl"

4. To add the capping residue to the C-terminus, select the Carbonyl carbon atom of the last residue by typing the following command (if you have followed the previous step, ensure that you have deselected all atoms before proceeding):

```
> select pk1, name c and resi <residue_number>
```

Alternatively, simply select the proper atom by clicking on it with the mouse. Then select from the PyMOL menu: "Build > Residue > N-Methyl"

5. Save the molecule by typing:
   ```
   > save <peptide_name>
   ```
   Or you could do it by clicking "File > Save Molecule".
   Repeat these steps to create all three starting conformations.

3.2 System Preparation for Running the MD Simulations with GROMACS 5.0.4

This protocol has been designed using the AMBER99SB-ILDN force field with periodic boundary conditions. To facilitate the combined analysis of the MD trajectories originating from different peptide conformations, it is recommended to make sure that every simulation contains the same number of water molecules. An example of a MD parameters file (*.mdp) suited for use with the AMBER99SB-ILDN force field is provided in the supplementary material. The commands described in the following subsections can also be performed by running the script *automd.sh* provided in the supplementary material (see the README section at the top of the script for instructions).

3.2.1 Determination of the Optimal Box Dimensions

We will perform the MD simulation in a rhombic dodecahedral box, to minimize the volume of the simulation cell. The dimensions of the box should be selected carefully to avoid interactions between neighboring periodic images. To determine the appropriate box dimensions, follow these steps:

1. Use the peptide in its extended conformation to determine the optimal box dimensions, considering that this represents the conformation with longest end-to-end distance.

2. Prepare the topology for the extended peptide:
   ```
   > gmx pdb2gmx -f protein.pdb -o protein.gro
   -ignh -ff amber99sb-ildn -water tip3p
   ```

3. Prepare a box that accommodates the peptide in its center and ensure that the minimum distance from the box edge to the peptide is at least half the nonbonded cutoff (as per the minimum image convention):
   ```
   > gmx editconf -f protein.pdb -o protein_
   pbc.gro -bt dodecahedron -d 1.0
   ```

4. Note the **"new box vectors"** value in the last lines of the output. This is the box vector that you must use with the "**-box**" flag during box preparation for the other peptides. An example is shown:
   ```
   system size : 2.108 2.098 0.985 (nm)
   diameter : 2.786(nm)
   center : 2.581 0.469 1.224 (nm)
   box vectors : 2.109 2.098 0.985 (nm)
   box angles : 90.00 90.00 90.00 (degrees)
   ```

```
box volume : 4.36 (nm^3)
shift : 1.009 3.120 0.468 (nm)
new center : 3.589 3.589 1.692 (nm)
new box vectors : 4.786 4.786 4.786 (nm)
new box angles : 60.00 60.00 90.00 (degrees)
new box volume : 77.51 (nm^3)
```

3.2.2 Determination of the Optimal Number of Water Molecules

Given the same box dimensions, the three peptides will accommodate a slightly different number of water molecules (to fill the box completely), depending on a variety of factors including surface areas and volumes. Therefore, it is necessary to determine this and use the smallest number of water molecules among the three systems (again to facilitate the combined analysis later—but not per se a requirement in principle). To determine this number of water molecules that the simulation system can accommodate, follow these steps:

1. Prepare the topology for the extended peptide:
   ```
   > gmx pdb2gmx -f protein.pdb -o protein.gro
   -ignh -ff amber99sb-ildn -water tip3p
   ```

2. Prepare the box, defining the box vector with the value obtained in Subheading 3.2.1:
   ```
   >gmx editconf -f protein.pdb -o protein_pbc.
   gro -bt dodecahedron -box 4.786
   ```

3. Run a first energy minimization in vacuum:
   ```
   >gmx grompp -f vacuum.mdp -c protein_pbc.
   gro -p topol.top -o protein_vac.tpr
   > gmx mdrun -v -deffnm protein_vac
   ```

4. Solvate the system:
   ```
   >gmx solvate -cp protein_vac.gro -cs spc216.
   gro -p topol.top -o protein_solvated.gro
   ```

5. Note the number of water molecules, referred to as "**SOL molecules**" that were added from the output:
   ```
   Output configuration contains 7605 atoms in
   2508 residues
   Volume : 77.518 (nm^3)
   Density : 983.104 (g/l)
   Number of SOL molecules: 2499
   ```

6. Repeat the above steps for all the peptides in all the conformations and note the lowest number of the three cases. Solvate the other two peptides (which are not the lowest) by using the "**-maxsol**" flag to set the optimal number of water molecules:
   ```
   > gmx solvate -cp protein_vac.gro -cs spc216.gro
   -p topol.top -o protein_solvated.gro -maxsol
   2499
   ```

3.2.3 System Preparation and Production MD

The following steps prepare the system for the production MD simulation. They must be repeated for all three peptide conformations. Since the initial energy minimization in vacuum and solvation has been done in the previous optimization steps, we can proceed with neutralizing the electrical charge of the system. Note that the various parameter files (.mdp) are provided in the supplementary material.

1. Add counter ions to the system to neutralize its electrical charge:

```
> gmx grompp -f ions.mdp -c protein_solvated.
gro -p topol.top -o protein_pp_ionize.tpr
> gmx genion -s protein_pp_ionize.tpr -p
topol.top -o protein_ionized.gro -neutral
```

If and when prompted to "Select a continuous group of solvent molecules," choose the group "SOL" or "non-Protein" or "Water" as they are all the same in our system.

2. Run energy minimization:

```
> gmx grompp -f ions.mdp -c protein_ionized.
gro -p topol.top -o protein_neutral_relaxed.
tpr
> gmx mdrun -v -deffnm protein_neutral_relaxed
```

3. Run a first equilibration under NVT conditions (constant volume):

```
> gmx grompp -f nvt.mdp -c protein_neutral_
relaxed.gro -p topol.top -o protein_nvt.tpr
> gmx mdrun -v -deffnm protein_nvt
```

4. Continue the equilibration under NPT conditions (constant pressure):

```
> gmx grompp -f npt.mdp -c protein_nvt.gro
-p topol.top -o protein_npt.tpr
> gmx mdrun -v -deffnm protein_npt -nt 48
-ntmpi 12 -pin on
```

5. Progressively release the position restraints during successive short MD runs:

```
> sed -e 's/1000   1000   1000/ 100    100
100/g' posre.itp > tmp.itp && mv tmp.itp
posre.itp
> gmx grompp -f npt.mdp -c protein_npt.gro
-p topol.top -o protein_npt_progrel100.tpr
> gmx mdrun -v -deffnm protein_npt_progrel100
> sed -e 's/100 100 100/ 10 10 10/g' posre.
itp > tmp.itp && mv tmp.itp posre.itp
> gmx grompp -f npt.mdp -c protein_npt_progrel100.
gro -p topol.top -o protein_npt_progrel10.tpr
> gmx mdrun -v -deffnm protein_npt_progrel10
```

6. Run a final short unrestrained equilibration MD step:

```
> gmx grompp -f unrestrained.mdp -c protein_
npt_progrel10.gro -p topol.top -o protein_
all_set.tpr
> gmx mdrun -v -deffnm protein_all_set
```

7. Run the production MD simulation:

```
> gmx grompp -f production.mdp -c protein_
all_set.gro -p topol.top -o protein_md.tpr
> gmx mdrun -v -deffnm protein_md
```

Please bear in mind that in all the steps involving energy minimization and MD simulations (which start with **gmx mdrun**), it is likely that GROMACS might use all of the processor(s) available to the user. It is possible that **gmx mdrun** may then give error messages if the system cannot be parallelized with the given conditions. In such cases, the user must set the number of threads and the PME ranks by using the -**nt** and -**ntmpi** flags with the **gmx mdrun** command. If one wishes to optimize parallelization, GROMACS offers a utility (**gmx tune_pme**) to find the optimal PME conditions for a given number of parallel processes (see the online GROMACS manual pages for more information, http://manual.gromacs.org/archive/5.0.4/programs/gmx-tune_pme.html).

3.3 Clustering by Dihedral Principal Component Analysis (dPCA)

The protocol to perform Dihedral PCA was adapted from the GROMACS documentation (http://www.gromacs.org/Documentation/How-tos/Dihedral_PCA). It consists of the following steps:

1. Concatenate the various trajectories (three in this particular case):

```
> gmx trjcat -f <path_trajectory_1> <path_
trajectory_2> <path_trajectory_3> -o all_
traj.xtc -settime
```

Ensure that you set the proper start time for the trajectories, which in our case would be 0, 50,000, and 100,000 in picoseconds (ps). This will not be done automatically. Hence, we use the -**settime** flag.

2. Make an index file containing all residues except the capping residues:

```
> gmx make_ndx -f protein_md.tpr -o index.
ndx
```

Follow the interactive menu to isolate the indices of the peptide residues without the capping groups.

3. Remove the capping groups from the trajectory:

```
> gmx trjconv -f all_traj.xtc -s protein_
md.tpr -n index.ndx -o protein_uncapped.xtc
```

Choose the index group corresponding to the peptide residues without the capping groups that were made in the previous step.

4. Create a topology for the peptide that excludes the capping residues:

```
> gmx convert-tpr -s protein_md.tpr -n index.
ndx -o protein_uncapped.tpr
```

Select the same index group as the previous step.

5. Now create a topology containing only the backbone atoms of all residues and also a trajectory file with the corresponding coordinates:

```
> gmx convert-tpr -s protein_uncapped.tpr
-o protein_bb.tpr
> gmx trjconv -f protein_uncapped.xtc -s
protein_uncapped -o protein_bb.xtc
```

When prompted, in both cases, choose the index group titled "Backbone," which by default would be option 4.

6. Create an index group for the backbone dihedral angles by creating an angle index file:

```
> gmx mk_angndx -s protein_bb.tpr -type dihedral
-o dangle.ndx
```

7. Open **dangle.ndx** with any text editor like, for example, vim or nano, and identify the index groups that correspond to the atoms forming the φ and ψ angles. Every four consecutive atom indices present in the index groups in the **dangle.ndx** file correspond to a single dihedral angle. For example, the first index group (Titled as [**Phi=180.0_2_10.46**]) in the text box shown below (which is a sample of **dangle.ndx** file), the atom indices 2 3 4 5 correspond to the first dihedral angle and 5 6 7 8 correspond to the second dihedral angle, and so on. The atoms that contribute to the φ angles are Co_i-N_{i+1}-$C\alpha_{i+1}$-Co_{i+1} and the ψ angles are N_i-$C\alpha_i$-Co_i-N_{i+1}. When your topology contains only the backbone atoms, like in our case here, the φ angles usually have the atom indices that start with "3," which represents the first backbone carbonyl carbon atom, and the ψ angles usually have atom indices that start with "1" that represents the first backbone nitrogen atom. In the example below, the atom indices that correspond to the φ angles are highlighted in bold and the ones that correspond to the ψ angles are highlighted in Italic:

```
[ Phi=180.0_2_10.46 ]
2 3 4 5 5 6 7 8 8 9 10 11
11 12 13 14 14 15 16 17 17 18 19 20
[Phi=0.0_2_1.13]
3 4 5 6 6 7 8 9 9 10 11 12
12 13 14 15 15 16 17 18 18 19 20 21
```

```
[Phi=0.0_3_1.76]
3 4 5 6 6 7 8 9 9 10 11 12
12 13 14 15 15 16 17 18 18 19 20 21
[Phi=180.0_1_1.88]
1 2 3 4 4 5 6 7 7 8 9 10
10 11 12 13 13 14 15 16 16 17 18 19
[ Phi=180.0_2_6.61 ]
1 2 3 4 4 5 6 7 7 8 9 10
10 11 12 13 13 14 15 16 16 17 18 19
[Phi=180.0_3_2.30]
1 2 3 4 4 5 6 7 7 8 9 10
10 11 12 13 13 14 15 16 16 17 18 19
```

8. Delete all the other index groups and combine the ones that correspond to φ and ψ angles that were identified, you can also rename the group name for convenience:

```
[Psi and Phi]
3 4 5 6 6 7 8 9 9 10 11 12
12 13 14 15 15 16 17 18 18 19 20 21
1 2 3 4 4 5 6 7 7 8 9 10
10 11 12 13 13 14 15 16 16 17 18 19
```

9. Extract the dihedral angles from the backbone trajectory file:
```
> gmx angle -f protein_bb.xtc -n dangle.ndx
-or dangle.trr -type dihedral
```

10. Note the number of atom positions that are filled with cos/sin of the angles:
```
There are 12 dihedrals. Will fill 8 atom po-
sitions with cos/sin
```
and create an index file (referred to in the following steps as "covar.ndx") containing indices from 1 to the number of positions that are filled with cos/sin:
```
[Covar]
1 2 3 4 5 6 7 8
```

11. Make a reference structure for constructing the covariance matrix using the dihedral angles, you can choose any frame for the "**-e**" flag:
```
> gmx trjconv -f dangle.trr -s protein_
bb.tpr -o resized.gro -n covar.ndx -e 100
```

12. Perform the covariance analysis:
```
> gmx covar -f dangle.trr -n covar.ndx -ascii
-xpm -nofit -nomwa -noref -nopbc -s resized.
gro
```

13. Since the first two eigenvectors (Principal Components) contain the largest variance in data, we will use them for the analysis by calculating the Potential of Mean Force (PMF) at every

time frame and projecting the vectors on each other to obtain a 2D free energy landscape.

```
> gmx anaeig -v eigenvec.trr -f dangle.
trr -s resized.gro -first 1 -last 2 -2d
2dproj_1_2.xvg
```

14. Convert the 2D plot into a 3D Gibbs free energy landscape by Bolzmann invertion:

```
> gmx sham -f 2dproj_1_2.xvg -notime
-bin bindex-1_2.ndx -lp prob-1_2.xpm -ls
gibbs-1_2.xpm -lsh enthalpy-1_2.xpm -g shamlog-1_2
-lss entropy-1_2.xpm
```

Any of the *.xpm files can be converted to *.eps using the GROMACS command "gmx xpm2ps"

15. As explained in the introduction of Subheading 3 (*see* also Fig. 1), our aim is to obtain 30 representative peptide conformations from the MD simulations. For this, we will use the 30 biggest bins (conformational states) identified by the PCA analysis. In order to identify the largest bins, check the "**shamlog-1_2.log**" file that contains the energy of each bin estimated by the Boltzmann Inversion method (by **gmx sham** in the previous step). Therefore, the size of the bin is inversely proportional to its energy. Bin indices sorted by their energies are listed after the "**Minima sorted after energy**" line in the "**shamlog-1_2.log**" file (with the lowest energy bin—the most populated one—having an energy of 0). This part is shown below where the energies are highlighted in bold and the bin indices are highlighted in Italic:

```
. . .

. . .

Minima sorted after energy
Minimum 0 at index 249 energy 0.000
Minimum 1 at index 766 energy 0.101
Minimum 2 at index 754 energy 0.884
Minimum 3 at index 918 energy 1.621
Minimum 4 at index 570 energy 1.669
Minimum 5 at index 594 energy 2.533

. . .

. . .
```

Note down the indices of the 30 lowest energy bins (Minimum 0 to 29).

16. From the "**bindex-1_2.ndx**" note down (any) one frame number from each of the 30 bins whose indices were sought in the previous step. In all our test cases, the first frame that is listed under the bin index was used. Samples from the "**bindex-1_2.ndx**" are shown below where the first three

structures in the biggest bins: 249, 766, and 570 are shown and the first frame in these respective bins is highlighted in bold:

```
. . .
. . .
[249]
232
233
235
. . .
. . .
[766]
121
122
159
. . .
. . .
[570]
141
149
151
. . .
. . .
```

17. Extract the selected frames from the combined trajectory. Please note that you should extract the frames from the trajectory that contains ALL atoms and not just the backbone atoms that were used to perform PCA. Also, the numbers noted in the previous steps are the frame numbers. These need to be multiplied by 50 (time frequency in ps at which coordinates were saved in the production MD—change this value if you have modified the save frequency in the parameter files) to obtain the time stamp. In the following example, we used frame number 219 as an example:

```
> gmx trjconv -f protein_nocap.xtc -s protein_
nocap.tpr -dump 10950 -o frame_219.pdb -pbc
mol
```

When prompted, choose the index group 1, containing all atoms.

Repeat this step to extract a total of 30 conformations (or less). We do not recommend using too many conformations for ensemble docking since it might lead to under-sampling of each conformation during the docking (a "dilution" problem), as the number of rigid-body docking models generated in HADDOCK is fixed.

Steps 9–17 can be automated by using the **dpca.sh** script that is given in the supplementary material. Ensure that the filenames are identical to what has been described in **steps 1–8** and the script should be run in the directory where all the files generated in these steps are present.

3.4 Docking and Scoring

3.4.1 Introduction to Docking Using HADDOCK v2.2

We provide here a brief introduction into the various docking and refinement stages of HADDOCK and its scoring functions. Before using a locally installed version of HADDOCK, it is recommended that the user also reads the HADDOCK manual (http://www.bonvinlab.org/software/haddock2.2/manual.html).

Initial Rigid Body Docking Stage (it0)

The initial stage of HADDOCK consists of a randomization of the starting orientations of the various molecules, followed by rigid body energy minimization (it0). During the randomization stage, the molecules are separated by a minimum of 25 Å and randomly rotated around their respective center of mass and translated within a 10 Å cube. During the following energy minimization, the molecules are treated as rigid bodies, i.e., all molecular bonds, angles, and internal rotations around bonds are frozen. The energy function being minimized contains the interaction restraints and the intermolecular van der Waals and electrostatics potentials. Typically, between 1000 (default) and 10,000 models are written to disk at this stage. The rigid body energy minimization is repeated multiple times internally (5 by default) with symmetrical solutions being sampled and minimized and only the best solution based on the HADDOCK score (see below) being written to disk. Typically, the top few hundred solutions (by default 200) are selected for further refinement.

Semi-flexible Refinement Stage (it1)

The second stage in HADDOCK consists of a semi-flexible refinement by high-temperature molecular dynamics in torsion angle space (it1), during which increasing flexibility is introduced in the interface of the complex. It consists of four stages:

1. High temperature rigid body dynamics (hot).
2. Rigid-body simulated annealing (cool1).
3. Semi-flexible simulated annealing with flexible side-chains at the interface (cool2).
4. Semi-flexible simulated annealing with fully flexible backbone and side-chains at the interface (cool3).

During this stage, the interface of the complex model is optimized. The flexible regions are defined by default automatically for all residues within 5 Å of a partner molecule (the user can, however, also define these manually). Usually, all the solutions at this stage are passed to the final refinement stage in explicit solvent.

Explicit Solvent Refinement Stage (water)

The last stage is a flexible refinement in explicit solvent (TIP3P water by default, but DMSO is also supported as a lipid environment mimic). The models from the previous stage are solvated in an 8.5 Å solvent shell and further refined using molecular dynamics simulations in Cartesian space, with weak position restraints on backbone atoms outside the interface, followed by a final energy minimization.

Scoring Functions
of HADDOCK

The HADDOCK score is a weighted sum of various energy and other scoring terms. It is used to rank the models at the various stages of the docking. The weight of various terms differs for each stage of the docking process. By default, the HADDOCK score is calculated as follows (a combination that has been shown to be successful for various types of systems, from protein-protein to protein-nucleic acid complexes):

$$it0: \quad E = 0.01E_{air} + 0.01E_{vdw} + 1.0E_{elec} + 1.0E_{desolv} - 0.01 \text{ BSA}$$

$$it1: \quad E = 0.1E_{air} + 1.0E_{vdw} + 1.0E_{elec} + 1.0E_{desolv} - 0.01 \text{ BSA}$$

$$\text{water}: \quad E = 0.1E_{air} + 1.0E_{vdw} + 0.2E_{elec} + 1.0E_{desolv}$$

where E_{air} is the ambiguous interaction restraints energy, E_{vdw} is the intermolecular van der Waals energy described by a 12-6 Lennard-Jones potential, E_{elec} is the intermolecular electrostatic energy described by a Coulomb potential, E_{desolv} is an empirical desolvation energy term [28], and BSA is the buried surface area in Å^2. The nonbonded energies are calculated using an 8.5 Å cutoff using the OPLS force field parameters [29]. Additional energy terms can be included in the scoring function if other restraint types are used.

3.4.2 Docking Protocol

As described in Subheadings 3.1–3.3, various peptide conformations have been obtained, either built in ideal conformations, or through MD simulations followed by clustering by dihedral PCA, from which we selected 30 representatives. This ensemble of peptide conformations (33 in total) along with the unbound receptor structure will be used for docking. The following protocol describes the steps to perform docking using a local version of HADDOCK. As for the MD part, we use the 1CZY case (*see* Table 1) as an example.

Defining Ambiguous
Interaction
Restraints (AIRs)

Since HADDOCK is an information-driven docking approach, it is necessary to define information about the putative interfaces (note that HADDOCK does offer various ab initio docking protocols, although these will not be described here). For a manual run, this means generating an ambiguous interaction restraints (AIRs) file (the webserver will simply take a comma-separated list of residue numbers). In order to generate AIRs, it is necessary to define active and passive residues at the interface of each molecule based on experimental data and/or bioinformatics predictions. In this example, we will assume that we know the peptide-binding site on the receptor—defining it from the known complex as all residues on the receptor that are within a 5 Å distance from the peptide. This represents of course an ideal situation. In case there is no available experimental information about interface residues, various bioinformatics predictors, including our webserver CPORT (http://haddock.science.uu.nl/services/CPORT/), can help predict interface residues. HADDOCK distinguishes between active and passive residues: Active residues must be in the interface (i.e., make contacts to either passive or active residues of the other molecule), otherwise an energy penalty will be paid; passive residues, in contrast, can be part of the interface, but are not penalized if otherwise. While active residues are typically selected based on high-quality experimental data or bioinformatics predictions, passive residues can be defined as the surface neighbors of active residues, or as a user-defined surface patch. In this protocol, we will define all residues of the peptide as passive. For our example 1CZY, the protein active residues are:

"41,42,44,60,62,66,67,68,77,114,115,116,117, 120,121,122,123,131,132,133,134,135,136,137," and the peptide passive residues:

"1,2,3,4,5,6,7."

After obtaining the list of active and passive residues for each molecule, the webserver for generating ambiguous interaction restraints can be used to generate the AIRs file required for HADDOCK. It can be found at http://www.bonvinlab.org/software/Haddock2.2/generate_air.html. A description of the format of the various restraints files in HADDOCK can be found in Box 4 of the original HADDOCK server article [17]. In the section "Active residues for 1st molecule" and "Passive residues for 2nd molecule," fill in the 1CZY protein active residues and peptide passive residues, respectively. In the section "Segid of 1st molecule:" and "Segid of 2nd molecule:", specify the segment IDs to use for each molecule during the docking (typically A and B). Then move to the bottom of the page, click on "Generate AIR restraints" to generate AIRs. You should be redirected to a new page that contains all the AIRs. Save this page as a restraint file "**restraints.tbl**" and copy it into your working directory. The "**restraints.tbl**" file will be used when setting up your new HADDOCK project.

Determining the Protonation State of Histidine Residues

In general, small charge differences can have a strong impact on the results of the docking. It is therefore advisable to define the protonation state of Histidine residues prior to docking (if not, by default Histidines will be considered positively charged). The HADDOCK webserver can do that automatically, using MolProbity [14] to guess the most plausible Histidine protonation states. Here, we provide a Python script that uses the "reduce" executable of MolProbity to assign the protonation state. Run the script on the pdb file(s) to determine the protonation states of Histidines:

```
> python ./molprobity.py ./protein.pdb
```

The *reduce* program will be used to add hydrogen atoms and perform basic validation checks and optimizations on the protein structure to generate a temporary new structure. Based on the presence and location of the hydrogen atoms in a Histidine residue, the script will determine its protonation state. For each histidine, if atom HD1 and HE2 exist, the doubly protonated, charged histidine HIS+ state is assigned, if HD1 exists but not HE2, a singly protonated histidine HISD is assigned, and if HE2 exists but not HD1, a singly protonated histidine HISE is assigned.

The output of the script is shown below for the unbound form of 1CZY:

```
## Executing Reduce to assign histidine pro-
tonation states
## Input PDB: protein.pdb
HIS (73) --> HISD
HIS (109) --> HISE
```

Save this information since it will be used later while editing the run parameters.

Starting a New Project

To start a new project, it is necessary to generate the **"new.html"** file that contains the information about all the required input data for the docking. An online tool to prepare this file is available on the HADDOCK webpage: http://www.bonvinlab.org/software/Haddock2.2/start_new.html.

Three sections need to be filled. In the first section "HADDOCK and project setup," you should define the path to your local HADDOCK installation under "Current HADDOCK program directory" and the path to where your project will be created in "Path of the new project." Then, define the "Run number," 1 in this case, and the "Number of molecules for docking (max. 6)," 2 in this particular case.

In the second section, "Define the various molecules for docking", set "PDB file of 1st molecule" to the absolute path of the protein structure, "PDB file of 2nd molecule" to the absolute path of the peptide (use one of the peptide PDB files for this), use the default values for "Segid of 1st molecule" and "Segid of 2nd molecule" (unless of course you changed those values when creating the AIR file). Since we will make use of an ensemble of peptide conformations for the docking, a list file containing all absolute (or relative) paths of those conformations (3 extreme + 30 MD conformations) should be created, and set "file list for 2nd molecule (opt.)" to the absolute path of this list file. Note that in principle, instead of absolute paths, relative paths can also be used, e.g., "./peptide.pdb."

In the last section "Define the various restraint files," the "AIR restraints" should be set to the path of the **"restraints.tbl"** file generated in Subheading "Defining Ambiguous Interaction Restraints (AIRs)."

Move to the bottom of the page and click on "Save updated parameters." You should see a new webpage containing the generated file. Save it as the "new.html" file, and then copy it into your working directory.

An example of the **"new.html"** file for 1CZY (also provided in supplementary material) is given below:

```
<html>
<head>
<title>HADDOCK - start</title>
</head>
<body bgcolor=#ffffff>
<h2>Parameters for the start:</h2>
<BR>
<!-- HADDOCK -->
HADDOCK_DIR=/home/software/haddock/
haddock2.2<BR>
N_COMP=2<BR>
PDB_FILE1=../pdb_files/protein.pdb<BR>
PDB_FILE2=../pdb_files/MD_conformations/md_1.
pdb<BR>
PDB_LIST2=../pdb_files/structures.list<BR>
PROJECT_DIR=./<BR>
PROT_SEGID_1=A<BR>
PROT_SEGID_2=B<BR>
AMBIG_TBL=./restraints.tbl<BR>
RUN_NUMBER=1<BR>
submit_save=Save updated parameters<BR>
</h4><!-- HADDOCK -->
</body>
</html>
```

Now go to your working directory, make sure that the **"new. html"** file is present and then type the command (which should be defined after proper installation of HADDOCK and sourcing of the configuration script):

```
> haddock2.2
```

A new directory "runX" (X is a number that is defined as "Run number" above) will be created, containing several subdirectories and the important **"run.cns"** parameter file, which will need to be edited in the next step. For details on the various subdirectories, please refer to the online HADDOCK manual at http://www. bonvinlab.org/software/Haddock2.2/start_new_help.html.

Setting the Docking Parameters

It is necessary to modify some docking parameters in the freshly generated **"run.cns"** file. For this, the user can use an online tool or edit the file directly with a text editor. The easiest way to modify this file is to use our online tool: In the http://www.bonvinlab. org/software/Haddock2.2/Haddock-start.html page (best viewed with Firefox or Google Chrome), upload your **"run.cns"** file and click on "Edit file." You will be redirected to a new webpage that contains all parameters to run HADDOCK. For details on all the parameters, see the online manual (http://www.bonvinlab.org/

software/Haddock2.2/run.html). Below, we will deal with only the parameters that should be adjusted to run our protein-peptide protocol.

Histidine patches: These parameters are used to define the protonation state of Histidines. By default, a Histidine is doubly protonated and thus positively charged in HADDOCK. The histidine parameters only need to be defined when a histidine should be singly protonated (HISD or HISE). The information on the protonation state of the various Histidines obtained in Subheading "Determining the Protonation State of Histidine Residues," will be used here. For our example 1CZY in that section, Histidine 73 should be in HISD state and Histidine 109 in HISE. In the section, "Patch to change doubly protonated HIS to singly protonated histidine (HD1)," set the first residue of "molecule (Protein) A" to 73. Then in "Patch to change doubly protonated HIS to singly protonated histidine (HE2)," set the first residue of "molecule (Protein) A" to 109. In this particular example, the peptide does not contain any Histidine; otherwise, the same procedure should be followed for the second molecule.

Definition of fully flexible segments: This section defines the segments that are defined as fully flexible during all stages of it1. In this protocol, because of the intrinsic high flexibility of peptides, we will define all residues of the peptide as fully flexible. Therefore, set the "Start Residue" to 1 and the "End Residue" to the residue number of the last residue in the peptide, here should be 7 for the case 1CZY.

Topology and parameter files: The linkage file in this section allows defining the charged states of the N- and C-termini of the protein and peptide. If the protein or peptide is a fragment of a larger protein or was capped in experiments for some specific reason, the N- and/or C-terminus should be uncharged. In HADDOCK, the default linkage file used to generate the topology is "protein-allhdg5-4.link," which produces charged N and C termini. For uncharged termini, the linkage file "protein-allhdg5-4-noter.link" should be used. For uncharged N-terminus and charged C-terminus, use "protein-allhdg5-4-noNter.link"; for charged N-terminus and uncharged C-terminus, use "protein-allhdg5-4-noCter.link."

For our particular example 1CZY, we will use "protein-allhdg 5-4-noter.link" for the peptide since the peptide is a fragment of a larger protein and capped in its N-terminus.

Number of structures to dock: Due to the flexibility of the peptide, the number of decoys to generate should be increased to improve the sampling of all conformations of the protein-peptide complex. Since we used 33 peptide conformations (3 extreme conformations + 30 MD cluster representatives) as initial ensemble for docking, we will change the "number of structures for rigid body docking" from

1000 to 9900 (in that way each starting conformation is used 300 times), and the "number of structures for refinement" from 200 to 400.

Number of MD steps in the docking protocol: To improve the sampling of protein-peptide interactions and in particular to allow the peptide to better sample its conformation in the context of the receptor, the number of MD steps for the it1 stages also needs to be increased: From 500/500/1000/1000 to 2000/2000/4000/4000 for the hot, cool1, cool2, and cool3 stages in it1, respectively.

Final explicit solvent refinement: Just like the number of structures to dock, the "number of structures for the explicit solvent refinement" is increased from 200 to 400.

Analysis and clustering: The "clustering method" is set to RMSD, and due to the smaller size of peptide the "RMSD cutoff for clustering" is decreased to 5 Å.

Parallel jobs: The user should specify the local "queue command" (e.g., simply csh if using a single computer, or a batch queue submission command for a cluster (e.g., qsub)), the absolute path of "cns executable" and define for "cpunumber" a number that matches the number of cores on the system (or the number of allocated slots in the queue in the batch system). The internal job dispatching routines will use this setting to limit the number of concurrent refinement jobs. Note that for rigid-body stage jobs, given the computational efficiency of this algorithm, several individual minimizations are bundled together in each job.

After updating the parameters above, click "Save updated file" on the page bottom, and then save the file as a new run.cns and copy it to the runX directory to replace the old one.

The following lines describe how these changes would look like if done manually simply in a text editor:

Histidine patches:
```
A_hisd_resid_1=73;
A_hise_resid_1=109;
```

The histidine protonation states are defined by these parameters.

Definition of fully flexible segments:
```
B_start_fle_1="1";
B_end_fle_1="7";
```

The parameters define the fully flexible segments in it1 step.

Topology and parameter files:
```
prot_link_B="protein-allhdg5-4-noter.link";
```

The parameter sets the charged states of N-terminus and C-terminus of the molecule. Here, it sets both termini uncharged for the peptide of 1CZY.

Number of structures to dock:

```
structures_0=9900;
structures_1=400;
```

These parameters set the number of structures to dock for it0 and it1.

DOCKING protocol:

```
initiosteps=2000;
cool1_steps=2000;
cool2_steps=4000;
cool3_steps=4000;
```

These parameters set the number of MD steps for hot, cool1, cool2, and cool3 stage in it1 step.

Final explicit solvent refinement:

```
waterrefine=400;
```

The parameter sets the number of structures to dock for the water step.

Analysis and clustering:

```
clust_meth="RMSD";
clust_cutoff=5;
```

These parameters set the clustering method and its cutoff.

Parallel jobs:

```
queue_1="csh";
cns_exe_1="/home/software/bin/cns";
cpunumber_1=50;
```

These parameters set the local queue command, path of cns program, and the number of parallel jobs.

Start the Docking

After ensuring that all parameters have been properly defined as explained in the previous steps, navigate to the runX directory and launch the docking by typing:

```
> haddock2.2 &>HADDOCK.log &
```

The docking should start in background and the information about the run will be written to the HADDOCK.log file. During the docking process, HADDOCK writes docking decoys in PDB format and outputs the ranked PDB files in file.cns, file.list and file.nam files at the end of each docking step in runX/structures/it0, runX/structures/it1 and runX/structures/it1/water directories, respectively. The file.cns, file.list and file.nam files contain a list of generated structures sorted on HADDOCK score. For it1 and water stages, the generated structures are automatically analyzed and the results are placed in the runX/structures/it1/analysis and runX/structures/it1/water/analysis directories, respectively.

As described above, the results of automatic analysis for it1 and water steps are placed in the analysis directories under each directory, respectively. Here, we will describe some relevant files containing useful information.

`fileroot_ave.pdb` and `fileroot_X.pdb`:

These are the models in PDB format. fileroot_ave.pdb is the average structure generated by superimposing the structures of docking solutions on the backbone atoms of interface residues, while the superimposed models are fileroot_N.pdb (N is a number that corresponds to the ranking of the model in the file.list file). The interface residues are automatically determined from an analysis of all generated models. Note that the average model might not be of much relevance in cases where very different solutions are sampled.

`fileroot_rmsd.disp`:

This file contains the pairwise RMSD matrix calculated over all models. The RMSD calculated here is the ligand interface RMSD, i.e., the structures are fitted on backbone atoms of interface residues of the first molecule and the RMSD is calculated on the interface backbone atoms of the second molecule. This file is used as input for the RMSD clustering. If the clustering method defined in run.cns is FCC (fraction of common contacts) instead, the name of this file will become fileroot_fcc.disp. and contain the fraction of common contacts between models [30].

`cluster.out`:

The file contains the list of clusters generated based on the matrix in the fileroot_rmsd.disp or fileroot_fcc.disp file, depending on the clustering method used. The clusters are numbered according to the size of the cluster, e.g., the largest cluster is cluster 1. The cluster.out file is used as input for analysis of clusters (ana_cluster.csh) described below.

`energies.disp, edesolv.disp` and `ene-reside.disp`:

These files contain various energy terms. The bonded and nonbonded energies and buried surface area for each structure are written to energies.disp, together with the average values over the ensemble. The empirical desolvation energies are contained in edesolv.disp. The ene-residue.disp file lists the per-residue intermolecular energies for all interface residues.

`hbonds.disp, ana_hbonds.lis` and `nbcontacts.disp, ana_nbcontacts.lis`:

The hbonds.disp file contains the intermolecular hydrogen bonds for each model, while the ana_hbonds.lis file lists all hydrogen bonds with their occurrence and average distance.

Similar information for intermolecular hydrophobic contacts is provided in nbcontacts.disp and ana_nbcontacts.lis.

`geom.disp:`

The file contains the averaged deviations from ideal covalent geometry (bonds, angles, impropers, and dihedrals) for each structure and averaged over all structures.

`noe.disp:`

The file contains the number of distance restraints violations per structure and averaged over the ensemble over all distance restraint classes and for each class (unambiguous, ambiguous, hbonds). Similar files are generated for dihedral angle restraints (dihedrals.disp), residual dipolar coupling restraints (sani.disp), intervector projection angle restraints (vean.disp), diffusion anisotropy restraints (dani.disp), and pseudo contact shifts restraints (pcs.disp).

`ana_XXX.lis:`

These files report restraint violations over the ensemble of models, giving the number of times various restraints are violated, the average distance, and the violation per restraint. The XXX can be dihed_viol, dist_viol_all, hbond_viol, noe_viol_all, noe_viol_ambig, and noe_viol_unambig.

Manual Analysis

Besides the automatic analysis, the user should also perform manual analysis of the models and clusters. For this purpose, a number of scripts are provided in the runX/tools directory.

1. *Collecting model statistics using ana_structures.csh*: This script extracts information from the header of the PDB files such as various energy terms, violation statistics, and buried surface area and calculates the overall backbone RMSD of each structure superimposed on the top ranking model. To run it type:

`>$HADDOCKTOOLS/ana_structures.csh`

in the runX/structures/it1 or runX/structures/it1/water directory.

It generates a number of "file.nam_XXX" and "structures_XXX-sorted.stat" files (XXX is energy term). The "file.nam_XXX" file contains the values of the respective energy (or other) term XXX for all structures. All of these terms are combined into one file and sorted in different ways, generating the corresponding "structures_XXX-sorted.stat" files. Of these, structures_haddock-sorted.stat is usually the most important, which corresponds to the HADDOCK score ranking.

Relationships between these energy terms can be checked by plotting. For this purpose the make_ene-rmsd_graph.csh script is provided. For example, the user can make a plot of the HADDOCK score as a function of the RMSD:

```
> $HADDOCKTOOLS/make_ene-rmsd_graph.csh 3 2
structures_haddock-sorted.stat
```

It will generate a ene_rmsd.xmgr file in xmgr format which can be displayed with xmgr or xmgrace:

```
> xmgrace ene_rmsd.xmgr
```

2. *RMSD clustering using cluster_struc*: This program is used in HADDOCK to perform clustering based on RMSDs. In the process of automatic analysis, if RMSD clustering was defined in run.cns, it has been run automatically in each analysis directory. However, the user can run it again to try different clustering cutoffs depending on the complex studied. It takes the fileroot_rmsd.disp file as input:

```
> $HADDOCKTOOLS/cluster_struc [-f] fileroot_
rmsd.disp cutoff min_cluster_size>cluster.
out
```

Here, the -f is an option for full linkage clustering algorithm (not used by default), the cutoff is the RMSD cutoff used to determine if two structures belong in the same cluster, and min_cluster_size is the minimum number of models to define a cluster.

The output in the cluster.out file looks like:

```
Cluster 1 -> 2 4 5 9 11 12 14 20 121 127 129
141 145 156 170
Cluster 2 -> 1 48 51 56 58 93 96 139 161 164
171 181 187
Cluster 3 -> 36 7 37 49 112 148
```

The numbering of the clusters is based on the size of the cluster, and the numbering of the structures corresponds to the position of the structure in the file.list file. The first structure of each cluster corresponds to the cluster center and the other structures are sorted according to their index.

3. *FCC clustering using cluster_fcc.py*: This Python script is used to perform clustering based on the fraction of common contacts (FCC) if FCC clustering was defined in run.cns. FCC is an alternative metric to measure the structural similarity between two docking models, based on the network of residue-residue interactions at the interface of the models. As for RMSD clustering, the user can choose to run it again to try different clustering cutoffs depending on the complex studied. It takes the fileroot_fcc.disp file as input:

```
> $HADDOCKTOOLS/cluster_fcc.py fileroot_fcc.
disp cutoff -c min_cluster_size > cluster.out
```

where cutoff is the FCC cutoff used to determine if two structures belong to the same cluster, and min_cluster_size is the minimum number of models to define a cluster.

4. *Analysis of clusters using ana_cluster.csh*: The ana_clusters.csh script calculates various statistics on a per cluster level. It takes the cluster.out file as input. To run it type:

```
>$HADDOCKTOOLS/ana_clusters.csh [-best #]
analysis/cluster.out
```

in the runX/structures/it1 or runX/structures/it1/water directory. The -best # is an optional argument to generate additional files with calculation only on the best # structures of a cluster, e.g., the top four structures of a cluster sorted on their HADDOCK score as done by default by the HADDOCK web server; this allows removing the dependency of the calculated averages on size of the various clusters.

Like the output of the ana_structures.csh script, the ana_clusters.csh script also generates a number of files containing values of different energy terms XXX but over models belonging to the same cluster (clustX), e.g., file.nam_clustX_XXX files, based on the list of models for each cluster stored in the file.nam_clustX files. The script also calculates averages of various energy terms for each cluster, which can be found in the various cluster_XXX.txt files. All these are combined and sorted in various ways in clusters_XXX-sorted.stat files. If the option "–best #" is used, additional files will be created containing the average values over the best # structures of each cluster, i.e., file.nam_clustX_best#, cluster_XXX.txt_best# and clusters_XXX-sorted.stat_best# files. Of all these files, clusters_haddock-sorted.stat and clusters_haddock-sorted.stat_best# are usually the most relevant.

5. *Rerunning automatic analysis on the basis of clusters*: After having performed the cluster-based analysis, it is possible to rerun the HADDOCK automatic analysis for a given cluster. For this the user needs to create cluster-specific files (e.g., file.cns_clustX_best#, file.list_clustX_best#, and file.nam_clustX_best#) and directory (e.g., analysis_clustX_best#). To simplify this process, the make_links.csh script is provided. To run it type:

```
>$HADDOCKTOOLS/make_links.csh clustX_best#
```

This will automatically move the original file.cns, file.list and file.nam files and analysis directory to new files and directory by adding a suffix _all, and then make links to cluster-specific files and directory, i.e.,

```
file.cns    -> file.cns_clustX_best#
file.list    -> file.list_clustX_best#
file.nam    -> file.nam_clustX_best#
analysis    -> analysis_clustX_best#
```

To rerun the analysis, go back to the runX directory and restart HADDOCK:

```
> haddock2.2
```

Once finished, the user will find a new directory analysis_clustX_best# that contains cluster-specific result files described in Subheading "Automatic Analysis."

3.5 Case Studies

Following the protocol described above, we performed unbound/unbound docking for six protein-peptide complexes from the protein-peptide benchmark [11], using a combination of ideal peptide conformations and MD cluster representatives. These six complexes (Table 1) correspond to two easy, two medium, and two hard docking cases (based on the classification of Trellet et al. [9]). The length of the peptides in these systems varies from 6 to 13 amino acids, while the proteins are much larger, varying from 74 to 214 residues.

We performed the docking using both the original three conformations (alpha-helix, polyproline-II, and extended) protocol (regular protocol) [9] and by adding 30 additional conformations sampled in MD simulations as described in this chapter (MD-based protocol). To assess the performance of the docking, the interface RMSD measure from the community-wide experiment CAPRI (Critical Assessment of PRedicted Interactions) [31, 32] is used as criteria, which is calculated on interface residues by superimposing the docking solutions to the crystal structure of bound complex. In the case of protein-peptide complexes, in CAPRI a docking solution is considered acceptable if its interface RMSD is less than 2 Å.

We summarized in Table 2, for both the regular protocol and the current MD-based protocol, the number of acceptable models out of the 400 water-refined models, together with the rank of the first acceptable model and the first acceptable cluster in the list of models or clusters sorted on HADDOCK score. This allows us to compare the docking performance of both protocols. For the easy cases, 1DDV and 1LVM, the MD-based protocol generated less acceptable models. This is due to the "dilution" problem mentioned above: with a larger number of starting conformations, only few will lead to acceptable models and accordingly the total number of acceptable models is expected to decrease depending on the information used to drive the docking. On the other hand, when large conformational changes are taking place, it seems that the MD-based protocol does improve the number of acceptable models (1CZY and 1NX1) and the ranking. However, both protocols fail for two cases, for long peptides (11 and 13 amino acids, for 1D4T and 1HC9, respectively), with rather large conformational changes. 1HC9 is especially challenging since the peptide forms a b-hairpin conformation in its bound form that is not sampled in the starting models. This clearly illustrates the challenges of

Table 2
Comparison of (A) unbound/unbound docking performance between the original three-conformations protocol and the MD-based protocol presented in this chapter and (B) interface RMSD of the best and first acceptable model using the MD-based protocol

(A)

Case difficulty	PDB ID complex	Number of acceptable[a] models		Rank[b] of first acceptable model		Rank[b] of first acceptable cluster[c]	
		Regular protocol	MD-based protocol	Regular protocol	MD-based protocol	Regular protocol	MD-based protocol
Easy	1DDV	39	30	11	3	5	1
	1LVM	176	92	1	1	1	1
Medium	1CZY	74	175	1	1	1	1
	1D4T	0	0	NA	NA	NA	NA
Hard	1HC9	0	0	NA	NA	NA	NA
	1NX1	58	62	6	3	3	2

(B)

Case difficulty	PDB ID complex	i-RMSD (Å)/rank of best model (Å)	i-RMSD (Å)/rank of first acceptable[a] model
Easy	1DDV	1.74/95	1.96/3
	1LVM	1.26/40	1.64/1
Medium	1CZY	0.93/42	1.31/1
	1D4T	2.31/3	NA
Difficult	1HC9	4.42/131	NA
	1NX1	1.28/43	1.59/3

The original protocol [9] uses three peptide conformations (alpha-helix, polyproline-II, and extended), while 30 additional conformations sampled in MD simulations were added in the MD-based protocol. Both protocols output 400 docking models at the end of the HADDOCK process
[a]A model is defined as acceptable if its interface RMSD (i-RMSD) from the reference is less than 2 Å according to the criteria of CAPRI. The i-RMSD is calculated on interface backbone atoms of docking models superimposed onto the crystal structure
[b]The ranking of the first acceptable model/cluster is the position of the first acceptable model/cluster in the list of models/clusters sorted on HADDOCK score
[c]A cluster is defined as acceptable when at least one model is acceptable within the top four models

protein-peptide docking. The best models for each case are shown in Fig. 2, superimposed onto the reference crystal structure.

In conclusion, the presented results, although taken from a limited number of cases, seem to indicate that the presented MD-based protocol is better at generating acceptable models with HADDOCK. This was however for an ideal case where the binding site on the receptor is well defined. Lack of proper information, high-flexibility, and large conformational changes still remain major challenges to be addressed in protein-peptide interaction modeling.

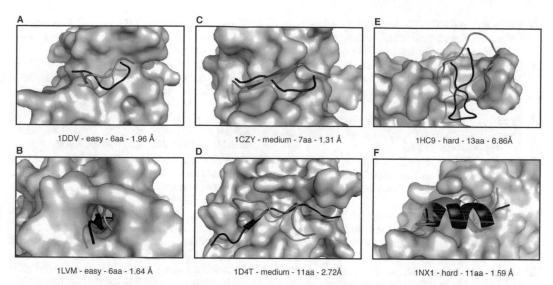

Fig. 2 View of the HADDOCK top-ranking first acceptable model for the six protein-peptide unbound/unbound docking cases using the MD-based protocol. The PDB ID as well as difficulty, peptide length and interface RMSD values are indicated for all six cases. The model selected for illustration is the acceptable model with the best rank for (**a**) 1DDV, (**b**) 1LVM, (**c**) 1CZY, and (**f**) 1NX1, and the model with the best HADDOCK score for (**d**) 1D4T and (**e**) 1HC9 (no acceptable models were generated for those two cases). The model peptide is shown in *orange* together with the reference peptide in the crystal structure of the complex in *black*. Docking model and crystal structure were superimposed on backbone atoms of the protein. The protein in the crystal structure is shown in surface representation. Figure generated with PyMOL [12]

Acknowledgments

C. Geng acknowledges financial support from the China Scholarship Council, grant NO. 201406220132. This protocol is adapted from a computer practical offered to our chemistry bachelor students [33].

References

1. Craik DJ, Fairlie DP, Liras S, Price D (2013) The future of peptide-based drugs. Chem Biol Drug Des 81:136–147

2. Otvos L (2008) Peptide-based drug design: here and now. Methods Mol Biol 494:1–8

3. Fischer E (1894) Einfluss der Configuration auf die Wirkung der Enzyme. Berichte der Dtsch Chem Gesellschaft 27:2985–2993

4. Koshland DE (1959) Enzyme flexibility and enzyme action. J Cell Comp Physiol 54: 245–258

5. Kumar S, Ma B, Tsai CJ, Sinha N, Nussinov R (2000) Folding and binding cascades: dynamic landscapes and population shifts. Protein Sci 9:10–19

6. Rubin MM, Changeux JP (1966) On the nature of allosteric transitions: implications of non-exclusive ligand binding. J Mol Biol 21: 265–274

7. Monod J, Wyman J, Changeux J-P (1965) On the nature of allosteric transitions: a plausible model. J Mol Biol 12:88–118

8. Vogt AD, Di Cera E (2012) Conformational selection or induced fit? A critical appraisal of the kinetic mechanism. Biochemistry 51: 5894–5902

9. Trellet M, Melquiond ASJ, Bonvin AMJJ (2013) A unified conformational selection and induced fit approach to protein-peptide docking. PLoS One 8:e58769

10. Diella F et al (2008) Understanding eukaryotic linear motifs and their role in cell signaling and regulation. Front Biosci 13:6580–6603

11. London N, Movshovitz-Attias D, Schueler-Furman O (2010) The structural basis of peptide-protein binding strategies. Structure 18:188–199

12. Pymol T, Graphics M, Schrödinger V. The PyMOL Molecular Graphics System, Version 1.5.0.4 Schrödinger, LLC. 5

13. Abraham MJ et al (2015) GROMACS: high performance molecular simulations through multi-level parallelism from laptops to super-computers. SoftwareX 1:19–25

14. Chen VB et al (2010) MolProbity: all-atom structure validation for macromolecular crystallography. Acta Crystallogr D Biol Crystallogr 66:12–21

15. Dominguez C, Boelens R, Bonvin AMJJ (2003) HADDOCK: a protein-protein docking approach based on biochemical or biophysical information. J Am Chem Soc 125:1731–1737

16. de Vries SJ et al (2007) HADDOCK versus HADDOCK: new features and performance of HADDOCK2.0 on the CAPRI targets. Proteins 69:726–733

17. de Vries SJ, van Dijk M, Bonvin AMJJ (2010) The HADDOCK web server for data-driven biomolecular docking. Nat Protoc 5:883–897

18. van Zundert GCP et al (2015) The HADDOCK2.2 Web Server: user-friendly integrative modeling of biomolecular complexes. J Mol Biol. doi:10.1016/j.jmb.2015.09.014

19. Brünger AT et al (1998) Crystallography & NMR system: a new software suite for macromolecular structure determination. Acta Crystallogr D Biol Crystallogr 54:905–921

20. Brunger AT (2007) Version 1.2 of the crystallography and NMR system. Nat Protoc 2:2728–2733

21. Mitternacht S (2016) FreeSASA 0.6.2: Solvent accessible surface area calculation. doi:10.5281/zenodo.44748

22. Rodrigues JPGLM, Bonvin AMJJ (2014) Integrative computational modeling of protein interactions. FEBS J 281:1988–2003

23. Karaca E, Bonvin AMJJ (2013) Advances in integrative modeling of biomolecular complexes. Methods 59:372–381

24. Melquiond ASJ, Bonvin AMJJ (2010) Data-driven docking: using external information to spark the biomolecular rendez-vous. In: Zacharias M (ed) Protein-protein complexes analysis, modeling and drug design. Imperial College Press, Munich, pp 183–209. doi:10.1142/9781848163409_0007

25. de Vries SJ, Bonvin AMJJ (2008) How proteins get in touch: interface prediction in the study of biomolecular complexes. Curr Protein Pept Sci 9:394–406

26. de Vries SJ, Bonvin AMJJ (2011) CPORT: a consensus interface predictor and its performance in prediction-driven docking with HADDOCK. PLoS One 6:e17695

27. Lindorff-Larsen K et al (2010) Improved side-chain torsion potentials for the Amber ff99SB protein force field. Proteins 78:1950–1958

28. Fernández-Recio J, Totrov M, Abagyan R (2004) Identification of protein-protein interaction sites from docking energy landscapes. J Mol Biol 335:843–865

29. Jorgensen WL, Tirado-Rives J (1988) The OPLS [optimized potentials for liquid simulations] potential functions for proteins, energy minimizations for crystals of cyclic peptides and crambin. J Am Chem Soc 110:1657–1666

30. Rodrigues JPGLM et al (2012) Clustering biomolecular complexes by residue contacts similarity. Proteins 80:1810–1817

31. Janin J (2005) Assessing predictions of protein-protein interaction: the CAPRI experiment. Protein Sci 14:278–283

32. Lensink MF, Wodak SJ (2013) Docking, scoring, and affinity prediction in CAPRI. Proteins 81:2082–2095

33. Rodrigues JPGLM, Melquiond ASJ, Bonvin AMJJ (2016) Molecular Dynamics Characterization of the Conformational Landscape of Small Peptides: A series of hands-on collaborative practical sessions for undergraduate students. Biochemistry and Molecular Biology Education 44:160–167

Chapter 9

Modeling Peptide-Protein Structure and Binding Using Monte Carlo Sampling Approaches: Rosetta FlexPepDock and FlexPepBind

Nawsad Alam and Ora Schueler-Furman

Abstract

Many signaling and regulatory processes involve peptide-mediated protein interactions, i.e., the binding of a short stretch in one protein to a domain in its partner. Computational tools that generate accurate models of peptide-receptor structures and binding improve characterization and manipulation of known interactions, help to discover yet unknown peptide-protein interactions and networks, and bring into reach the design of peptide-based drugs for targeting specific systems of medical interest.

Here, we present a concise overview of the Rosetta FlexPepDock protocol and its derivatives that we have developed for the structure-based characterization of peptide-protein binding. Rosetta FlexPepDock was built to generate precise models of protein-peptide complex structures, by effectively addressing the challenge of the considerable conformational flexibility of the peptide. Rosetta FlexPepBind is an extension of this protocol that allows characterizing peptide-binding affinities and specificities of various biological systems, based on the structural models generated by Rosetta FlexPepDock. We provide detailed descriptions and guidelines for the usage of these protocols, and on a specific example, we highlight the variety of different challenges that can be met and the questions that can be answered with Rosetta FlexPepDock.

Key words Peptide-protein interactions, Peptide binding, Peptide specificity, Peptide docking, Peptide modeling, Rosetta FlexPepDock, Rosetta FlexPepBind

1 Introduction

1.1 The Importance of Peptide-Mediated Interactions

Protein-protein interactions are main components of various critical biological processes in living cells [1, 2]. A significant fraction of these interactions (15–40%) are mediated by short linear interacting motifs (SLIMs) embedded inside disordered regions of a protein [1, 3]. Peptidic stretches may also appear in-between domains [4], or as flexible loops that bulge out of structured domains and mediate a protein-protein interaction (e.g., [5]). Thus, even among domain-domain interactions it is often one linear segment that contributes most of the binding affinity [6, 7]. Isolated peptides also play critical regulatory roles. Many have been recently found to be encoded and transcribed from short

Ora Schueler-Furman and Nir London (eds.), *Modeling Peptide-Protein Interactions: Methods and Protocols*, Methods in Molecular Biology, vol. 1561, DOI 10.1007/978-1-4939-6798-8_9, © Springer Science+Business Media LLC 2017

open reading frames [8]. Others are generated posttranslationally by proteolytic cleavage [9, 10]. Finally, peptides are also an important source of drugs used for medical purposes [11].

1.2 The Challenges of Peptide Docking

The flexibility of the peptide partner prior to binding presents considerable challenges for peptide docking [12, 13]. Even when the binding site on the receptor is known, and even if information about the approximate location and conformation of the peptide is available, identifying the correct conformation in this huge sampling space can still be a formidable challenge that has been met only recently. By efficiently dividing peptide docking into separate, predominantly independent modules, and by focusing on relevant conformations, this space can be considerably reduced. On the peptide side, the *conformational sampling* model states that peptides fluctuate in equilibrium between a more or less restricted number of preferred conformational states, and binding results in an equilibrium-shift toward pre-existing bound conformations [14–16], while simultaneous binding and folding upon encountering the receptor (i.e., the *induced fit* model) is less prevalent [4, 14, 17, 18]. Therefore, mapping out the conformational ensemble can significantly reduce the internal degrees of freedom to be sampled for the peptide during docking. Consequently, the peptide backbone conformation space can be reduced to a significant extent by biasing to specific conformation(s), such as helix and extended poly-proline [19], or a limited set of initial plausible peptide backbone conformations generated using molecular dynamics simulations [20] or other folding algorithms [21], or fragments extracted from existing protein structures [22]. On the receptor side, the search can be restricted to promising binding sites identified by mapping of the receptor surface using, e.g., solvent mapping [23] or mapping of individual amino acids [24]. Thus, while the generation of accurate structures of peptide-protein complexes is still challenging, it has become doable, as demonstrated by the range of different approaches that have been developed recently, many of them described in this book.

Given the simplifications described above, initial models of a peptide-receptor complex can be generated efficiently. However, for practical use, they often need to be further refined to high quality. This last step of docking and optimization may result in significant conformational changes of the peptide, in particular if *induced fit* upon binding is considerable and the peptide folds and binds simultaneously. Therefore, the docking protocol needs to be able to sample the required conformational changes so that near-native conformations are sampled and well refined to low-energy models.

The protocols described in this chapter are aimed at the refinement to high resolution of a peptide with known approximate location in a binding site.

1.3 The Rosetta FlexPepDock Peptide-Receptor Modeling Suite

Rosetta FlexPepDock [22, 25] (Fig. 1; Subheadings 3.1–3.3) was one of the first approaches that overcame the challenge of the vast conformational space of peptide conformations: It accurately modeled peptides beyond five residues into their receptor-binding site, using a combination of docking (rigid body optimization) and folding (extensive backbone optimization of the peptide).

A. FlexPepDock Refinement: **Full-atom refinement of coarse models**

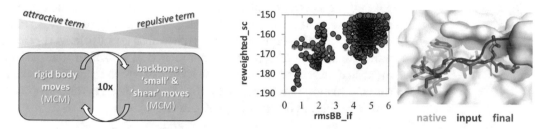

B. FlexPepDock *ab-initio*: **Coarse-grained peptide folding and docking into given binding site (fold & dock + refinement = final model)**

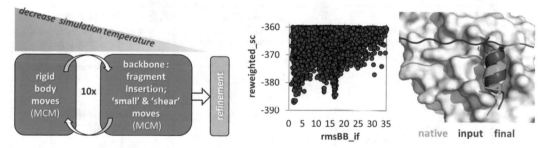

Fig. 1 Overview of the Rosetta FlexPepDock protocols described in this chapter. (**A**) Rosetta FlexPepDock Refinement. *Left panel*: Outline of the Refinement protocol. The starting structure is refined by iterative cycles of rigid body and peptide backbone optimizations. These cycles start with reduced Van der Waals (VdW) repulsive and increased VdW attractive terms, which are gradually ramped back to their original value in the energy function. *Middle panel*: Output models from $n=1000$ independent simulations are ranked using the Rosetta energy score and the top-scoring model (*magenta circle*) is chosen. *Right panel*: Slam tail peptide bound to the SH2 domain of the XLP protein SAP (pdb id: 1d4t [80]). Starting from an initial coarse model (*red*), refinement produces a final model (*magenta*) that is very similar to the solved structure (*green*). (**B**) Rosetta FlexPepDock ab-initio peptide docking. *Left panel*: Outline of ab-initio peptide docking. The starting structure (in an arbitrary, extended conformation) is refined in iterative cycles of rigid body and peptide backbone optimizations, using a coarse-grained model with side chains represented by a centroid. These cycles start at a high simulation temperature that is gradually ramped back to normal. Each resulting model is further refined to a full-atom model using FlexPepDock refinement (*orange box*; see **a**). *Middle panel*: Top scoring output models from $n=50,000$ independent simulations are ranked using the Rosetta energy score (*lower right*) and clustered to select several representative models (*magenta circles*). Note the larger sampled range. *Right panel*: The native, input and final folded and docked model of a peptide derived from Nuclear receptor coactivator 1 bound to Mineralocorticoid receptor (pdb id: 2a3i [81]; coloring as above). *Abbreviations*: MCM Monte Carlo with Minimization, *rmsBB_if* Interface backbone RMSD (i.e., the root mean square deviation of the peptide backbone atoms of residues at the interface), *reweighted_sc* reweighted score, energy score with up-weighted contribution of the peptide (*see* Table 3 for more details)

This protocol is embedded within the Rosetta framework [26–28] that is based on efficient Monte-Carlo sampling of relevant conformations, using an energy function well calibrated on a range of different structural modeling and design applications.

1.4 Use of Structural Models to Study Peptide-Binding Affinity and Specificity

Accurate models of peptide-protein complexes can be used as starting point to further characterize the details of an interaction. Understanding peptide-binding specificity of a given receptor protein can in principle be addressed by varying amino acids at different peptide sites and evaluating their compatibility with the binding site, to determine a sequence profile [29–32]. Alternatively, a set of different protein sequences can be threaded and optimized onto a template peptide-receptor complex, and new substrates can be identified based on energies associated with the resulting structures. This strategy is used in the *Rosetta FlexPepBind* protocol [33–35] (Fig. 2; Subheadings 3.4 and 3.5).

A. FlexPepBind framework: calibration on a set of known substrates and non substrates & application of calibrated protocol to detect novel substrates

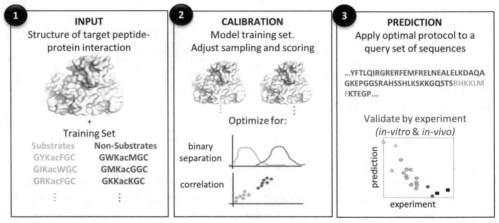

B. Biological systems studied using FlexPepBind framework

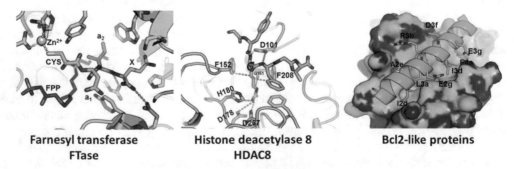

Farnesyl transferase
FTase

Histone deacetylase 8
HDAC8

Bcl2-like proteins

Fig. 2 Rosetta FlexPepBind. (**A**) Overview of the Rosetta FlexPepBind approach. Initially, a set of known substrates and non-substrates is used to calibrate the protocol to optimize the distinction between substrates and non-substrates. The calibrated protocol is then used to identify novel substrates. (**B**) Details of the substrate binding sites for three biological systems studied using the FlexPepBind protocol that are described in the text

While traditional approaches have characterized binding specificity based on sequence profiles derived from known binders (compiled, e.g., in the comprehensive Eukaryotic Linear Motif, ELM, database [36, 37]), a sequence-profile derived from structural consideration may be less biased to already known binders and therefore can provide more general features. Consequently, new, nonstandard substrates may be identified.

In this chapter, we provide a detailed description of Rosetta FlexPepDock and its derived application FlexPepBind. Our hope is that this will give readers appetite, and provide practical details, to apply these tools to their biological systems of interest.

2 Materials

2.1 Rosetta

FlexPepDock is part of the Rosetta Macromolecular molecular modeling package [26–28]. Detailed instructions for downloading and installing Rosetta can be found on the RosettaCommons webpage at www.rosettacommons.org

2.2 Documentation for FlexPepDock and FlexPepBind

1. Online documentation that includes detailed description of runline options and optimal parameter settings can be accessed at the following web address:
   www.rosettacommons.org/docs/latest/application_documentation/docking/flex-pep-dock

2. In the downloaded Rosetta package, located in the *rosetta/demos/public/* directory, the subdirectories:

   (a) *refinement_of_protein_peptide_complex_using_FlexPepDock*

   (b) *abinitio_fold_and_dock_of_peptides_using_FlexPepDock*

   (c) *peptide_specificity_using_FlexPepBind*
       include detailed instructions and example files for running the different protocols.

3. A protocol capture (*rosetta/demos/protocol_capture/ FlexPepDock_AbInitio* directory of the downloaded Rosetta package): describes how to set up, run, and postprocess the result for the ab-initio FlexPepDock protocol.

Queries regarding the various FlexPepDock protocols can be addressed to the Rosetta Forum webpage (www.rosettacommons.org/forum). In addition, a dedicated web server allows submission of FlexPepDock refinement runs online and to obtain results quickly without the need to manage command lines: (flexpepdock.furmanlab.cs.huji.ac.il) [38].

3   Methods

In this section, we first describe the functionalities of the different FlexPepDock protocols (FlexPepDock: Fig. 1 and Subheadings 3.1–3.3; FlexPepBind: Fig. 2 and Subheadings 3.4 and 3.5). Then we demonstrate on an example study, namely the generation and mode of action of the bactericidal EDF peptide [39, 40], how the use of a combination of these protocols has significantly advanced our understanding of the structural details of mode of action in this system (Fig. 3 and Subheading 3.6).

A. Modeling of the structural details of the EDF – MazF toxin interaction

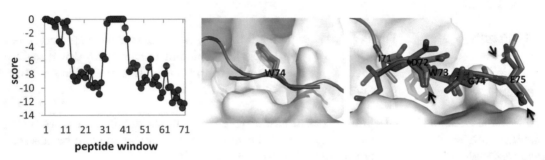

B. ClpP mediated generation of EDF precursor from glucose 6-phosphate dehydrogenase

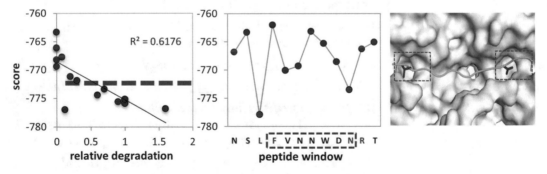

Fig. 3 Example application of the Rosetta FlexPepDock suite: Structure-based characterization of the bacterial external death factor (EDF) activity and generation. (**A**) Mechanism of action of EDF. EDF competes with the MazE antitoxin for toxin MazF, and by this removes toxin inhibition. *Left panel*: identification of binding site of EDF: the EDF pentapeptide sequence was threaded onto each overlapping MazE pentapeptide stretch, and each structure was optimized. The best-scoring complex identifies the location of mimicry and competition, namely the c-terminal MazE residues 71–75. *Middle* and *right panels*: Comparison of the solved structure of MazE bound to MazF to the model of EDF bound to MazF highlights the sophisticated mimicry of EDF: The critical interaction with the central tryptophan side chain is conserved (*middle panel*), and several hydrogen bonds between the peptide backbone and side chain atoms in MazF are retained (*highlighted by arrows*). (**B**) Mechanism of EDF generation. The *E. coli* protease ClpP acts on Zwf to generate EDF. *Left panel*: calibration of known ClpP substrates and non-substrates: FlexPepBind predictions are strongly correlated with experimentally determined cleavage propensities, allowing the definition of a threshold for subsequent substrate prediction (*red dashed line*). *Middle panel*: screening of overlapping six residue windows of Zwf predicts two putative, consequent ClpP cleavage sites that suggests the generation of a heptapeptide intermediate (*highlighted by a dashed red rectangle*). *Right panel*: Model of the ClpP-Zwf interaction, highlighting the two consecutive P1 cleavage sites bound to adjacent subunits in the homo-multimeric ClpP (in *red dashed rectangles*)

In contrast to protein-protein docking, where both partners are predominantly pre-folded, the challenge in peptide-protein docking lies in the need to sample both the rigid body orientation of the peptide relative to its receptor, as well as its internal flexibility. Starting from the RosettaDock framework of protein-protein docking [41–43], Rosetta FlexPepDock was optimized to sample in addition to these internal peptide degrees of freedom.

3.1 Refinement of Peptide-Protein Structures Using Rosetta FlexPepDock Refinement (Fig. 1a)

The Rosetta FlexPepDock Refinement protocol [25] was developed with the purpose of refining a coarse peptide-protein complex, obtained, e.g., from a homolog complex, to a near-native model. Starting with this coarse model, the FlexPepDock Refinement protocol refines all of the peptide's degrees of freedom (i.e., its rigid body orientation, as well as backbone dihedral angles) using a Monte-Carlo-Minimization based approach, in which each perturbation is followed by extensive minimization to find the local energy minimum, before the Metropolis-criterion is applied to accept/reject the new structure. All side chains at the interface (both on the peptide and receptor side) are continuously readjusted during the refinement, by allowing their replacement from a rotamer library that includes also the native receptor side chain conformation. The refinement starts with an energy function up-weighted for contacts (VdW attractive term) and down-weighted for clashes (VdW repulsive term); these are gradually ramped back to the default energy function. This allows initial efficient sweeping of the highly rugged energy landscape that is followed by accurate refinement into the energy minima.

FlexPepDock Refinement was benchmarked on a set of perturbed peptide-protein structures, and can successfully refine to near-native models (peptide backbone [bb]-RMSD <2.0 Å) structures with initial peptide bb-RMSD of up to 5.5 Å, with an effective range of up to 50° between phi-psi angles (*see* **Note 8**).

3.2 Simultaneous Docking and De-Novo Peptide Folding Using FlexPepDock Ab-Initio (Fig. 1b)

In many cases, no prior information about the peptide conformation is known, and therefore sampling needs to be increased beyond local refinement. Toward this goal, we developed the extended Rosetta FlexPepDock ab-initio protocol [22] for simultaneous docking and de-novo folding of peptides on a specific region on a receptor. An initial approximate description of a peptide-protein interaction site can be obtained either from experiments or from prediction (e.g., with PeptiMap [23], PepSite [44, 45], PEP-SiteFinder [21]).

Starting with a coarse-grained structural representation of the peptide and the receptor, FlexPepDock ab-initio efficiently samples the space of possible peptide backbone conformations and rigid-body orientations using the Rosetta fragments library and various torsional sampling moves [22, 46]. The resulting coarse-grained models are subsequently refined into high-resolution models using the FlexPepDock Refinement protocol [25]. Models are selected after clustering of the top-scoring models and selecting best scoring cluster representatives (using the reweighted score measure, *see* Table 3).

In a benchmark, this protocol identifies near-native peptide conformations among the top ten clusters in 7/14 cases [22].

The FlexPepDock ab-initio and refinement protocols have significantly impacted the docking field: By extending the scope of state-of-the-art methods for high-resolution peptide modeling, they have motivated the development of a range of new approaches for peptide-protein docking (reviewed in [13], and presented in other chapters in this book). Together, these have enabled detailed, structure-based studies and manipulation of many more interactions.

The FlexPepDock protocols have been applied by users to a large variety of different applications, among them to model MHC-peptide antigen complexes for assessment of tumorigenic mutations and other applications [47, 48], modeling of protease substrates [49–52] and the design of protease inhibitors [53], to study host-pathogen interactions [54, 55], and a variety of other interactions between peptides and peptide-binding domains (e.g., [56–61]). As an example, we used this protocol, together with loop modeling, to characterize the activation mechanism of myosin II heavy chain kinase A of *Dictyostelium*: We first modeled how the c-terminal tail binds to the kinase active site for autophosphorylation, and then how this phosphorylated tail binds into a putative allosteric site to subsequently activate the kinase [62]. A similar approach was used by others to model the phosphorylation-dependent interaction of the α-synuclein tail with Rab8a [63]. FlexPepDock was also used to generate a plausible starting model for a molecular dynamics simulation of a glycogen synthase kinase 3β kinase/substrate peptide interaction [64]. In Subheading 3.6 we describe in detail a specific example, the study of the peptide EDF (Fig. 3).

3.3 How to Run FlexPepDock Refinement and FlexPepDock Ab-Initio Protocols

Below we provide a detailed description about the various input files needed and runline command for running refinement and ab-initio modeling using FlexPepDock. The various command line options and scoring measures are summarized in Tables 2 and 3, respectively. We refer the reader also to the extensive documentation available online (*see* Subheading 2).

3.3.1 Input Files Needed to Run FlexPepDock Simulations

1. **Starting structure**: An initial approximate structure of a peptide-protein complex with or without side-chain coordinates. In ab-initio mode, the starting backbone conformation of the peptide may be arbitrary (e.g., extended). The exact way in which the starting conformation is created may vary depending on the specific application (*see* **Notes 1** and **5**). **Multichain receptors**: In order for FlexPepDock to correctly handle multichain receptors, the PDB file must contain first the receptor chains in a consecutive manner, followed by the peptide chain and ligand chains that come last.

2. **Native structure (optional)**: This is a reference structure for RMSD comparisons and statistics of final models, in case a native structure is available. If a native structure is not supplied, the starting structure is used for reference instead.

3. **Constraint file (optional)**: This file specifies conditions to be met for final models. Constraints are incorporated as additional terms into the Rosetta energy function during the simulation (for detailed description on the Rosetta resfile format visit www.rosettacommons.org/docs/latest/rosetta_basics/file_types/constraint-file for more information). They are mainly used to generate a starting model of a peptide in a given binding site for further FlexPepDock refinement, and to restrict sampling to relevant conformations for FlexPepBind (*see* Subheadings 3.4, 3.5 and Table 1A).

4. **Fragment files (for ab-initio docking)**: 3-mer, 5-mer, and 9-mer Rosetta fragment files should be provided for the peptide sequence when using the ab-initio protocol (*see* Subheading 3.3.3). If the peptide length is smaller than 9, use 3-mer and 5-mer libraries.

3.3.2 Steps to Run FlexPepDock Refinement and Analyze the Results

Below we list the runline commands of the different steps. For more details about the different parameters used, *see* Table 2.

1. Prepack your initial complex to optimize the side-chains of each monomer according to the Rosetta energy function (*see* **Note 2**):

```
$FlexPepDocking.{ext} -database ${ro-
setta_db} -s start.pdb -native native.pdb
-flexpep_prepack -ex1 -ex2aro [-unboundrot
unbound_receptor.pdb]
```

Here start.pdb is the input structure. This will generate start_0001.pdb, which is the input for refinement or ab initio modeling in the next step. The last three parameters define the receptor side chain flexibility in this and the following simulation steps (*see* **Note 3c**).

2. Refine the structure using FlexPepDock refinement (*see* **Note 9**): Generate 100 (or more) models with the -lowres_preoptimize flag (*see* **Note 3b**), and additional 100 models (or more) without this flag, by two separate runs:

```
$FlexPepDocking.{ext} -database ${ro-
setta_db} -s start_0001.pdb -native na-
tive.pdb -out:file:silent decoys.silent
-out:file:silent_struct_type binary -pep_re-
fine -ex1 -ex2aro -use_input_sc -unboundrot
unbound_receptor.pdb -nstruct 100 [-lowres_
preoptimize]
```

Table 1
Details of the FlexPepBind approach for the structure-based identification of peptide substrates, applied to three systems: FTase, HDAC8, and Bcl-2-like proteins. (A) The different steps of FlexPepBind calibration. Comparison of the optimal parameters for different systems reveals that with small changes FlexPepBind can be applied successfully to a range of different systems, including both modifying enzymes as well as noncatalytic binders. (B) Summary of calibration, validation, and application for each of the systems studied, including the main findings and lessons learned

(A)

System	FTase–CX$_1$X$_2$X$_3$'	HDAC8–XX$_1$KacX$_2$X$_3$X	Bcl/Mcl–BH3
Step 1: Generation of template (pdb id)	FTase–CNIQ (c-terminus of Rap2a)–farnesyl analog (1tn6 [82])	Optimized structure of HDAC8–GYKacFGC (best substrate in calibration set) starting from HDAC8–p53-derived peptide structure (2v5w [73])	BH3 region of BIM (EIWIAQELRRIGDEFNAYYAR) bound to: • Bcl-xL (BIM 3a L->F mutant) (3io8:CD [83]) • Mcl-1 (2pqk [84]) • Bcl-2 (2xa0 [85]); BH3 region of BAD replaced by BH3 region of BIM (from 2pqk)
Step 2: Constraints receptor-peptide (hydrogen bonds: HB; Zn coordination: Zn; Stacking: St)	HB: R202β–X$_2$ O HB: Q167α–X$_3$' O Zn: Zn^{2+}–Cys[SH]	HB: D101–Kac N and X$_2$ N St: F152 and F208–Kac Zn: metal binding D178, H180, D267–Kac	NA
Step 3: • Sampling: FlexPepDock refinement/ minimization only • Scoring: energy function used	• Minimization only • Score12	• Minimization only • Modified Score12: – Increased penalty for burial of carboxyl oxygen (Lazaridis-Karplus [86] ΔG_{free} OC = −13.5) – + Coulombic term (short-range; weight = 0.5)	• FlexPepDock Refinement (choose best among n =1000 models) • Modified Score12 (see HDAC8)
Step 4: Ranking criterion	Peptide_score without reference energy (pep_sc_noref)	Interface score (I_sc)	Reweighted score (reweighted_sc)

(B)

System	FTase–CX$_1$X$_2$X$_3$'	HDAC8–XX$_1$KacX$_2$X$_3$X	BCL/MCL–BH3

Calibration	CX₁X₂X₃: 77/51 substrates/non-substrates (multiple-turnover conditions) [69]	361 GX₁KacX₂GC peptides (X₁, X₂ = all amino acids except C) (fraction deacetylated) [35]	• TRAIN1 (360 peptides): all combinations of mutations @ positions 2d:IAF, 3a:LIPA, 3b:RD, 3d:IFDNA, 4a: FVN • TRAIN2 (181 peptides): single mutations to all amino acids (except C,M) @ 2d,2e,2g,3a,3b,3d,3e,3f,3g,4a (SPOT arrays, assayed with Bcl-xL & Mcl-1) [87]
Validation	1. 29/15 CXXX substrates/non-substrates [69] 2. 72 c-terminal sequences of known FTase substrates [69] 3. 24/17 CXXL substrates/non-substrates [70]	• 361 GRKacX₂X₃GC peptides [88] • 8 targets from proteome screen [74]	• 40/33/17 peptides that bind only Bcl-xL/ only Mcl-1/both Bcl-xL and Mcl-1 (yeast surface display) [87] • Prediction of specificity pattern of human BH3 only-derived peptides to Bcl-xL/Mcl-1
Application	Best-scoring CXXX' peptides 1. Top-scoring (13) 4. Top-scoring new candidates from the human proteome (16) **Outcome: 26/29 are substrates**	Candidates identified in PhosphoSite screens (hexapeptides in proteins) • Screen 1 (26) • Screen 2 (19) **Outcome: excellent correlation; detect strongest currently known substrates**	TRAIN2 (181) assayed for Bcl-2 [34] **Outcome: accurate prediction of Bcl-2 binding affinity**
Findings and Lessons learned	• Binding in a catalysis-competent conformation is the bottleneck of farnesylation -> binding assessment can be used to identify enzyme substrates • Novel class of FTase substrates identified: CXX[DE]'	• Strongest HDAC8 substrates identified so far • Excellent correlation between predicted binding (in catalysis-competent conformation) and substrate activity • Significant extension of the deacetylome	• Longer peptide substrate, no constraints • Helical flexibility is crucial for successful prediction • Needs considerably more sampling • Specificity-determining residues identified • Bcl-xL and Bcl-2 substrates well identified • Mcl-1 substrate distinction needs improvement

Table 2
Runline options for FlexPepDock simulations. (A) Common FlexPepDock flags; (B) Relevant Common Rosetta flags; and (C) Expert flags

(A) Common FlexPepDock flags		
Flag	**Description**	**Default**
`-receptor_chain`	Chain id of receptor protein	First chain
`-peptide_chain`	Chain id of peptide protein	Second chain
`-lowres_abinitio`	Low-resolution ab-initio folding and docking mode	False
`-pep_refine`	Refinement mode	False
`-lowres_preoptimize`	Perform a preliminary round of centroid mode optimization before Refinement	False
`-flexpep_prepack`	Prepacking mode. Optimize the side-chains of each monomer separately (no docking)	False
`-flexpep_score_only`	Read in a complex, score it and output interface statistics	False
`-flexPepDocking MinimizeOnly`	Minimization mode: Perform only a short minimization of the input complex	False
`-ref_startstruct`	Alternative start structure for scoring statistics (useful as reference for rescoring previous runs with the -flexpep_score_only flag)	N/A
`-peptide_anchor`	Set the peptide anchor residue manually. Only recommended if one strongly suspects the critical region for peptide binding to be remote from its center of mass	Residue nearest to the peptide center of mass
(B) Relevant Common Rosetta flags		
Flag	Description	
`-in::file::s` `-in:file:silent`	Specify starting structure (PDB or silent format, respectively)	
`-in::file::silent_ struct_type` `-out::file::silent_ struct_type`	Format of silent file to be read in/out. For silent output, use the binary file type since other types may not support ideal form Models can be extracted using `extract_pdbs`	
`-native`	Specify the native structure for which to compare in RMSD calculations. When the native is not given, the starting structure is used as reference	
`-nstruct`	Number of models to create in the simulation	
`-unboundrot`	Add the position-specific rotamers of the specified structure to the rotamer library (usually used to include rotamers of unbound receptor)	

(continued)

Table 2
(continued)

(A) Common FlexPepDock flags		
Flag	**Description**	**Default**
-use_input_sc	Include rotamer conformations from the input structure during side-chain repacking. Unlike the -unboundrot flag, not all rotamers from the input structure are added each time to the rotamer library, only those conformations accepted at the end of each round are kept and the remaining conformations are lost	
-ex1/-ex1aro -ex2/-ex2aro ex3 ex4	Adding extra side-chain rotamers (highly recommended). The -ex1 and -ex2aro flags are recommended as default values	
-database	The Rosetta database	
-frag3 -flexPepDocking:frag5 -frag9	3mer/5mer/9mer fragments files for ab-initio peptide docking (9mer fragments for peptides longer than 9)	
(C) Expert flags		
Flag	Description	Default
-rep_ramp_cycles	The number of outer cycles for the protocol. In each cycle, the repulsive energy of Rosetta is gradually ramped up and the attractive energy is ramped down, before inner-cycles of Monte-Carlo with Minimization (MCM) are applied	10
-mcm_cycles	Number of inner-cycles for both rigid-body and torsion-angle Monte-Carlo with Minimization (MCM) procedures	8
-smove_angle_range	Defines the perturbation size of small/sheer moves	6.0
-extend_peptide	Start the protocol with the peptide in extended conformation (neglect original peptide conformation; extend from the anchor residue)	false
-frag3/5/9_weight	Relative weight of different fragment libraries in ab-initio fragment insertion cycles	1.0/0.25/0.1

Note that in this and following runs we used a binary silent output to save disk space. Selected models can be extracted from the decoys.silent file using the following command:

```
$extract_pdbs.{ext} -database ${ro-
setta_db} -in:file:silent decoys.silent
-in:file:fullatom -in:file:tags decoy_tag
```

Table 3
Description of the various scoring and quality assessment measures

Total_score[a]	Total score of the complex
reweighted_sc	Reweighted score of the complex, in which interface residues are given double weight, and peptide residues are given triple weight
I_bsa	Buried surface area of the interface
I_hb	Number of hydrogen bonds across the interface
I_pack	Packing statistics of the interface
I_sc	Interface score (sum over energy contributed by interface residues of both partners)
pep_sc	Peptide score (sum over energy contributed by the peptide to the total score; consists of the internal peptide energy and the interface energy)
pep_sc_noref	Peptide score without an amino acid dependent reference energy term E_{aa}, originally introduced to bias for natural protein sequences during protein design
I_unsat	Number of buried unsatisfied HB donors and acceptors at the interface
rms (ALL/BB/CA)	RMSD between output model and the native structure, over all peptide (heavy/backbone/C-alpha) atoms
rms (ALL/BB/CA)_if	RMSD between output model and the native structure, over all peptide interface (heavy/backbone/C-alpha) atoms
startRMS(all/bb/ca)	RMSD between start and native structures, over all peptide (heavy/backbone/C-alpha) atoms

For the common Rosetta scoring terms, please also see www.rosettacommons.org/docs/latest/rosetta_basics/scoring/score-types
[a]For all interface terms, the interface residues are defined as those whose C-beta atoms (C-alpha for Glycines) are up to 8 Å away from any corresponding atom in the partner protein

where the Rosetta executable extract_pdbs is used, and decoy_tag is the tag(s) of the desired decoy(s) to be extracted (e.g., -in:file:tags start_0001 start_0002).

3. Select model(s): sort the score files of both runs (score.sc by default) and choose the top-scoring models as candidate models for further inspection (sort according to reweighted score or interface score; *see* **Note 4** and Table 3 for how to choose the scoring term).

3.3.3 Steps to Run FlexPepDock Ab-Initio and Analyze the Results

1. Prepack your initial complex (*see* Subheading 3.3.2, **step 1**).

2. Generate fragment files needed for ab initio sampling of peptide conformations.

   ```
   $perl make_fragments.pl -verbose -id
   xxxxx xxxxx.fasta
   ```

   where xxxxx is the name of the peptide (needs to be five digits) and the fasta file contains the sequence. This will generate the input files for the fragment picker below (xxxxx.psipred_ss2, xxxxx.checkpoint along with other files):

```
$fragment_picker.{ext} -data-
base ${rosetta_db} -in:file:vall vall.
jul19.2011 -in:file:checkpoint xxxxx.check-
point -frags:describe_fragments frags.
fsc -frags:frag_sizes 3 5 9    -frags:n_
candidates 2000 -frags:n_frags 500
-frags:ss_pred    xxxxx.psipred_ss2
psipred -frags:scoring:config psi_L1.cfg
-frags:bounded_protocol true
```

where vall.jul19.2011 is the Rosetta fragment database, the psi_L1.cfg file provides the fragment scoring weights (e.g., for different secondary structure and profile scores). For more details about the other command line parameters, see www.rosettacommons.org/docs/latest/application_documentation/utilities/app-fragment-picker

The fragment files should be re-indexed to the positions of the peptide in the structure (using the script *scripts/frags/shift.sh* in the demo *abinitio_fold_and_dock_of_peptides_using_FlexPepDock;* e.g., for a receptor of 100 residues, the first residue index in the fragment file for the peptide should be 101). Fragments are not required for the receptor. *See* **Note 6** for tips for reinforcing a specific secondary structure of the peptide.

3. Optimize the structure using FlexPepDock ab-initio: Generate 50,000 models (*see* **Note 3a** for how to determine this number, and Table 2 for a description of the individual runline parameters):

```
$FlexPepDocking.{ext}   -database ${roset-
ta_db}  -s start_0001.pdb -native native.pdb
-out:file:silent decoys.silent -out:file:silent_
struct_type binary -lowres_abinitio -pep_re-
fine -ex1 -ex2aro -use_input_sc -unboundrot
unbound_receptor.pdb   -frag3   <frag3   file>
-flexPepDocking:frag5 <frag5 file> -frag9 <frag9
file> -nstruct 50000
```

4. Cluster the top-scoring models: To maximize the diversity of the final models, and to estimate the size of the local energy basin, we suggest clustering the results (using the Rosetta clustering application (for detailed clustering command line options visit www.rosettacommons.org/docs/latest/application_documentation/utilities/cluster). We recommend clustering the top-1 % models (either based on reweighted score or interface score) and choosing the clusters with lowest-energy representatives. We have found that a clustering radius of 2.0 Å (peptide CA atoms only) provides a diverse and representative

set of conformations, among which good solutions are often found within the top 1–10 clusters. Note that the RMSD values need to be modified to reflect RMSD of the whole complex as the clustering is performed on the whole complex and in that context the clustering radius will change. Use the clustering scripts provided in the demo "*abinitio_fold_and_dock_of_peptides_using_FlexPepDock*" that will automatically adjust the RMSD (alternatively, it is possible to exclude residues during the clustering process). *See* **Note 4** for additional model selection options after clustering.

3.3.4 Steps to Run FlexPepDock Minimization

Local minimization finds a similar structure in the nearby minimum of the Rosetta Energy landscape. This protocol is embedded inside the refinement and the ab initio sampling cycles. As stand-alone, this protocol is used in the FlexPepBind framework (*see* Subheading 3.5.2). It simply minimizes all the peptide dihedral angles and its rigid-body orientation, as well as the receptor side-chains (the receptor backbone is not changed).

1. Prepack your initial complex (*see* Subheading 3.3.2, **step 1**; make sure to include the side chain conformations of the initial complex; not needed if you start from a model; *see* **Note 2**).

2. Optimize the structure using FlexPepDock minimization:
   ```
   $FlexPepDocking.{ext} -database ${rosetta_
   db} -s start_001.pdb -native native.pdb -flex-
   PepDockingMinimizeOnly -ex1 -ex2aro -unboundrot
   unbound_receptor.pdb
   ```

3.4 Modeling Peptide-Binding Specificity Using FlexPepBind

One of the major applications of FlexPepDock is the extension from the modeling of peptide-protein complex *structures* to the modeling of peptide-protein *binding*. While a structure of a protein complex, whether solved by experiment or modeled by an accurate protocol, can provide important insights into the interactions, unfortunately, despite community-wide attempts and progress, the development of general protocols for the prediction of binding and binding affinity based on structure is only at its beginning [65, 66]. Luckily however, for specific cases, we and others have made considerable progress in defining the range of peptide substrates for a given receptor [33–35, 67] (*see* Fig. 2, Table 1, and other chapters in this book).

The Rosetta FlexPepBind protocol is based on the observation that structure-based prediction of binding is particularly successful when critical features of an interaction with a given receptor can be identified and reinforced. The core strategy of Rosetta FlexPepBind for the prediction of peptide-binding specificity for a given protein receptor is to use an existing template structure of a peptide-receptor complex, onto which we model each of the investigated peptide sequences with constraints that reinforce defined critical features, such as conserved hydrogen bonds between specific side chains in the receptor and the peptide backbone. These models are then used to distinguish binders from nonbinders based on their energies. This section contains an overview of FlexPepBind; the following Subheading 3.5 provides the details of how to run the protocol and analyze the obtained data.

Rosetta FlexPepBind consists of the following steps (Table 1 and Fig. 2, upper panel): *Step 1—Prepare a starting template structure of a peptide-receptor complex:* The binding of each of a list of peptides of interest will be modeled onto that template structure (*see Step3* below) (*see* **Note** 7); *Step 2: Derive features that are critical for bindi*ng: Conformational sampling will be biased toward relevant conformations, by implementation of constraints that reproduce these conserved structural features; *Step 3: Model each peptide sequence on the peptide-receptor complex template:* Each peptide sequence of interest is threaded onto the template and optimized under the constraints defined in *Step2*. Depending on the system, either simple minimization [33, 35] or extensive optimization using the FlexPepDock refinement protocol might be needed [34]. *Step 4: Rank the optimized peptide-receptor models based on binding energy and identify binders:* different scoring measures may be used to rank the optimized peptide-receptor complex structures, to select the peptides predicted to bind the strongest (*see* Table 3 and Subheading 3.5.3).

Rosetta FlexPepBind needs to be tailored to each specific system to work well [33–35] (*see* Table 1 and Subheading 3.5.3). However, these system-specific changes are rather minor. In order to adapt the protocol to the specific system, the protocol is first calibrated on a training set with known binders and nonbinders. This involves selection of the optimal template structure (*Step 1* above), and the identification of conserved features specific to that system, which are then defined as constraints (*Step 2* above) (*see* Fig. 2a, left panel). Furthermore, we test which optimization performs best—if minimization only works well, we opt for this option as it is much faster and makes it easy to apply the final protocol on large scale (*Step 3* above). Finally, we investigate which scoring function performs best for the identification of binders (binary distinction), as well as for predictions that correlate best with experimental data (quantitative predictions) (*Step 4* above).

The protocol can be applied to characterize peptide-binding specificity, as well as to understand peptide substrate specificity of enzymes. For the latter, the underlying assumption is that the ability of a peptide substrate to bind to the catalytic site in a catalysis-competent conformation is critical, and that therefore binding can be taken as a useful approximate to substrate strength. The catalysis-competent conformation is enforced using a set of constraints identified in *Step 2* above.

After rounds of calibration (Fig. 2a, middle panel), the optimized protocol can be applied to novel potential substrates (Fig. 2a, right panel). These are then tested by experiment. If necessary, feedback from these validations can be used to further improve the protocol.

We have applied Rosetta FlexPepBind to investigate several biological systems, including the (a) discovery of novel prenylation targets of Farnesyltransferase (FTase) [33], based on the peptide sequence of the four c-terminal residues of a protein; (b) identification of Histone Deacetylase 8 (HDAC8) non-histone substrates [35]; and (c) elucidation of BH3 binding specificity toward Bcl-2 like proteins [34]. A general overview of these applications, including details about the specific modification of FlexPepBind for each system, is presented in Table 1 and Fig. 2b, and the background and main findings for each are summarized below.

3.4.1 Discovery of Novel Prenylation Targets of Farnesyltransferase (FTase) [33]

Background: The addition of a farnesyl moiety by FTase to a cysteine residue near the c-terminus is critical for the function of many proteins involved in signal transduction, e.g., by targeting the modified protein substrates to the membrane. Until recently, it was believed that a C-terminal CaaX' motif is required for farnesylation (i.e., the modified cysteine is followed by two aliphatic residues and an additional c-terminal residue). However, farnesylation experiments have suggested broader substrate diversity [69, 70]. Sequence-based methods such as PrePS allowed indeed to generate a more general profile of FTase substrates, but were restricted to the pool of sequences of known substrates [71]. By calibrating FlexPepBind for FTase substrate detection, we were able to discover a wide range of new in vitro FTase peptide substrates that do not confer to previously reported sequence motifs [33]. Identification of many potential substrates in pathogenic proteomes suggests that pathogens hijack the host farnesylation machinery for their needs [72].

Results: Table 1, left column provides details of FTase FlexPepBind calibration, assessment on known substrates, and application for the detection of new substrates. This study validated our assumption that the assessment of binding ability in a catalysis-competent conformation is a good approximation for substrate strength, allowing modeling of substrates based on binding, without the need to explicitly model the catalysis. Importantly, prospective in vitro experiments validate that 26/29 predicted novel substrate peptides indeed undergo farnesylation. Other than the discovery of putative novel farnesylation targets in the human genome, as well as possible inhibitors, we provide insights into the main determinants of farnesylation.

3.4.2 Structure-Based Identification HDAC8 Non-histone Substrates [35]

Background: Histone Deacetylases (HDAC) catalyze the deacetylation of specific acetyl lysine residues in histones. Recent studies however have identified a range of additional non-histone substrate proteins. By calibrating FlexPepBind for HDAC8 substrate detection, we were able to discover a wide range of new in vitro HDAC8 peptide substrates, defining an expanded HDAC8 acetylome.

Results: The results are summarized in Table 1, middle column. In this study, the challenge was to obtain an appropriate template structure that could be used to model efficiently the binding of different peptide substrates. After an initial calibration round, we chose as template a structural model of the strongest substrate in the training set, GYKacFGC, bound to HDAC8. This structure was generated using Rosetta FlexPepDock, starting from a solved structure of HDAC8 bound to a p53-derived peptide [73]. Calibration and validation involved first optimal binary distinction of substrates/non-substrates, and later correlation between predicted binding affinity and experimental substrate efficiency. We identified in this study the strongest HDAC8 peptide substrate yet reported (FGKacFSW, $k_{cat}/K_M = 4800$ M^{-1} s^{-1}), and highlight the relevance of structure-based substrate identification and its complementarity to sequence-based approaches and proteomic experiments [74] toward a comprehensive picture of the HDAC8 deacetylome.

3.4.3 Characterization of BH3–Bcl-2 Like Proteins Binding Specificity [34]

Background: Interactions between Bcl-2-like proteins and BH3 domains play a key role in the regulation of apoptosis. Despite the overall structural similarity of their interaction with helical BH3 domains, Bcl-2-like proteins exhibit an intricate spectrum of binding specificities whose underlying basis is not well understood. We used the FlexPepBind protocol to predict the BH3-only peptide-binding specificity Bcl-2 like proteins.

Results: Table 1, right column provides details of Bcl FlexPepBind calibration, assessment on known binders, and application for the detection of new binders. This study characterized the distinct patterns of binding specificity of BH3-only derived peptides for the Bcl-2 like proteins Bcl-xL, Mcl-1, and Bcl-2 and provided insight into the structural basis of determinants of specificity. Since this involved the modeling of a longer, helical peptide, without any constraints, considerable more sampling was necessary to obtain good correlation between prediction and experiment.

The successful application of Rosetta FlexPepBind to three very different systems (summarized in Table 1 and Fig. 2b) provides a framework for the extension to additional systems characterized by a few known substrates/non-substrates and a solved template structure of a peptide-protein complex. It also highlights specific features that need to be taken into account, depending on the system studied.

3.5 How to Run FlexPepBind

Below we provide a detailed description about the various input files needed and runline command for running the FlexPepBind protocol. We refer the reader also to the extensive documentation available online (*see* Subheading 2).

1. **Template structure**: A solved structure of the receptor bound to a substrate is required. In cases where only the structure of the receptor is available, a template can be created by extensive modeling of a substrate peptide (using FlexPepDock protocols described in Subheadings 3.1–3.4). The template structure has to be prepacked to remove internal clashes (*see* **step 1** in Subheading 3.3 and **Note 2** on preparing the structure). If the template is an enzyme then it is important that the active site residues will not move much during repacking.

2. **Constraint file (optional)**: Constraints critical for binding of the peptide to the receptor can be extracted from the solved structure to bias docking simulations.

3. **Calibration and Validation sets**: A known set of substrates and non-substrates can be used to calibrate and validate the protocol before it can be applied at large scale.

1. Thread various peptide sequences onto the template to generate a starting structure for each of the peptides to be tested. We use the Rosetta fixbb protocol [68]. The fixbb design protocol can be run as

   ```
   $fixbb.{ext} -database ${rosetta_db} -s
   template.pdb -resfile peptide_resfile -ex1-
   ex2aro -use_input_sc -unboundrot unbound_re-
   ceptor.pdb -nstruct 1
   ```

   Where template.pdb is the template peptide-protein complex; peptide_resfile contains instruction regarding threading of the peptide sequence onto the template peptide backbone (for detailed description of the resfile format, visit www.rosettacommons.org/docs/latest/rosetta_basics/file_types/resfiles). Detailed run line options of the Rosetta fixbb design protocol can be found here: www.rosettacommons.org/docs/latest/application_documentation/design/fixbb.

2. Optimize each starting structure (using one of the FlexPepDock protocols). In our previously studied systems we have used both FlexPepDock refinement (Subheading 3.3.2) for extensive optimization of the threaded peptide, as well as simple FlexPepDock minimization only (Subheading 3.3.4).

Given an initial training and test set of substrates and non-substrates (i.e., binders and nonbinders), the aim is to optimize the protocol for optimal distinction of the substrates. This is done by running a (simplified) version of FlexPepBind on the training set using different parameter combinations.

The following parameters can be calibrated toward this aim (*see* Table 1 for a concise summary of which parameter combination has worked best in previous applications [33–35, 40]):

1. Template structure: If multiple template structures are available, we recommended first selecting the best template by a quick initial assessment of performance on the calibration set, or a small set of peptides, using simple FlexPepDock minimization only (Subheading 3.3.4).

2. Constraints: The careful definition of binding constraints will reduce the False Positive Rate by preventing the sampling of nonrelevant conformations that would nevertheless score well. In addition, it will speed up the protocol by efficiently restricting sampling to the relevant conformation space. Different sets of constraints can be used to calibrate the optimum set of constraints (*see* Table 1 for different type constraints for different systems [33–35, 40]).

3. Ranking the peptides: In our previous studies we found that for different biological systems under investigation, different scoring measures provide best distinction of binders (*see* Table 3 for a description of the scoring measures, and Table 1 for the scoring measures selected for different systems). To select the best performing term for a specific system, it is therefore advised to assess the performance of these scoring measures on a calibration set.

4. Sampling: Either simple minimization (*see* Subheading 3.3.4) or extensive refinement (*see* Subheading 3.3.2) might be necessary after threading the peptide sequences into the template backbone. It is recommended to check both in the calibration step.

3.6 Study of Diverse Aspects of a Peptide Using Rosetta FlexPepDock Tools: Bacterial External Death Factor (EDF) as Example

The *E. coli* external death factor, EDF, was first detected as a substance secreted to the medium that leads to bacterial death, and then characterized as a short peptide with the sequence NNWNN [39]. Below we detail how the modeling of the structure and binding of peptide-protein interactions has contributed to the elucidation of some of the details of the controlled generation of this peptide, as well as its mode of action (Fig. 3). These mechanistic details help also to shed more light on the functional importance of EDF.

The mode of action of EDF involves the manipulation of the MazF-MazE toxin-antitoxin interaction, and by this the regulation of bacterial cell death. In turn, the generation of EDF is regulated by the MazF toxin, a specific RNase. A series of experiments have shed light on the details of these mutual regulations. We describe below how these have been assisted and guided by FlexPepDock-based modeling tools.

The MazF-MazE toxin-antitoxin system consists of a bicistronic joint operon encoding for an unstable MazE antitoxin that binds to and thereby inhibits the stable RNAse toxin MazF. Toxin-antitoxin pairs have been implied in programmed cell death that can be invoked by different types of stress, and they play crucial regulatory roles within a bacterial community [75]. One of the main regulators of the toxin-antitoxin interaction is EDF. We describe here how we generated a detailed structural model of the interaction of EDF with MazF using Rosetta FlexPepDock, and what this model has taught us about its form of action [39] (Fig. 3a).

Our starting point was the solved structure of the MazE/ MazF complex [PDB id: 1ub4 [76]], which shows how the MazE antitoxin wraps around a MazF toxin dimer. Our underlying assumption was that EDF would use a local mimic of the MazE antitoxin to outcompete its binding to the toxin. We therefore searched for the EDF-binding site by scanning each overlapping penta-peptide derived from MazE in the complex structure as a candidate target-binding site of EDF. For each such pentapeptide backbone "window," we generated a starting structure by changing the sequence to that of EDF (NNWNN), and subsequently optimized this starting structure using FlexPepDock refinement. The MazE c-terminal pentapeptide sequence, IDWGE 71–75, produced the model with the lowest interface energy and was selected as predicted binding site (Fig. 3a, left panel).

Our model suggests how EDF can mimic MazE and outcompete its interaction with MazF (Fig. 3a, middle and right panels): Crucial interactions of this region in the MazE-MazF complex are mimicked by EDF, including (1) the conserved buried tryptophan: W3 is buried in the hydrophobic pocket of MazF that accommodates corresponding MazE residue W73, and (2) conserved hydrogen bonds and electrostatic interactions: In particular, rearrangement of the peptide allows superposition of the side chain tip of N5 in EDF to the tip of E75 in MazE, conserving a hydrogen bond with MazF R86. EDF residues N2 and N4 do not interact with their side chains with MazF. This model explains the effect of various EDF point mutations on EDF activity: While mutants NNGNN and GNWNG are predicted to affect binding most significantly, single N replacements, or a double mutant NGWGN do not significantly affect binding. These results strongly support a possible mechanism in which EDF displaces MazE by mimicking its binding to MazF.

In addition to the study of the mode of action of the EDF peptide, models generated using FlexPepDock have also helped to elucidate part of the process of EDF generation (Fig. 3b).

EDF generation is very complex and tightly regulated. In short, EDF is derived from the gene encoding for glucose 6-phosphate dehydrogenase, *zwf*. Under specific conditions, Zwf mRNA is cut by the MazF toxin RNase to produce a leader-less mRNA specifically translated under stress conditions. The EDF precursor peptide is cleaved out from this translation product (generating peptide NNWDN from its positions 199–203, which is subsequently modified to NNWNN). While the ClpP/ClpA protease was known to be involved, no structural details about the cleavage step were available.

We describe here how we used Rosetta FlexPepDock to generate a detailed structural model of the interaction of the Zwf region that contains the EDF precursor peptide with the ClpP protease, and what this model has taught us [40]. Our underlying assumption was that the EDF-spanning region of Zwf contains strong signals for ClpP protease cleavage, and therefore, that we should be able to identify these cleavage sites based on a structural model of the protein-ClpP interaction. This is based on our working hypothesis that binding estimates can be used as predictor for enzymatic activity on a given substrate.

3.6.2.1 Calibration of FlexPepBind ClpP

In a first step, we calibrated FlexPepBind to successfully identify known ClpP substrates. Based on the solved structure of ClpP [covalently bound to a peptide Chloromethyl Ketone at the active site, pdb id: 2fzs [77]], we predicted substrates among a set of peptides with reported catalytic activity for ClpP/ClpA, namely the known ClpP substrate MAPMALPV and several mutants of this sequence [78]. To generate an initial template structure for further modeling of mutants, we first modeled the structure of ClpP bound to the MAPMALPV peptide: we added peptide residues not resolved in the template structure in an extended conformation, and optimized the whole peptide by constraining the cleaved residues to the adjacent active sites during optimization using the FlexPepDock ab initio protocol [22]. Active site residue S97 was mutated to G97 (to avoid severe steric clash, since we did not model the covalent bond of the peptide to the protease in the template structure) and a distance constraint of 3.6 Å between the C atom of the residue preceding the cleaving bond and the Cα atom of active site residue 97 was used to enforce catalytic C=O bond position at the active site. The standard setup of ab initio FlexPepDock was used to generate a starting structure to be used as template for optimization (see Subheading 3.2; in brief, out of 50,000 structures generated using the FlexPepDock ab initio protocol, the top ranking 1 % were clustered with a 2.0 Å RMSD cutoff, and the best scoring model in the largest cluster was selected; the reweighted score was used for ranking). We modeled hexamer peptide sequences (P3-P2-P1-P1'-P2'-P3'), assuming that the influence of more distant residues is negligible. The excellent correspondence between good predicted score and high experimental rate/amount of degradation allowed us to derive a score threshold to be used to predict ClpP cleavage sites (Fig. 3b, left panel).

3.6.2.2 Identification of Potential ClpP Cleavage Site in Zwf Using ClpP FlexPepBind

With the calibrated protocol in hand, we proceeded to predict potential ClpP cleavage site(s) in the region of interest of Zwf, ^{192}FANSLFV*NNWDN*RTIDH^{208} (which contains the EDF precursor sequence, highlighted in italics). For each overlapping hexapeptide, we optimized a structural model based on the template generated above, and ranked different peptides according to their predicted binding energy. The results suggest that L_{196} (peptide NSL-FVN) and N_{203} (peptide WDN-RTI) are strong ClpP cleavage sites (Fig. 3b, middle panel). ClpP cleavage is thus predicted to generate a heptapeptide encompassing residues 197–203.

Given the distance of approximately seven residues between the consecutive active sites in the solved crystal structure, and the processive nature of ClpP-mediated cleavage [79], we propose that L_{196} (L_{196}FVNNWDNR) is anchored in the first active site, allowing the C-terminal N_{203} (NNWD**N203**R) to be located in the next active site in the proteolytic chamber (Fig. 3b, right panel). The resulting heptamer FVNNWDN would then be further cleaved either by ClpP, or other exo-proteases, to generate the NNWDN pentamer EDF precursor.

4   Notes

Even though we have made every effort to generate with FlexPepDock a general protocol that is broadly applicable to a range of different systems, performance and sample size are still dependent on the number of degrees of freedom of a given system and should be evaluated with care. Below are several tips that we have found to work well in a wide variety of different systems that we and others have analyzed using FlexPepDock and the related FlexPepBind application.

1. **How to generate a starting structure for FlexPepDock.** (*a*) *from a structure of a homolog complex:* If similar structures exist (this is common for peptide binders with multiple specificity, as for example for PDZ domains and many domains that bind signaling peptides), the initial structure can be constructed from a homology model of a similar structure using the Rosetta tool for comparative modeling, or any other homology modeling tool. (*b*) *given a binding site:* If only the binding site is known, an initial peptide chain conformation (either extended or alpha-helical) can be created from a FASTA file, using the BuildPeptide Rosetta utility (for detailed description on this protocol, visit www.rosettacommons.org/docs/latest/application_documentation/utilities/build-peptide), or using external tools such as PyMOL (www.pymol.org) Builder. The chain can then be positioned manually in the vicinity of the binding site using external tools such as PyMOL and Chimera (www.cgl.ucsf.edu/chimera). Alternatively, the peptide may be

positioned manually in an arbitrary orientation relative to the receptor protein, and guided to the binding site with Rosetta FlexPepDock (or RosettaDock), by enforcing appropriate distance constraints to the binding site (as specified in a constraint file; *see* Subheading 3.3.1).

2. **Prepack the receptor to remove internal clashes**: Always prepack the receptor structure (using the prepacking mode; *see* **step 1** in Subheading 3.3.2) before running a docking protocol, to remove potential internal clashes in the structure and generate a uniform energy background in non-interface regions. This removes irrelevant energy differences between models and allows better comparison and ranking of different models. Prepacking is not needed if the simulation is started from a model generated in a previous FlexPepDock run using the same parameters.

   For FlexPepBind applications, it is especially important to make sure that for an enzyme the active site residues do not rearrange too much (e.g., Histidine side-chain flipping). Resfiles can be provided to prevent repacking of selected residues (for the description of the resfile format, see www.rosettacommons. org/docs/latest/rosetta_basics/file_types/resfiles).

   In certain cases, it might be advisable to also pre-optimize the backbone of the receptor (using, e.g., the Rosetta Relax protocol; for detailed instructions on how to prepare an input structure using the Relax protocol, see www.rosettacommons. org/docs/latest/rosetta_basics/preparation/preparing-structures). In this case, it is recommended to use both backbone and side-chain constraints to restrict the extent of structural rearrangement.

3. **Selection of sampling parameters**

   (a) **Determination of adequate amount of sampling needed**: The amount of sampling (e.g., number of models generated) depends on the complexity of the problem. The optimal parameters of the different FlexPepDock protocols depend on the amount of available information about the interaction studied. If no accurate information about the binding mode is available, then it is advised to use FlexPepDock ab-initio to generate a considerably larger number of models (e.g., 50,000) to sample a larger space, and to guarantee convergence of the protocol. This helps to identify reliable low-energy conformations in the energy landscape. In turn, when information about the binding mode is available, the refinement protocol can be used to generate a relatively smaller number of model (e.g., 200–1000), and even better, when constraints can be defined that characterize peptide interactions with a given receptor, less models need to be generated, and the results for, e.g., binding predictions will be more reliable. It is important to note that the refinement protocol can refine models that are close to the correct solution both in terms of Cartesian, as well as dihedral φ/ψ distance. Even if the peptide is placed in the correct rigid body orientation, but needs to undergo significant dihedral changes (e.g., transition from extended conformation to helix), it is advised to use the ab-initio protocol that will use fragments to sample large dihedral changes more efficiently.

(b) **Boost local refinement with the -lowres_preoptimize flag**: This flag will add a preemptive centroid-mode optimization step, before performing full-atom, high-resolution refinement. As a rule of thumb, it is recommended to use this flag when the quality of the initial starting structure is less well defined (roughly more than 3A peptide backbone-RMSD), and thus sampling an extended range makes sense.

(c) **Receptor side chain flexibility—not too much but not too rigid**: *(a) Include information about the interface side chain conformations in the free receptor structure* (-unbound-rot). In many cases, the unbound receptor (or peptide) may contain side-chain conformations that are more similar to the final bound structure than those in the rotamer library. In order to save this useful information, it is possible to specify a structure whose side-chain conformations will be appended to the rotamer library during prepacking or docking, and may improve the chances of a well-scoring near-native result. This option was originally developed for the RosettaDock protocol. (*b*) *Add extra rotamers:* It is highly recommended to use the Rosetta extra rotamer flags that increase the number of rotamers used for repacking (we recommend using at least the -ex1 and -ex2aro flags). It is important to choose the same parameters for both pre-packing and model production runs.

4. **Selection of the final model**: The final model should be selected based on score (and cluster size). For refinement it is recommended to inspect the top 5–10 scoring models in addition to the best-ranked model. For ab-initio peptide docking, we recommend first clustering the results and then selecting top-scoring representatives. The clusters can be ranked either by energy or size; though we recommend going for the energy criteria for most applications.

   Ranking of models: Different scoring measures may be used to rank and select models (*see* Table 3). For structure prediction, ranking according to the reweighted score (reweighted_sc) provides, at least for ab-initio peptide docking, in our hands best results. We therefore suggest using this term for ranking in structure prediction, unless indicated otherwise. For binding prediction using the FlexPepBind protocol, the optimal scoring measure for the identification of binders is determined for each system separately (*see* Subheading 3.5.3 and Table 1A).

5. **Do NOT use FlexPepDock for fully blind docking (i.e., without a specified binding site)**: Neither of the provided protocols is intended for fully blind docking. The ab-initio protocol assumes that the peptide is located in the vicinity of the binding site, but does not assume anything about the initial peptide backbone conformation. The Refinement protocol is more restricted—it is intended for obtaining high-resolution peptide models given a coarse-grain starting structure, which should resemble the native solution to some extent (up to 5A backbone-RMSD for the native peptide, even though in some cases, the protocol works well for starting structures with up to 12A bb-RMSD from the native). *See* **Note 1** for a description of how to generate an appropriate starting structure.

6. **Include information about the secondary structure of the peptide in FlexPepDock simulations**: The FlexPepDock Refinement protocol is designed to allow substantial peptide backbone flexibility, but not enough to switch well between secondary structures (e.g., from strand to helical conformation). Hence, it may be useful to initially assign a canonical secondary structure to the peptide based on prior information (e.g., from secondary structure prediction algorithms, from homolog structures, or experimental information, such as CD experiments, etc.). Formally, FlexPepDock does not require initial secondary structure assignment, and can therefore be applied in cases where no confident information on the secondary peptide structure is available. However, the ab-initio FlexPepDock protocol may benefit implicitly from accurate secondary structure prediction when building the fragment libraries.

7. **The receptor backbone is not moved in the current FlexPepDock implementations**: Our protocol allows full receptor side-chain flexibility, and was shown to perform quite well when docking to unbound receptors or to alternative conformations. However, it is assumed that the receptor backbone does not change too much at the interface, as we do not yet model receptor backbone flexibility. To overcome this current restriction, it is advised to use different available structural models of the receptor and select models from different simulations.

   For FlexPepBind, where the binding site is known, a short precalibration step can identify the template receptor structure that will provide the best signal to distinguish substrates from non-substrates.

8. **Peptide length**: Our benchmarks consist of peptides of length 5–15, and our protocols performed well on this benchmark. Specific larger peptides have also been modeled successfully, but we do not have elaborate benchmark results for these.

9. **Use a cluster to obtain results quickly**: In our tests, producing 200 models with the Refinement protocol typically takes 3–10 CPU hours (approximately 1–3 min per model). Substantial speedup gain is obtained by running parallel processes using appropriate job-distributor flags (e.g., for using MPI; for details visit https://www.rosettacommons.org/docs/latest/rosetta_basics/Rosetta-Basics). The ab-initio protocol requires a much larger number of models (we experimented with 50,000 models), but the running time per model is similar (3–4 min). The running time may increase or decrease, depending mainly on the receptor size (for peptides of length 5–15).

References

1. Petsalaki E, Russell RB (2008) Peptide-mediated interactions in biological systems: new discoveries and applications. Curr Opin Biotechnol 19:344–350

2. Pawson T, Nash P (2003) Assembly of cell regulatory systems through protein interaction domains. Science 300:445–452

3. Neduva V, Linding R, Su-Angrand I, Stark A, de Masi F, Gibson TJ, Lewis J, Serrano L, Russell RB (2005) Systematic discovery of new recognition peptides mediating protein interaction networks. PLoS Biol 3:e405

4. Vacic V, Oldfield CJ, Mohan A, Radivojac P, Cortese MS, Uversky VN, Dunker AK (2007) Characterization of molecular recognition features, MoRFs, and their binding partners. J Proteome Res 6:2351–2366

5. Gamble TR, Vajdos FF, Yoo S, Worthylake DK, Houseweart M, Sundquist WI, Hill CP (1996) Crystal structure of human cyclophilin A bound to the amino-terminal domain of HIV-1 capsid. Cell 87:1285–1294

6. London N, Raveh B, Movshovitz-Attias D, Schueler-Furman O (2010) Can self-inhibitory peptides be derived from the interfaces of globular protein-protein interactions? Proteins 78(15):3140–3149

7. London N, Raveh B, Schueler-Furman O (2013) Druggable protein-protein interactions—from hot spots to hot segments. Curr Opin Chem Biol 17:952–959

8. Andrews SJ, Rothnagel JA (2014) Emerging evidence for functional peptides encoded by short open reading frames. Nat Rev Genet 15:193–204

9. Heemels MT, Ploegh H (1995) Generation, translocation, and presentation of MHC class I-restricted peptides. Annu Rev Biochem 64:463–491

10. Zhou A, Webb G, Zhu X, Steiner DF (1999) Proteolytic processing in the secretory pathway. J Biol Chem 274:20745–20748

11. Over B, Wetzel S, Grutter C, Nakai Y, Renner S, Rauh D, Waldmann H (2013) Natural-product-derived fragments for fragment-based ligand discovery. Nat Chem 5:21–28

12. London N, Movshovitz-Attias D, Schueler-Furman O (2010) The structural basis of peptide-protein binding strategies. Structure 18:188–199

13. London N, Raveh B, Schueler-Furman O (2013) Peptide docking and structure-based characterization of peptide binding: from knowledge to know-how. Curr Opin Struct Biol 23:894–902

14. Wright PE, Dyson HJ (2009) Linking folding and binding. Curr Opin Struct Biol 19:31–38

15. Kjaergaard M, Teilum K, Poulsen FM (2010) Conformational selection in the molten globule state of the nuclear coactivator binding domain of CBP. Proc Natl Acad Sci U S A 107:12535–12540

16. Rosal R, Pincus MR, Brandt-Rauf PW, Fine RL, Michl J, Wang H (2004) NMR solution structure of a peptide from the mdm-2 binding domain of the p53 protein that is selectively cytotoxic to cancer cells. Biochemistry 43:1854–1861

17. Sugase K, Dyson HJ, Wright PE (2007) Mechanism of coupled folding and binding of an intrinsically disordered protein. Nature 447:1021–1025

18. Fuxreiter M, Tompa P, Simon I (2007) Local structural disorder imparts plasticity on linear motifs. Bioinformatics 23:950–956

19. Trellet M, Melquiond AS, Bonvin AM (2015) Information-driven modeling of protein-peptide complexes. Methods Mol Biol 1268:221–239

20. Ben-Shimon A, Niv MY (2015) AnchorDock: blind and flexible anchor-driven peptide docking. Structure 23:929–940

21. Saladin A, Rey J, Thevenet P, Zacharias M, Moroy G, Tuffery P (2014) PEP-SiteFinder: a tool for the blind identification of peptide binding sites on protein surfaces. Nucleic Acids Res 42:W221–W226

22. Raveh B, London N, Zimmerman L, Schueler-Furman O (2011) Rosetta FlexPepDock ab-initio: simultaneous folding, docking and refinement of peptides onto their receptors. PLoS One 6:e18934

23. Lavi A, Ngan CH, Movshovitz-Attias D, Bohnuud T, Yuch C, Beglov D, Schueler-Furman O, Kozakov D (2013) Detection of peptide-binding sites on protein surfaces: the first step toward the modeling and targeting of peptide-mediated interactions. Proteins 81:2096–2105

24. Ben-Shimon A, Eisenstein M (2010) Computational mapping of anchoring spots on protein surfaces. J Mol Biol 402:259–277

25. Raveh B, London N, Schueler-Furman O (2010) Sub-angstrom modeling of complexes between flexible peptides and globular proteins. Proteins 78:2029–2040

26. Leaver-Fay A, Tyka M, Lewis SM, Lange OF, Thompson J, Jacak R, Kaufman K, Renfrew PD, Smith CA, Sheffler W et al (2011) ROSETTA3: an object-oriented software suite for the simulation and design of macromolecules. Methods Enzymol 487:545–574

27. Das R, Baker D (2008) Macromolecular modeling with rosetta. Annu Rev Biochem 77:363–382

28. Kaufmann KW, Lemmon GH, Deluca SL, Sheehan JH, Meiler J (2010) Practically useful: what the Rosetta protein modeling suite can do for you. Biochemistry 49:2987–2998

29. King CA, Bradley P (2010) Structure-based prediction of protein-peptide specificity in Rosetta. Proteins 78:3437–3449

30. Yanover C, Petersdorf EW, Malkki M, Gooley T, Spellman S, Velardi A, Bardy P, Madrigal A, Bignon JD, Bradley P (2011) HLA mismatches and hematopoietic cell transplantation: structural simulations assess the impact of changes in peptide binding specificity on transplant outcome. Immunome Res 7:4

31. Smith CA, Kortemme T (2010) Structure-based prediction of the peptide sequence space recognized by natural and synthetic PDZ domains. J Mol Biol 402:460–474

32. Grigoryan G, Reinke AW, Keating AE (2009) Design of protein-interaction specificity gives selective bZIP-binding peptides. Nature 458:859–864

33. London N, Lamphear CL, Hougland JL, Fierke CA, Schueler-Furman O (2011) Identification of a novel class of farnesylation targets by structure-based modeling of binding specificity. PLoS Comput Biol 7:e1002170

34. London N, Gulla S, Keating AE, Schueler-Furman O (2012) In silico and in vitro elucidation of BH3 binding specificity toward Bcl-2. Biochemistry 51:5841–5850

35. Alam N, Zimmerman L, Wolfson NA, Joseph CG, Fierke CA, Schueler-Furman O (2016) Structure-based identification of HDAC8 non-histone substrates. Structure 24:458–468

36. Dinkel H, Michael S, Weatheritt RJ, Davey NE, Van Roey K, Altenberg B, Toedt G, Uyar B, Seiler M, Budd A et al (2012) ELM—the database of eukaryotic linear motifs. Nucleic Acids Res 40:D242–D251

37. Dinkel H, Van Roey K, Michael S, Kumar M, Uyar B, Altenberg B, Milchevskaya V, Schneider M, Kuhn H, Behrendt A et al (2016) ELM 2016-data update and new functionality of the eukaryotic linear motif resource. Nucleic Acids Res 44:D294–D300

38. London N, Raveh B, Cohen E, Fathi G, Schueler-Furman O (2011) Rosetta FlexPepDock web server—high resolution modeling of peptide-protein interactions. Nucleic Acids Res 39:W249–W253

39. Belitsky M, Avshalom H, Erental A, Yelin I, Kumar S, London N, Sperber M, Schueler-Furman O, Engelberg-Kulka H (2011) The Escherichia coli extracellular death factor EDF induces the endoribonucleolytic activities of the toxins MazF and ChpBK. Mol Cell 41:625–635

40. Kumar S, Kolodkin-Gal I, Vesper O, Alam N, Schueler-Furman O, Moll I, Engelberg-Kulka H (2016) Escherichia coli quorum-sensing EDF, a peptide generated by novel multiple distinct mechanisms and regulated by trans-translation. MBio 7

41. Gray JJ, Moughon S, Wang C, Schueler-Furman O, Kuhlman B, Rohl CA, Baker D (2003) Protein-protein docking with simultaneous optimization of rigid-body displacement and side-chain conformations. J Mol Biol 331:281–299

42. Wang C, Schueler-Furman O, Baker D (2005) Improved side-chain modeling for protein-protein docking. Protein Sci 14:1328–1339

43. Wang C, Bradley P, Baker D (2007) Protein-protein docking with backbone flexibility. J Mol Biol 373:503–519

44. Petsalaki E, Stark A, Garcia-Urdiales E, Russell RB (2009) Accurate prediction of peptide binding sites on protein surfaces. PLoS Comput Biol 5:e1000335

45. Trabuco LG, Lise S, Petsalaki E, Russell RB (2012) PepSite: prediction of peptide-binding sites from protein surfaces. Nucleic Acids Res 40:W423–W427

46. Rohl CA, Strauss CE, Misura KM, Baker D (2004) Protein structure prediction using Rosetta. Methods Enzymol 383:66–93

47. Yadav M, Jhunjhunwala S, Phung QT, Lupardus P, Tanguay J, Bumbaca S, Franci C, Cheung TK, Fritsche J, Weinschenk T et al (2014) Predicting immunogenic tumour mutations by combining mass spectrometry and exome sequencing. Nature 515:572–576

48. Arber C, Feng X, Abhyankar H, Romero E, Wu MF, Heslop HE, Barth P, Dotti G, Savoldo B (2015) Survivin-specific T cell receptor targets tumor but not T cells. J Clin Invest 125:157–168

49. Solomonson M, Huesgen PF, Wasney GA, Watanabe N, Gruninger RJ, Prehna G, Overall CM, Strynadka NC (2013) Structure of the mycosin-1 protease from the mycobacterial ESX-1 protein type VII secretion system. J Biol Chem 288:17782–17790

50. Eckhard U, Huesgen PF, Brandstetter H, Overall CM (2014) Proteomic protease specificity profiling of clostridial collagenases reveals their intrinsic nature as dedicated degraders of collagen. J Proteomics 100:102–114

51. Marino G, Huesgen PF, Eckhard U, Overall CM, Schroder WP, Funk C (2014) Family-wide characterization of matrix metalloproteinases from Arabidopsis thaliana reveals their distinct proteolytic activity and cleavage site specificity. Biochem J 457:335–346

52. Barre O, Dufour A, Eckhard U, Kappelhoff R, Beliveau F, Leduc R, Overall CM (2014) Cleavage specificity analysis of six type II transmembrane serine proteases (TTSPs) using PICS with proteome-derived peptide libraries. PLoS One 9:e105984

53. Velarde-Salcedo AJ, Barrera-Pacheco A, Lara-Gonzalez S, Montero-Moran GM, Diaz-Gois A, Gonzalez de Mejia E, Barba de la Rosa AP (2013) In vitro inhibition of dipeptidyl peptidase IV by peptides derived from the hydrolysis of amaranth (Amaranthus hypochondriacus L.) proteins. Food Chem 136:758–764

54. Lamarque MH, Roques M, Kong-Hap M, Tonkin ML, Rugarabamu G, Marq JB, Penarete-Vargas DM, Boulanger MJ, Soldati-Favre D, Lebrun M (2014) Plasticity and redundancy among AMA-RON pairs ensure host cell entry of Toxoplasma parasites. Nat Commun 5:4098

55. Srinivasan P, Ekanem E, Diouf A, Tonkin ML, Miura K, Boulanger MJ, Long CA, Narum DL, Miller LH (2014) Immunization with a functional protein complex required for erythrocyte invasion protects against lethal malaria. Proc Natl Acad Sci U S A 111:10311–10316

56. Lesniewska K, Warbrick E, Ohkura H (2014) Peptide aptamers define distinct EB1- and EB3-binding motifs and interfere with microtubule dynamics. Mol Biol Cell 25:1025–1036

57. Poplawski A, Hu K, Lee W, Natesan S, Peng D, Carlson S, Shi X, Balaz S, Markley JL, Glass KC (2014) Molecular insights into the recognition of N-terminal histone modifications by the BRPF1 bromodomain. J Mol Biol 426: 1661–1676

58. Ozawa K, Horan NP, Robinson A, Yagi H, Hill FR, Jergic S, Xu ZQ, Loscha KV, Li N, Tehei M et al (2013) Proofreading exonuclease on a tether: the complex between the E. coli DNA polymerase III subunits alpha, epsilon, theta and beta reveals a highly flexible arrangement of the proofreading domain. Nucleic Acids Res 41:5354–5367

59. Doray B, Misra S, Qian Y, Brett TJ, Kornfeld S (2012) Do GGA adaptors bind internal DXXLL motifs? Traffic 13:1315–1325

60. Thomas S, Rai J, John L, Schaefer S, Putzer BM, Herchenroder O (2013) Chikungunya virus capsid protein contains nuclear import and export signals. Virol J 10:269

61. Shi X, Betzi S, Lugari A, Opi S, Restouin A, Parrot I, Martinez J, Zimmermann P, Lecine P, Huang M et al (2012) Structural recognition mechanisms between human Src homology domain 3 (SH3) and ALG-2-interacting protein X (Alix). FEBS Lett 586:1759–1764

62. Crawley SW, Gharaei MS, Ye Q, Yang Y, Raveh B, London N, Schueler-Furman O, Jia Z, Cote GP (2011) Autophosphorylation activates Dictyostelium myosin II heavy chain kinase A by providing a ligand for an allosteric binding site in the alpha-kinase domain. J Biol Chem 286:2607–2616

63. Yin G, Lopes da Fonseca T, Eisbach SE, Anduaga AM, Breda C, Orcellet ML, Szego EM, Guerreiro P, Lazaro DF, Braus GH et al (2014) alpha-Synuclein interacts with the switch region of Rab8a in a Ser129 phosphorylation-dependent manner. Neurobiol Dis 70:149–161

64. Buch I, Fishelovitch D, London N, Raveh B, Wolfson HJ, Nussinov R (2010) Allosteric regulation of glycogen synthase kinase 3beta: a theoretical study. Biochemistry 49: 10890–10901

65. Vangone A, Bonvin AM (2015) Contacts-based prediction of binding affinity in protein-protein complexes. Elife 4:e07454

66. Vreven T, Moal IH, Vangone A, Pierce BG, Kastritis PL, Torchala M, Chaleil R, Jimenez-Garcia B, Bates PA, Fernandez-Recio J et al (2015) Updates to the integrated protein-protein interaction benchmarks: docking benchmark Version 5 and affinity benchmark Version 2. J Mol Biol 427:3031–3041

67. Chaudhury S, Gray JJ (2009) Identification of structural mechanisms of HIV-1 protease specificity using computational peptide docking: implications for drug resistance. Structure 17:1636 1648

68. Kuhlman B, Baker D (2000) Native protein sequences are close to optimal for their structures. Proc Natl Acad Sci U S A 97:10383–10388

69. Hougland JL, Hicks KA, Hartman HL, Kelly RA, Watt TJ, Fierke CA (2010) Identification of novel peptide substrates for protein farnesyl-transferase reveals two substrate classes with distinct sequence selectivities. J Mol Biol 395:176–190

70. Krzysiak AJ, Aditya AV, Hougland JL, Fierke CA, Gibbs RA (2010) Synthesis and screening of a CaaL peptide library versus FTase reveals a surprising number of substrates. Bioorg Med Chem Lett 20:767–770

71. Maurer-Stroh S, Eisenhaber F (2005) Refinement and prediction of protein prenylation motifs. Genome Biol 6:R55

72. Al Quadan T, Price CT, London N, Schueler-Furman O, AbuKwaik Y (2011) Anchoring of bacterial effectors to host membranes through host-mediated lipidation by prenylation: a common paradigm. Trends Microbiol 19:573–579

73. Vannini A, Volpari C, Gallinari P, Jones P, Mattu M, Carfi A, De Francesco R, Steinkuhler C, Di Marco S (2007) Substrate binding to histone deacetylases as shown by the crystal structure of the HDAC8-substrate complex. EMBO Rep 8:879–884

74. Olson DE, Udeshi ND, Wolfson NA, Pitcairn CA, Sullivan ED, Jaffe JD, Svinkina T, Natoli T, Lu X, Paulk J et al (2014) An unbiased approach to identify endogenous substrates of "histone" deacetylase 8. ACS Chem Biol 9:2210–2216

75. Ramisetty BC, Natarajan B, Santhosh RS (2015) mazEF-mediated programmed cell death in bacteria: "what is this?". Crit Rev Microbiol 41:89–100

76. Kamada K, Hanaoka F, Burley SK (2003) Crystal structure of the MazE/MazF complex: molecular bases of antidote-toxin recognition. Mol Cell 11:875–884

77. Szyk A, Maurizi MR (2006) Crystal structure at 1.9A of E. coli ClpP with a peptide covalently bound at the active site. J Struct Biol 156:165–174

78. Thompson MW, Maurizi MR (1994) Activity and specificity of Escherichia coli ClpAP protease in cleaving model peptide substrates. J Biol Chem 269:18201–18208

79. Jennings LD, Lun DS, Medard M, Licht S (2008) ClpP hydrolyzes a protein substrate processively in the absence of the ClpA ATPase: mechanistic studies of ATP-independent proteolysis. Biochemistry 47:11536–11546

80. Poy F, Yaffe MB, Sayos J, Saxena K, Morra M, Sumegi J, Cantley LC, Terhorst C, Eck MJ (1999) Crystal structures of the XLP protein SAP reveal a class of SH2 domains with extended, phosphotyrosine-independent sequence recognition. Mol Cell 4:555–561

81. Li Y, Suino K, Daugherty J, Xu HE (2005) Structural and biochemical mechanisms for the specificity of hormone binding and coactivator assembly by mineralocorticoid receptor. Mol Cell 19:367–380

82. Reid TS, Terry KL, Casey PJ, Beese LS (2004) Crystallographic analysis of CaaX prenyltransferases complexed with substrates defines rules of protein substrate selectivity. J Mol Biol 343:417–433

83. Lee EF, Czabotar PE, Yang H, Sleebs BE, Lessene G, Colman PM, Smith BJ, Fairlie WD (2009) Conformational changes in Bcl-2 pro-survival proteins determine their capacity to bind ligands. J Biol Chem 284:30508–30517

84. Fire E, Gulla SV, Grant RA, Keating AE (2010) Mcl-1-Bim complexes accommodate surprising point mutations via minor structural changes. Protein Sci 19:507–519

85. Ku B, Liang C, Jung JU, Oh BH (2011) Evidence that inhibition of BAX activation by BCL-2 involves its tight and preferential interaction with the BH3 domain of BAX. Cell Res 21:627–641

86. Lazaridis T, Karplus M (1999) Effective energy function for proteins in solution. Proteins 35:133–152

87. Dutta S, Gulla S, Chen TS, Fire E, Grant RA, Keating AE (2010) Determinants of BH3 binding specificity for Mcl-1 versus Bcl-xL. J Mol Biol 398:747–762

88. Gurard-Levin ZA, Kilian KA, Kim J, Bahr K, Mrksich M (2010) Peptide arrays identify isoform-selective substrates for profiling endogenous lysine deacetylase activity. ACS Chem Biol 5:863–873

Part III

Prediction and Design of Peptide Binding Specificity

Flexible Backbone Methods for Predicting and Designing Peptide Specificity

Noah Ollikainen

Abstract

Protein–protein interactions play critical roles in essentially every cellular process. These interactions are often mediated by protein interaction domains that enable proteins to recognize their interaction partners, often by binding to short peptide motifs. For example, PDZ domains, which are among the most common protein interaction domains in the human proteome, recognize specific linear peptide sequences that are often at the C-terminus of other proteins. Determining the set of peptide sequences that a protein interaction domain binds, or it's "peptide specificity," is crucial for understanding its cellular function, and predicting how mutations impact peptide specificity is important for elucidating the mechanisms underlying human diseases. Moreover, engineering novel cellular functions for synthetic biology applications, such as the biosynthesis of biofuels or drugs, requires the design of protein interaction specificity to avoid crosstalk with native metabolic and signaling pathways. The ability to accurately predict and design protein–peptide interaction specificity is therefore critical for understanding and engineering biological function. One approach that has recently been employed toward accomplishing this goal is computational protein design. This chapter provides an overview of recent methodological advances in computational protein design and highlights examples of how these advances can enable increased accuracy in predicting and designing peptide specificity.

Key words Protein interactions, Protein design, Sampling algorithms, Backbone flexibility, Conformational ensembles, Specificity prediction, Designing specificity

1 Computational Protein Design Predicts Sequences that Are Optimal for a Given Structure

In general, the goal of computational protein design is to predict protein sequences that fold into a particular three-dimensional structure and perform a desired function. For example, one might want to design an enzyme to catalyze a chemical reaction or design a protein therapeutic to bind to and inhibit the function of a viral protein [1–3]. While a variety of computational protein design methods exist, they generally share two basic components: (1) a search algorithm that will sample different protein sequences and (2) an energy function that will evaluate the energy of each protein

Ora Schueler-Furman and Nir London (eds.), *Modeling Peptide-Protein Interactions: Methods and Protocols*, Methods in Molecular Biology, vol. 1561, DOI 10.1007/978-1-4939-6798-8_10, © Springer Science+Business Media LLC 2017

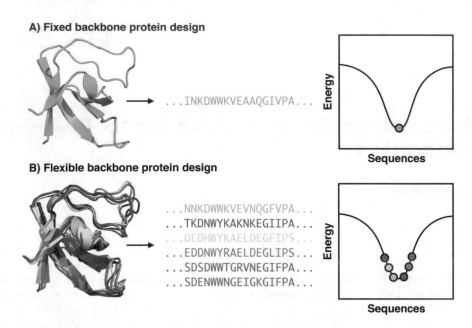

Fig. 1 Computational protein design methods take a protein backbone structure as input and return low-energy sequences for that structure. Fixed backbone protein design (**a**) does not allow the input backbone structure to change, whereas flexible backbone protein design (**b**) allows slight conformational changes in the input structure while maintaining the same overall fold

sequence when modeled onto the desired target structure. In early computational protein design work, the goal was typically to identify the lowest energy protein sequence for a fixed protein backbone structure [4] (Fig. 1a). The lowest energy sequence for a given structure is assumed that is most likely to fold into that structure. The problem of finding the lowest energy sequence can be challenging due to the enormous number of possible protein sequences of a given length. For example, there are 20^{100} (or $\sim 10^{130}$) possible sequences for a protein that is 100 amino acid residues long. Moreover, most amino acid residues have flexible side chains that are often modeled as discrete conformations called "rotamers" [5]. If, for example, 10 rotamers are used for each amino acid side chain, then there are 200^{100} (or $\sim 10^{230}$) possible side-chain conformations for a 100 amino acid protein.

To identify the lowest energy amino acid sequence and side-chain conformation in such a massive search space, an approach called Dead End Elimination (DEE) can be used to eliminate side-chain rotamers from the search space if they are provably not part of the global minimum energy conformation [6]. After pruning the search space with DEE, the remaining conformations can then be enumerated to find the protein sequence with the global minimum energy conformation. While deterministic approaches like DEE were typically employed in early computational protein design work [4], recent efforts have taken advantage of stochastic approaches such as [7, 8]. These methods start with a random protein sequence and side-chain conformation, iteratively make random changes to the sequence and conformation called "moves," and then accept or reject these moves based on whether or not they result in a decrease in energy. While stochastic methods do not guarantee finding the global minimum energy sequence, they are often much faster than deterministic methods and therefore can be run numerous times to identify many different low-energy sequences [9].

2   Beyond the Global Minimum: Predicting Sequence Tolerance

The ability to find a large number of low-energy protein sequences for a desired structure, rather than just the global minimum energy sequence, can be tremendously useful for many protein design applications. One reason for this is that the energy functions used to evaluate designed proteins are not perfect. Consequently, the probability that any given designed protein will adopt its desired structure or perform its desired function can be quite low for challenging protein design applications such as enzyme design or the design of protein–protein interactions [1–3]. Fortunately, it is becoming increasingly feasible to experimentally screen hundreds or even thousands of designed proteins for a given function, which increases the likelihood of obtaining at least one successful design. As a result, the goal of computational protein design is beginning to toward the identification of a large set of sequences that will likely adopt the desired structure, as the set of "tolerated" sequences for a given structure. Both deterministic and stochastic methods can be applied to predict sequence tolerance, since deterministic approaches can be extended to enumerate all sequences within a given energy of the global minimum sequence [10, 11] and stochastic approaches can be run many times to identify diverse local minima sequences [12].

In addition to aiding protein design applications, the accurate identification of tolerated sequences for a protein structure is critical for applying computational protein design to predict the set of peptide sequences that a protein can bind with high affinity, as a protein's "peptide specificity." For example, one approach for predicting peptide specificity is to obtain a crystal structure of a protein–peptide complex and then perform computational protein design on the peptide. The resulting designed peptide sequences will represent a set of peptides that are likely to have high affinity interactions with the protein. One major challenge with this approach is that proteins are often promiscuous and capable of binding a highly diverse set of peptide sequences [13]. Traditional computational protein design methods make the assumption that the protein backbone remains fixed in its 3D structure; however, this assumption is not valid for flexible peptides with diverse sequences that can adopt a variety of backbone conformations when bound to a protein. Additionally, the region of the protein that interacts with the peptide may undergo changes in protein backbone conformation that could be important for peptide binding [14]. Consequently, modeling protein backbone flexibility during computational protein design can be crucial for accurately predicting and designing peptide specificity [15].

3   Modeling Backbone Flexibility Can Improve the Accuracy of Computational Protein Design

Computational protein design methods that model protein backbone flexibility are referred to as "flexible backbone design" methods in contrast to traditional "fixed backbone design" methods that do not allow the protein backbone to change conformation (Fig. 1). One major advantage of flexible backbone design methods is their ability to identify a diverse set of tolerated sequences for a given protein structure (Fig. 1b). This is because changes in backbone conformation can allow the protein to achieve energetically favorable amino acid side-chain conformations that would otherwise be incompatible with the initial backbone conformation. On a large scale by directly comparing fixed and flexible backbone design methods performed on 40 diverse protein folds. The sequences designed with flexible backbone methods were much more diverse and exhibited greater similarity to naturally occurring protein sequences with the corresponding protein folds compared to sequences generated with fixed backbone design. This study revealed instances where backbone movements were necessary to enable optimal packing of hydrophobic residues or precise side-chain geometries required for hydrogen bonding.

To model backbone flexibility, flexible backbone design methods perform moves that alter the protein backbone conformation in addition to the traditional moves that change amino acid sequence and side chain conformations. For example, one type of backbone move, called a "backrub," moves a local segment of the backbone as a rigid body by rotating it about an axis defined by the first and last atoms of the segment (Fig. 2a). This type of motion was originally described by the Richardson group and initially observed via alternate conformations in ultra-high resolution protein crystal structures [17]. Since its initial observation, this move has been implemented into protein modeling and design tools [18, 19], enabling improvements in the ability to recapitulate properties of natural proteins. For example, backrub moves were implemented into the Rosetta protein structure prediction and design software [20] and demonstrated to improve the accuracy of predicting point mutant side-chain conformations relative to fixed

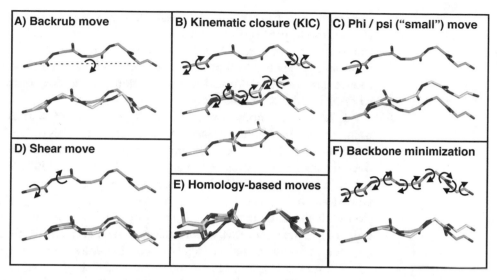

Fig. 2 Many different types of backbone moves can be used to model protein backbone flexibility. These include backrub moves (**a**), kinematic closure (KIC) moves (**b**), random phi/psi perturbations called "small" moves (**c**), shear moves (**d**), homology-based moves like fragment insertion (**e**), and all-atom minimization of backbone phi and psi torsions (**f**)

backbone sampling alone [18]. These moves have also been shown to allow protein modeling and design methods to better recapitulate natural properties of proteins including conformational heterogeneity, sequence entropy, and amino acid covariation [16, 21–23]. Finally, a recent study described a method to couple backrub moves to changes in amino acid sequence and side-chain conformation and showed that this approach significantly improved both the prediction of ligand binding site sequence tolerance and the prediction of mutations that alter enzyme substrate specificity compared to fixed backbone design [24].

Backrub moves provide a simple way to make small, subtle backbone motions, but other types of backbone motions can be helpful to model larger conformational changes. For example, a robotics-inspired move called kinematic closure (or KIC) was used to achieve sub-angstrom accuracy in modeling protein loop conformations (Fig. 2b) [25, 26]. While originally applied to protein loops, KIC moves can be performed on any segment of the backbone that is at least three residues long. Once a segment has been defined by starting and ending residues, a KIC move can be performed with two steps: (1) randomize a set of phi and psi backbone torsions between the start and end residues, thus creating a "chain break" in the protein backbone, (2) select six of the phi/psi torsions and analytically solve for a set of angles that will close the chain [27]. Because KIC moves allow the simultaneous change of many backbone degrees of freedom, they enable a potentially larger range of motion that is particularly useful for modeling backbone regions with high intrinsic flexibility like protein loops. In addition to previous work showing that KIC moves can improve the accuracy of protein loop modeling compared to other approaches, flexible backbone design methods that employ KIC moves have been shown to increase sequence tolerance compared to fixed backbone design [16].

Like the backrub motion, KIC moves only perturb a local segment of the backbone between a start and end residue and are therefore useful for modeling and design applications where the overall protein fold must be maintained. Similarly, although not strictly a local move, "shear" moves randomly perturb a phi torsion angle followed by a compensating rotation of the preceding psi torsion with the same magnitude (Fig. 2c) [28]. In addition to these types of localized motions, other backbone moves can be performed to achieve more global changes in conformation. For example, "small" moves rotate a phi or psi torsion of a single residue by a random small angle, resulting in a potentially large downstream change in structure (Fig. 2d) [28]. Additionally, homology-based approaches like fragment insertion can be used to sample backbone conformations from protein structures with homologous sequences (Fig. 2e) [29]. Finally, torsion angle minimization can be used to globally relax a protein structure into a local minimum conformation by simultaneously altering all torsion angles while minimizing the total energy of the protein (Fig. 2f) [30, 31].

4   Predicting Peptide Specificity Using Conformational Ensembles

Given the variety of different types of backbone motions, integrating them into a protein design algorithm that must also sample amino acid sequences and side-chain conformations can be challenging. A simple approach for incorporating backbone flexibility into design involves first generating a "conformational ensemble" [21]. Given an input structure, this approach performs many Monte Carlo simulations that iteratively apply backbone moves throughout the structure, resulting in an ensemble of structures that closely resemble the input structure but have differing backbone conformations (Fig. 1b). Computational protein design can then be performed on each structure in the conformational ensemble to identify low-energy sequences. Conformational ensembles generated using backrub moves, called "backrub ensembles," were shown to recapitulate the native conformational heterogeneity of ubiquitin, and sequences designed using these backrub ensembles had higher similarity to ubiquitin family sequences compared to sequences designed with fixed backbone design [21]. Since these observations, many studies have used backrub moves to generate conformational ensembles for the purpose of modeling or design [15, 16, 21, 32, 33], and the Kortemme lab provides a web server that allows users to generate backrub ensembles for a given input structure [34].

One key aspect of computational protein design approaches that use conformational ensembles is that they are capable of generating highly diverse sequences with variable backbone conformations (Fig. 1b), making them well suited for the problem of predicting protein–peptide specificity. Prediction of protein–peptide specificity can be formulated as the problem of identifying the set of tolerated sequences for a peptide in its protein bound state. By generating backrub ensembles for peptide-bound PDZ domains and then using computational protein design to sample and score approximately 10,000 peptide sequences for each backbone structure in the ensemble. These sequences were weighed by their scores and used to create a "specificity profile" for each PDZ domain, which reflects the likelihood of observing each amino acid at each position in the peptide sequence (Fig. 3a). The authors first used this protocol to predict specificity profiles for 17 human PDZ domains that each had (1) a crystal structure with a bound peptide available from the [35] to use as input and (2) a specificity profile that was previously determined by phage display experiments to use as validation [36]. Overall, the predicted specificity profiles

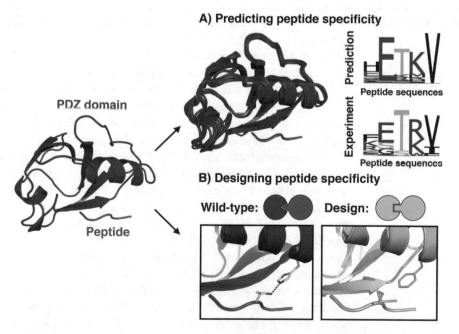

Fig. 3 Computational protein design methods can predict and design the peptide specificity of a protein given its peptide-bound structure. Predicting peptide specificity (**a**) involves generating a conformational ensemble of the complex and then performing computational protein design on the peptide to identify a set of low-energy peptide sequences. Designing peptide specificity (**b**) involves predicting a mutation in the protein that destabilizes the interaction with the wild-type peptide and then predicting a compensating mutation on the peptide that stabilizes the interaction with the mutant protein

were highly similar to the experimentally determined specificity profiles, demonstrating successful predictions of PDZ peptide binding specificity.

In addition to predicting the specificity profiles of wild-type PDZ domains, predict the specificity of 153 mutants of the Erbin PDZ domain, including 92 point mutants [36] and 61 mutants with 4–10 mutations [37]. While the prediction performance for Erbin point mutants was better on average than for the wild-type proteins (since the wild-type Erbin specificity is well predicted and most point mutations did not significantly alter specificity), prediction of specificity profiles for PDZ domains with 4–10 mutations was less accurate. One possible explanation for why these additional mutations make predicting specificity more difficult is that they introduce changes into the PDZ domain structure that are not accurately captured by the backrub ensemble, which is based on the wild-type crystal structure. This highlights a caveat with this approach for predicting specificity, which is that it requires a crystal structure of the protein–peptide complex as input. However, this structural information may not always be available, and therefore additional methods may be required to first build a model of the protein–peptide complex.

5 Modeling Protein–Peptide Interactions When Structural Information Is Missing

While predicting a protein's peptide specificity without knowing the exact structure of the bound peptide is challenging, a computational approach has been developed to specifically address this problem. That only needs an approximate location of a single residue on the peptide to predict peptide specificity [38]. In this method, the remainder of the peptide is assembled by a Monte Carlo simulation that iteratively appends residues with backbone conformations sampled from the PDB until the desired peptide length is reached. Once the peptide backbone is built, computational protein design can then be used to predict tolerated peptide sequences. This approach was further extended to predict peptide specificity for proteins without any available structural information. To accomplish this, generated homology models using Modeller [39] and then performed the *pepspec* protocol to predict peptide specificity starting from these models. To evaluate the accuracy of their predictions, the authors applied their protocol using varying amounts of structural information on four proteins with experimentally determined peptide specificity profiles. Overall, the authors found that but even with limited structural information they were able to recapitulate the major features of the specificity profiles for three of the four proteins using their homology model and peptide assembly approach.

The above methods for predicting specificity use computational protein design to sample and score different peptide sequences and conformations in complex with a bound protein. An alternative approach formulates peptide specificity prediction as the problem of screening a predefined set of peptide sequences to distinguish binders from. In this strategy, specific peptide sequences are docked onto the protein and then scored to estimate binding affinities. Docking of each peptide sequence can be performed using an approach such as FlexPepDock, which optimizes the rigid-body orientation of the peptide relative to the protein while incorporating peptide backbone flexibility using small and shear moves [40]. This technique has been used to successfully identify both known and novel peptide substrates of the enzyme Farnesyltransferase [41] and to predict the specificity of helical peptides for the apoptosis regulating proteins Bcl-xL, Mcl-1, and Bcl-2 [42]. FlexPepDock has been further extended to perform ab initio docking, which simultaneously folds and docks a peptide onto the surface of the protein and can therefore be used in cases where no information is available about the peptide backbone conformation when bound to the protein [43]. The Furman lab provides a web server that allows users to perform their own FlexPepDock simulations to model protein–peptide interactions [44].

6   Redesigning Peptide Specificity Using the Second-Site Suppressor Approach

Given the development of computational methods that predict the binding partners for a given peptide-binding domain, a natural extension of these methods is the redesign of protein–protein interaction specificity for synthetic biology applications. An example application is the engineering of metabolic or signaling pathways that do not interfere with the function of native proteins. The design of novel, "orthogonal" protein–protein interactions that avoid cross-talk with existing proteins has been of considerable interest in the field of protein design [45–48]. A common way to accomplish this is to use the "computational second-site suppressor" approach. Given an existing protein–protein interaction, this strategy aims to predict mutations in both proteins that would destabilize the interaction when only one protein is mutated, but stabilize the interaction if both proteins are mutated simultaneously (Fig. 3b).

The computational second-site suppressor technique has previously been applied to redesign the specificity of a number of protein–protein interactions [45–48], and recently it was used to redesign the peptide specificity of a PDZ domain [49]. To redesign PDZ specificity, used computational protein design to predict a single mutation in PDZ domain of α1-Syntrophin that altered its peptide specificity. This mutation changed a histidine on the PDZ domain, which formed a hydrogen bond to a serine or threonine on the peptide, into a phenylalanine (Fig. 3b). This new PDZ domain exhibited both decreased affinity with serine/threonine peptides and increased affinity with methionine peptides, demonstrating a switch in its peptide specificity. To determine how transferable this specificity switch mutation was between homologous domains, the authors made the same mutation in five other PDZ domains and measured peptide-binding preferences of the wild-type and mutant PDZ domains using fluorescence anisotropy. Despite the high structural similarity between the homologous domains, the specificity switch could only be transferred to three of the five PDZ domains tested. However, the authors were able to accurately recapitulate the energetic effects of the mutations in all homologous domains by performing flexible backbone simulations that use minimization to simultaneously optimize backbone and side-chain degrees of freedom. These simulations were subsequently used to predict mutations that further improved the specificity of the initial design. This study demonstrates both the challenge of generalizing the determinants of peptide specificity across related proteins and the success in computational models to capture the energetic consequences of subtle structural differences in protein–peptide interactions between homologous domains.

7   Designing Novel Peptide Ligands Using Flexible Backbone Design

Another useful application of computational methods that can model and predict peptide specificity prediction is the rational design of peptides that bind a given protein target. For example, one might want to design a therapeutic peptide or peptide-mimetic drug to inhibit a disease-associated protein–protein interaction. To accomplish this goal, Roberts et al. [50] applied computational protein design to engineer a peptide that competitively inhibits the interaction between the cystic fibrosis transmembrane conductance regulator (CFTR) and the PDZ domain of the CFTR-associated ligand (CAL). In cystic fibrosis patients, CFTR is mutated and forms a variant called ΔF508-CFTR, which is rapidly degraded via a lysosomal pathway involving its interaction with CAL. Inhibiting the interaction between ΔF508-CFTR and CAL could therefore potentially rescue its activity in cystic fibrosis patients and be used as a treatment. To design peptides that bind the PDZ domain of CAL and thus block CFTR binding, used an algorithm called K*, which first prunes rotamers using minDEE [51] that are not part of the lowest energy conformation for a given sequence and then performs a branch-and-bound search called A* [52] to enumerate the remaining low-energy conformations. To account for peptide flexibility, minimization was performed on the peptide backbone, side chains and rigid body orientation, generating an ensemble of conformations for each sequence. Finally, these conformational ensembles were Boltzmann-weighted and used to compute an approximate binding constant for each sequence. The best designed peptide, called kCAL01, bound to CAL with 170-fold higher affinity than CFTR and was shown to increase ΔF508-CFTR activity in human cells to a similar extent as an inhibitor drug identified experimentally by high-throughput screening [53].

In the previous example, the goal was to identify a peptide with the highest binding affinity to the target protein. However, some applications may require the design of a diverse set of peptide sequences that can form interactions across a wide range of affinities. For example, engineering novel metabolic or signaling pathways could necessitate designing proteins with transient interactions rather than stable, obligate interactions. To design diverse peptides that bound the apoptosis regulating protein Bcl-xL, Fu et al. [54] used a conformational ensemble-based approach that took advantage of two observations: (1) Bcl-xL binds peptides with α-helical secondary structures and (2) normal mode analysis of α-helical structures from the PDB showed that two bend modes and one twist mode can model alpha helical backbone flexibility [55]. used normal mode calculations to deform the backbone of the helical peptide bound to Bcl-xL in a crystal structure complex and used these backbone conformations as templates for design. Seventeen designed peptides with diverse sequences were selected to test experimentally, and among these, eight peptides bound strongly to Bcl-xL and four others showed weak but detectable binding. This study demonstrates the utility of flexible backbone approaches for designing diverse peptide sequences that exhibit a broad range of binding affinities.

8 Future Directions

Significant progress has been made during the past two decades in computational protein design. Early work focused on identifying low-energy sequences and amino acid side-chain conformations on a fixed protein backbone, and more recent work has demonstrated the substantial benefit of modeling protein backbone flexibility and the number of techniques for sampling different backbone conformations has grown dramatically. There are now more options for incorporating backbone flexibility into modeling and design than ever before; however, this comes with the cost of uncertainty in deciding which method should be used for a given application. It is likely that the types of backbone motions that work best in one context might not be the best for another context. For example, peptides bound to PDZ domains adopt β-strand conformations and therefore may exhibit different backbone dynamics compared to α-helical peptides bound to Bcl-2 family proteins. Understanding which types of motions work best in specific contexts will require computational benchmarks that evaluate the accuracy of many different flexible backbone methods using standardized datasets [56].

Modeling backbone flexibility can be helpful for predicting and designing peptide specificity, but there is still much room for improvement in how backbone flexibility is incorporated into computational protein design methods. In the conformational ensemble-based approaches described here, two separate simulations are performed: one where backbone moves are applied to generate an ensemble, and one where computational protein design is performed to identify low-energy sequences. In this scheme, changes in backbone conformation and changes in amino acid sequence occur independently of each other. However, in reality, these changes are coupled because mutations will alter a protein's backbone conformation. Although conformational ensembles are useful for increasing the diversity of sequences sampled by computational protein design, they may provide inaccurate predictions in cases where sequence-dependent changes to the backbone conformation occur. This may explain why predicting the specificity of PDZ domains with 4–10 mutations using the wild-type structure to generate ensembles was the most challenging test case Smith and Kortemme [15]. To overcome this challenge, backbone moves could be coupled to changes in side-chain conformation and changes in amino acid sequence. This type of "coupled move" has already been applied to couple backrub motions with amino acid mutations and this approach enabled the successful prediction of mutations that alter protein–ligand specificity [24]. Coupling other types of backbone motions to changes in side-chain conformation and amino acid sequence may further improve accuracy in predicting and designing peptide specificity in future studies.

References

1. Röthlisberger D, Khersonsky O, Wollacott AM, Jiang L, DeChancie J et al (2008) Kemp elimination catalysts by computational enzyme design. Nature 453:190–195. doi:10.1038/nature06879

2. Jiang L, Althoff EA, Clemente FR, Doyle L, Röthlisberger D et al (2008) De novo computational design of retro-aldol enzymes. Science 319:1387–1391. doi:10.1126/science.1152692

3. Fleishman SJ, Whitehead TA, Ekiert DC, Dreyfus C, Corn JE et al (2011) Computational design of proteins targeting the conserved stem region of influenza hemagglutinin. Science 332:816–821. doi:10.1126/science.1202617

4. Dahiyat BI (1997) De novo protein design: fully automated sequence selection. Science 278:82–87. doi:10.1126/science.278.5335.82

5. Ponder JW, Richards FM (1987) Tertiary templates for proteins. J Mol Biol 193:775–791. doi:10.1016/0022-2836(87)90358-5

6. Desmet J, Maeyer MD, Hazes B, Lasters I (1992) The dead-end elimination theorem and its use in protein side-chain positioning. Nature 356:539–542. doi:10.1038/356539a0

7. Kuhlman B, Baker D (2000) Native protein sequences are close to optimal for their structures. Proc Natl Acad Sci U S A 97:10383–10388. doi:10.1073/pnas.97.19.10383

8. Kuhlman B, Dantas G, Ireton GC, Varani G, Stoddard BL et al (2003) Design of a novel globular protein fold with atomic-level accuracy. Science 302:1364–1368. doi:10.1126/science.1089427

9. Voigt CA, Gordon DB, Mayo SL (2000) Trading accuracy for speed: a quantitative comparison of search algorithms in protein sequence design. J Mol Biol 299:789–803. doi:10.1006/jmbi.2000.3758

10. Gordon DB, Mayo SL (1999) Branch-and-Terminate: a combinatorial optimization algorithm for protein design. Structure 7:1089–1098. doi:10.1016/S0969-2126(99)80176-2

11. Ollikainen N, Sentovich E, Coelho C, Kuehlmann A, Kortemme T (2009) SAT-based protein design. Proc of ICCAD. pp 128–135

12. Saunders CT, Baker D (2005) Recapitulation of protein family divergence using flexible backbone protein design. J Mol Biol 346:631–644. doi:10.1016/j.jmb.2004.11.062

13. Castagnoli L (2004) Selectivity and promiscuity in the interaction network mediated by protein recognition modules. FEBS Lett 567:74–79. doi:10.1016/S0014-5793(04)00491-0

14. Münz M, Hein J, Biggin PC (2012) The role of flexibility and conformational selection in the binding promiscuity of PDZ domains. PLoS Comput Biol 8:e1002749. doi:10.1371/journal.pcbi.1002749

15. Smith CA, Kortemme T (2010) Structure-based prediction of the peptide sequence space recognized by natural and synthetic PDZ domains. J Mol Biol 402:460–474. doi:10.1016/j.jmb.2010.07.032

16. Ollikainen N, Kortemme T (2013) Computational protein design quantifies structural constraints on amino acid covariation. PLoS Comput Biol 9:e1003313. doi:10.1371/journal.pcbi.1003313

17. Davis IW, Arendall WB III, Richardson DC, Richardson JS (2006) The backrub motion: how protein backbone shrugs when a sidechain dances. Structure 14:265–274. doi:10.1016/j.str.2005.10.007

18. Smith CA, Kortemme T (2008) Backrub-like backbone simulation recapitulates natural protein conformational variability and improves mutant side-chain prediction. J Mol Biol 380:742–756. doi:10.1016/j.jmb.2008.05.023

19. Georgiev I, Keedy D, Richardson JS, Richardson DC, Donald BR (2008) Algorithm for backrub motions in protein design. Bioinformatics 24:i196–i204. doi:10.1093/bioinformatics/btn169

20. Leaver-Fay A, Tyka M, Lewis SM, Lange OF, Thompson J et al (2011) Rosetta3: an object-oriented software suite for the simulation and design of macromolecules. Methods Enzymol 487:545–574. doi:10.1016/B978-0-12-381270-4.00019-6

21. Friedland GD, Linares AJ, Smith CA, Kortemme T (2008) A simple model of backbone flexibility improves modeling of side-chain conformational variability. J Mol Biol 380:757–774. doi:10.1016/j.jmb.2008.05.006

22. Ollikainen N, Smith CA, Fraser JS, Kortemme T (2013) Flexible backbone sampling methods to model and design protein alternative conformations. Methods Enzymol 523:61–85. doi:10.1016/B978-0-12-394292-0.00004-7

23. Jackson EL, Ollikainen N, Covert AW III, Kortemme T, Wilke CO (2013) Amino-acid site variability among natural and designed proteins. PeerJ 1:e211. doi:10.7717/peerj.211

24. Ollikainen N, de Jong RM, Kortemme T (2015) Coupling protein side-chain and backbone flexibility improves the re-design of protein-ligand specificity. PLoS Comput Biol 11:e1004335. doi:10.1371/journal.pcbi.1004335

25. Mandell DJ, Coutsias EA, Kortemme T (2009) Sub-angstrom accuracy in protein loop reconstruction by robotics-inspired conformational sampling. Nat Methods 6:551–552. doi:10.1038/nmeth0809-551

26. Stein A, Kortemme T (2013) Improvements to robotics-inspired conformational sampling in rosetta. PLoS One 8:e63090. doi:10.1371/journal.pone.0063090

27. Coutsias EA, Seok C, Jacobson MP, Dill KA (2004) A kinematic view of loop closure. J Comput Chem 25:510–528. doi:10.1002/jcc.10416

28. Rohl CA, Strauss CEM, Misura KMS, Baker D (2004) Protein structure prediction using rosetta. Methods Enzymol 383:66–93. doi:10.1016/S0076-6879(04)83004-0

29. Simons KT, Kooperberg C, Huang E, Baker D (1997) Assembly of protein tertiary structures from fragments with similar local sequences using simulated annealing and bayesian scoring functions. J Mol Biol 268:209–225. doi:10.1006/jmbi.1997.0959

30. Tyka MD, Keedy DA, André I, DiMaio F, Song Y et al (2011) Alternate states of proteins revealed by detailed energy landscape mapping. J Mol Biol 405:607–618. doi:10.1016/j.jmb.2010.11.008

31. Kellogg EH, Leaver-Fay A, Baker D (2011) Role of conformational sampling in computing mutation-induced changes in protein structure and stability. Proteins 79:830–838. doi:10.1002/prot.22921

32. Humphris EL, Kortemme T (2008) Prediction of protein-protein interface sequence diversity using flexible backbone computational protein design. Structure 16:1777–1788. doi:10.1016/j.str.2008.09.012

33. Smith CA, Kortemme T (2011) Predicting the tolerated sequences for proteins and protein interfaces using RosettaBackrub flexible backbone design. PLoS One 6:e20451. doi:10.1371/journal.pone.0020451

34. Lauck F, Smith CA, Friedland GF, Humphris EL, Kortemme T (2010) RosettaBackrub—a web server for flexible backbone protein structure modeling and design. Nucleic Acids Res 38:W569–W575. doi:10.1093/nar/gkq369

35. Berman HM, Westbrook J, Feng Z, Gilliland G, Bhat TN et al (2000) The protein data bank. Nucleic Acids Res 28:235–242. doi:10.1093/nar/28.1.235

36. Tonikian R, Zhang Y, Sazinsky SL, Currell B, Yeh J-H et al (2008) A specificity map for the PDZ domain family. PLoS Biol 6:e239. doi:10.1371/journal.pbio.0060239

37. Ernst A, Sazinsky SL, Hui S, Currell B, Dharsee M et al (2009) Rapid evolution of functional complexity in a domain family. Sci Signal 2:ra50–ra50. doi:10.1126/scisignal.2000416

38. King CA, Bradley P (2010) Structure-based prediction of protein-peptide specificity in rosetta. Proteins 78:3437–3449. doi:10.1002/prot.22851

39. Fiser A, Sali A (2003) Modeller: generation and refinement of homology-based protein structure models. Methods Enzymol 374:461–491. doi:10.1016/S0076-6879(03)74020-8

40. Raveh B, London N, Schueler-Furman O (2010) Sub-angstrom modeling of complexes between flexible peptides and globular proteins. Proteins 78:2029–2040. doi:10.1002/prot.22716

41. London N, Lamphear CL, Hougland JL, Fierke CA, Schueler-Furman O (2011) Identification of a novel class of farnesylation targets by structure-based modeling of binding specificity. PLoS Comput Biol 7:e1002170. doi:10.1371/journal.pcbi.1002170

42. London N, Gullá S, Keating AE, Schueler-Furman O (2012) In Silicoand in VitroElucidation of BH3 binding specificity toward Bcl-2. Biochemistry 51:5841–5850. doi:10.1021/bi3003567

43. Raveh B, London N, Zimmerman L, Schueler-Furman O (2011) Rosetta FlexPepDock ab-initio: simultaneous folding, docking and refinement of peptides onto their receptors. PLoS One 6:e18934. doi:10.1371/journal.pone.0018934

44. London N, Raveh B, Cohen E, Fathi G, Schueler-Furman O (2011) Rosetta FlexPepDock web server—high resolution modeling of peptide-protein interactions. Nucleic Acids Res 39:W249–W253. doi:10.1093/nar/gkr431

45. Kortemme T, Joachimiak LA, Bullock AN, Schuler AD, Stoddard BL et al (2004) Computational redesign of protein-protein interaction specificity. Nat Struct Mol Biol 11:371–379. doi:10.1038/nsmb749

46. Joachimiak LA, Kortemme T, Stoddard BL, Baker D (2006) Computational design of a new hydrogen bond network and at least a 300-fold specificity switch at a protein–protein interface. J Mol Biol 361:195–208. doi:10.1016/j.jmb.2006.05.022

47. Kapp GT, Liu S, Stein A et al (2012) Control of protein signaling using a computationally designed GTPase/GEF orthogonal pair. Proc Natl Acad Sci U S A 109:5277–5282. doi:10.1073/pnas.1114487109

48. Sammond DW, Eletr ZM, Purbeck C, Kuhlman B (2010) Computational design of second-site suppressor mutations at protein-protein interfaces. Proteins 78:1055–1065. doi:10.1002/prot.22631/full

49. Melero C, Ollikainen N, Harwood I, Karpiak J, Kortemme T (2014) Quantification of the transferability of a designed protein specificity switch reveals extensive epistasis in molecular recognition. Proc Natl Acad Sci U S A 111:15426–15431. doi:10.1073/pnas.1410624111

50. Roberts KE, Cushing PR, Boisguerin P, Madden DR, Donald BR (2012) Computational design of a PDZ domain peptide inhibitor that rescues CFTR activity. PLoS Comput Biol 8:e1002477. doi:10.1371/journal.pcbi.1002477

51. Georgiev I, Lilien RH, Donald BR (2008) The minimized dead-end elimination criterion and its application to protein redesign in a hybrid scoring and search algorithm for computing partition functions over molecular ensembles. J Comput Chem 29:1527–1542. doi:10.1002/jcc.20909

52. Leach AR, Lemon AP (1998) Exploring the conformational space of protein side chains using dead-end elimination and the A* algorithm. Proteins 33:227–239. doi:10.1002/(SICI)1097-0134(19981101)33:2<227::AID-PROT7>3.0.CO;2-F

53. Pedemonte N, Lukacs GL, Du K, Caci E, Zegarra-Moran O et al (2005) Small-molecule correctors of defective DeltaF508-CFTR cellular processing identified by high-throughput screening. J Clin Investig 115:2564–2571. doi:10.1172/JCI24898

54. Fu X, Apgar JR, Keating AE (2007) Modeling backbone flexibility to achieve sequence diversity: the design of novel α-helical ligands for Bcl-xL. J Mol Biol 371:1099–1117. doi:10.1016/j.jmb.2007.04.069

55. Emberly EG, Mukhopadhyay R, Wingreen NS, Tang C (2003) Flexibility of α-helices: results of a statistical analysis of database protein structures. J Mol Biol 327:229–237. doi:10.1016/S0022-2836(03)00097-4

56. ÓConchúir S, Barlow KA, Pache RA, Ollikainen N, Kundert K et al (2015) A web resource for standardized benchmark datasets, metrics, and rosetta protocols for macromolecular modeling and design. PLoS One 10:e0130433. doi:10.1371/journal.pone.0130433

Simplifying the Design of Protein-Peptide Interaction Specificity with Sequence-Based Representations of Atomistic Models

Fan Zheng and Gevorg Grigoryan

Abstract

Computationally designed peptides targeting protein-protein interaction interfaces are of great interest as reagents for biological research and potential therapeutics. In recent years, it has been shown that detailed structure-based calculations can, in favorable cases, describe relevant determinants of protein-peptide recognition. Yet, despite large increases in available computing power, such accurate modeling of the binding reaction is still largely outside the realm of protein design. The chief limitation is in the large sequence spaces generally involved in protein design problems, such that it is typically infeasible to apply expensive modeling techniques to score each sequence. Toward addressing this issue, we have previously shown that by explicitly evaluating the scores of a relatively small number of sequences, it is possible to synthesize a direct mapping between sequences and scores, such that the entire sequence space can be analyzed extremely rapidly. The associated method, called Cluster Expansion, has been used in a number of studies to design binding affinity and specificity. In this chapter, we provide instructions and guidance for applying this technique in the context of designing protein-peptide interactions to enable the use of more detailed and expensive scoring approaches than is typically possible.

Key words Interaction specificity, Computational protein design, PDZ-peptide interactions, Cluster expansion, Flexible peptide docking

1 Introduction

It is estimated that a large fraction of cellular protein-protein interactions is mediated by peptide-recognition domains (PRDs) interacting with short amino-acid stretches on partner proteins [1]. Many families of PRDs are known [2, 3], with domains belonging to the same family closely related in sequence and structure but often with divergent functions. Selective inhibition of PRD-peptide interactions by means of designed reagents (e.g., inhibitor peptides) is an attractive strategy for the targeted functional modulation of cellular processes [4]. However, the achievement of selectivity in such systems—i.e., interaction with the

Ora Schueler-Furman and Nir London (eds.), *Modeling Peptide-Protein Interactions: Methods and Protocols*, Methods in Molecular Biology, vol. 1561, DOI 10.1007/978-1-4939-6798-8_11, © Springer Science+Business Media LLC 2017

desired PRD (target) and not similar domains of divergent function (competitors)—appears to be a fundamental challenge, due to the similarity shared by PRD family members.

The experimental engineering of PRD-targeting peptides is complicated in several ways: (1) for each considered peptide sequence, the binding to multiple PRDs (the target and competitors) should ideally be evaluated; (2) selective sequences are likely much rarer than simply those with appreciable binding affinities to the target, so the sequence space should be investigated rigorously; and (3) high-throughput experimental approaches detect positive interactions, but often cannot confirm noninteractions. For these reasons, it is highly desirable to have robust computational means of designing selective PRD-peptide recognition, which, in turn, requires effective models for quantifying PRD-peptide binding.

A range of computational methods for predicting PRD-peptide binding strengths has been proposed in recent years: reduced models emergent from training on high-throughput experimental data [5, 6], methods based on structural sampling and energy calculations [7–13], or hybrid approaches [14, 15]. Structure-based methods are particularly attractive due to their potential to generalize across different PRDs. However, the use of detailed atomistic modeling in designing specific recognition is severely limited by the need to consider a large number of sequence candidates. To mitigate this problem, we have previously suggested that when using a computationally expensive method of sequence evaluation in protein design, it is not necessary to repeat the calculation from scratch for every sequence considered. Instead, upon performing the calculation on a relatively small number of sequences, it is possible to effectively parameterize the computed property directly as a simple function of sequence, making further estimations of the property for new sequences many orders of magnitude faster. The associated method, called cluster expansion (CE) [16, 17], has been used in a number of design studies [18–21]. In the case of PRD-peptide recognition, the CE framework states that for a fixed PRD the binding score (no matter how it is calculated) is a function of the peptide sequence only and, in agreement with physico-chemical intuition, should be the result of contributions from individual amino acids and amino-acid groups. Thus, if $E(\vec{\sigma})$ represents the predicted binding score for a peptide sequence $\vec{\sigma}$ and a given domain, it can be expressed as a sum of contributions from constellations of amino acids at clusters of peptide positions (i.e., cluster functions or CFs):

$$E(\vec{\sigma}) = C + \sum_{\substack{i=1 \\ \sigma_i \neq \rho_i}}^{L} f_i(\sigma_i) + \sum_{\substack{i=1 \\ \sigma_i \neq \rho_i}}^{L-1} \sum_{\substack{j=i+1 \\ \sigma_j \neq \rho_j}}^{L} f_{ij}(\sigma_i, \sigma_j) + \ldots$$

where L is peptide length, $\vec{\rho}$ is a reference peptide sequence, σ_i and ρ_i are the amino acids in the ith position of $\vec{\sigma}$ and $\vec{\rho}$, respectively. $f_i(\sigma_i)$ is the point CF capturing the effective contribution of amino acid σ_i at position i, $f_{ij}(\sigma_i, \sigma_j)$ is the pair CF capturing the additional joined contribution due to having σ_i at position i and σ_j at position j simultaneously, and all higher-order CFs (similarly defined) are also present in the summation. The significance of the reference sequence is that for a given $\vec{\sigma}$, the expression only sums over those clusters (i.e., combinations of positions) for which all of the occupying amino acids are different from the corresponding ones in $\vec{\rho}$. Thus, C represents the binding score of $\vec{\rho}$ (the reference CF), whereas the remaining terms indicate the additional contributions due to the amino acids in $\vec{\sigma}$ differing from those in $\vec{\rho}$. Although the expansion is exact when all CFs are included, the importance of higher-order terms is expected to drop dramatically for most physical systems, so that an appropriately truncated CE, with only selected lower-order CFs (e.g., all up to pair), can effectively balance speed of calculation and accuracy. The particular set of clusters to include, and the optimal values of involved CFs, can be determined after the full structure-based calculation is explicitly run on a set of training sequences to derive their scores, by means of data fitting methods. The result of the CE framework is a simple expression that drastically speeds up the further evaluation of sequences, while retaining close agreement with the underlying explicit structure-based method. This, in turn, enables the efficient consideration of the specificity design problem, identifying optimal tradeoff between the affinity to the targeted PRD and the selectivity against competitors. Further, the linear dependence of CE-computed scores on sequence variables makes powerful algorithms like integer linear programming (ILP) applicable toward sequence optimization [18], further simplifying the search over large sequence spaces using potentially complex objective functions. While the details of the framework have been published and reviewed previously [22], below we outline practical aspects of applying it toward the design of selective PRD-peptide recognition.

2 Materials

The following resources are needed to apply our framework:

1. A linear algebra engine (e.g., the proprietary MathWorks MATLAB or the open-source GNU Octave).

2. A structure-based simulation tool (e.g., the macromolecular modeling suite Rosetta [23]).

3. Highly desirable: access to a high-performance computing cluster that is able to perform hundreds of jobs in parallel.

3   Methods

In this section, we outline how the task of designing PRD-binding peptides can be simplified using our CE framework, enabling the application of more sophisticated (and potentially more accurate) structure-based scoring models than is otherwise easily possible. Although we will frequently refer to our experience with designing PDZ-targeting peptides [19], where Rosetta FlexPepDock ab initio [24] was used as the underlying scoring method, the presented guidelines should be applicable to a wide variety of complex simulation techniques and model systems. In what follows, we assume that the goal is to design a peptide that interacts with a particular targeted domain T, and avoids interactions with n undesired domains U_1 through U_n. It is further assumed that a structure-based scoring method (SSM) exists, one that is possibly quite complex, that is known (or expected) to perform reasonably well at quantifying interactions between T, U_1 ... U_n and arbitrary peptides. The existence of such an SSM generally means that experimental structures (or high-quality homology models, *see* refs. 19, 25 for more info) must be available for T, U_1 ... Un (or sufficiently close homologues). A peptide with optimal affinity and selectivity for T is then desired, but the direct application of the SSM toward design is impossible/difficult in light of its computational complexity and the sequence space to consider.

3.1   Define a Set of Allowed Amino Acids for Each Peptide Position: i.e., the Design Alphabet

Much of the subsequent effort is reduced when fewer amino-acid options are allowed. It is thus practical to reduce choices based on any strong positional preferences in peptides known to bind the PRD family of interest or T specifically. However, unnecessarily limiting the design alphabet can make it difficult to design specificity, so only preferences with strong experimental evidence should be considered. In our PDZ study, we used previously reported phage display data [26] to define a "permissive" sequence space with 2–8 amino acids allowed at six peptide positions. This was sufficient to achieve the desired level of selectivity in our application [19], but other scenarios may require a more liberal choice of amino-acid options.

3.2   Frame the CE Model

As mentioned earlier, a truncated CE model (e.g., one with only constant, self, and pair CFs) can drastically reduce computational complexity while retaining accuracy relative to the underlying SSM being expanded. For a more succinct model, we can further restrict pair clusters to those pairs of peptide positions that are likely to influence each other's amino-acid identity—e.g., by interacting directly or through a common site on the PRD. For example, in the case of designing PDZ-binding peptides, since the peptide binds as a β-strand, we only considered $i-i+2$ pairs that map onto the same side of the binding interface [19]. Each cluster to be included gives rise to a numbers CFs, each representing a unique combination of non-reference amino acids at the corresponding positions. The total number of CFs is thus related to the size of the design alphabet. For example, if N_1 point and N_2 pair clusters are included, and n amino acids are allowed per site, then the number of CFs is $1 + (n-1) \cdot N_1 + (n-1)^2 \cdot N_2$. Note that not all of these considered CFs will end up in the final CE model (see below), so this initial set is referred to as candidate CFs.

3.3 Generate a CE Training Set by Randomly Drawing Sequences from the Design Alphabet

Each sequence in the training set will be subjected to the SSM calculation, so to keep the time to train a CE model manageable, it is important to keep the training set only as large as necessary. We recommend building a training set that (1) has at least twice as many sequences as the number of candidate CFs (N_{CF}^0), and (2) reasonably covers each of the candidate CFs (e.g., includes at least three instances of each CF). Thus, a good strategy is to first generate a set of $2\,N_{CF}^0$ random sequences, then check for the coverage of each candidate CF, and add sequences containing the under-represented CFs (but otherwise random) as necessary.

3.4 Run the SSM for All Sequences in the Training Set in Complex with T, U_l ... U_n

Extract the final binding score from each simulation. It is important to select a predictive simulation technique and optimize any parameters/options in the protocol for accuracy on the PRD of interest (see detailed discussion in [25]). Though tradeoffs between speed and accuracy are still relevant here (as in typical design scoring functions), it is important to remember that the SSM need only to be fast enough to enable the scoring of the training set in an acceptable amount of time. Additionally, scoring the training set is embarrassingly parallel, as every sequence can be treated independently, so the availability of a computer cluster drastically simplifies this task and further expands the possible field of practically admissible SSMs. For example, in the Rosetta FlexPepDock ab initio method of domain-peptide interaction modeling, the accuracy is associated with the degree of sampling [24], which is proportional to the invested running time. In our PDZ-peptide study [19], the preferred protocol took approximately 400 CPU hours for each PRD-peptide—this produced an acceptable level of accuracy, but the running time was far beyond what would be feasible in a design calculation. On the other hand, given that our CE training set was on the order of ~100 sequences, and the availability of a 1000-core cluster, the method was perfectly suitable for our CE-based approach.

3.5 Train the CE Model to Find Optimal Values for Each CF

Given a training set of sequences with precomputed SSM scores, a CE model can be easily derived for each domain using the freely available package CLEVER—a rigorous statistical framework we previously described [21, 22]. To demonstrate the training procedure transparently, here we describe a simplified implementation within a linear algebra engine (e.g., MATLAB or Octave), which we have found to work well in PRD-peptide binding designs (the advantages of statistical rigor offered by CLEVER become evident in more complex cases with larger numbers of candidate CFs and possibly higher-order CFs). First, create an $m \times n$ matrix M $(m > n)$, where m is the number of training sequences, and n is the number of candidate CFs, such that $M(i, j)$ is 1 if the ith sequence contains the amino acids specified by the jth CF at the corresponding positions, and 0 otherwise (note, in systems with symmetry it is possible for a single CF to be present within a sequence more than once, in which case $M(i, j)$ can take on integer values above 1). Create also an $m \times 1$ vector E, in which the ith element is the SSM score of the ith sequence calculated in Subheading 3.4. Optimal CF values can be attained by finding an $n \times 1$ vector b that minimizes the mean square difference between $\hat{E} = M \cdot b$ (CE-predicted scores) and E, where the jth element in b represents the value of the jth CF (we refer to this as training the CE model). This least-square solution can be easily calculated using the method of pseudo-inverse, as $\left(M^{\mathrm{T}}M\right)^{-1} M^{\mathrm{T}} E$. In MATLAB or Octave, this is simply:

```
b = (M'*M)^(-1)*M'*E;
```

(also, function calls *regress(E, M)* and *ols(E, M)* in Matlab and Octave, respectively, perform analogous calculations more efficiently for large matrices). To reduce overtraining, rather than including all candidate CFs into M at once, we recommend the following previously described procedure that selects a subset of the most statistically justifiable CFs [16]. First train a CE model with all candidate CFs (reference, point, and pair), the all-inclusive model, which is likely over-trained. Then train another model only including the reference and point CFs—the current model. Next, consider pair CFs, one at a time, in the decreasing order of their magnitudes in the all-inclusive model. For a given pair CF, train a new model in which the CF is added to the current model, and compare the cross-validation root-mean square errors (CV-RMS) between these two models. CV-RMS is the average error of predicting the score of each sequence when it is left out of the training set. If the new model has a lower CV-RMS, then update the current model to include the pair CF; otherwise, discard the pair CF. Move onto the next candidate pair CF until all are exhausted. Importantly, CV-RMS can be efficiently computed in a closed form:

$$\sqrt{\frac{1}{n}\sum_{i=1}^{n}\left(\frac{E_i - \hat{E}_i}{1 - M_i\left(M^{\mathrm{T}}M\right)^{-1} M_i^{\mathrm{T}}}\right)^2}$$

where M_i represents the ith row of matrix M, and E_i and $\hat{E}_i$ are the ith elements of vectors E and $\hat{E}$, respectively. In MATLAB or Octave, this corresponds to the following expression:

```
sqrt(sum(((E-M*b)./(1-sum(M.*(M*((M'
*M)^(-1))'),2))).^2)/length(E))
```

3.6 Randomly Generate a Test Set Containing Sequences Not in the Training Set

Again, we recommend drawing from the design alphabet randomly. This set need only be sufficiently large to enable a reliable estimation of true CE error. We thus recommend for the test set to be 2–5 times smaller than the training set. Run the same SSM protocol for these sequences (for each PRD). Estimate true CE error as the root-mean-square difference between the SSM and CE-predicted scores (test-set RMS). In an acceptable CE model this error, which is usually slightly higher than CV-RMSD estimated from the training set, should be at least lower than the typical score differences that discriminate between binders and nonbinders (Fig. 1a). Importantly, if the underlying SSM is stochastic, one should generally expect the CE error to be bound from below by the associated random error in the SSM. On the other hand, we have previously observed CE to reduce the effect of noise in the SSM due implicit averaging in the training procedure (Fig. 1b) [19]. Though we do not expect this to be the case with PRD-peptide systems, if the estimated CE error is nevertheless deemed too high, the following can be attempted to improve it: (1) consider whether important clusters are missing from the list of candidates and add them, repeating the training procedure; consider adding select higher-order CFs (e.g., triplet CFs); (2) if many CFs you believe to be important are discarded in the training

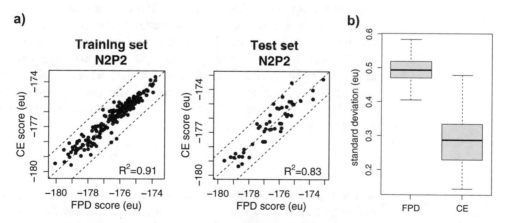

Fig. 1 CE model robustly estimates results of the SSM. Here, we show an example of a CE model from our previous study [19], which estimates the scores of Rosetta FlexPepDock ab initio for NHERF-2 PDZ2 with training and test peptide sequences. (**a**) CE predictions correlate well with Rosetta scores for peptide sequences in both training and test sets. For nearly all sequences, the prediction error is below 1.5 Rosetta energy units (eu), which is the typical score difference that discriminated between binder and nonbinder peptides for a given PDZ domain [19]. (**b**) CE reduces the effect of noise in the SSM calculations. The Rosetta scores of the sequences in the training set were perturbed by a normally distributed noise with zero mean and standard deviation of 0.5 eu, and the perturbed scores were used to train a new CE model. The standard deviation of CE-computed scores from 100 independent trials is significant lower than that of the Rosetta scores used to derive to CE model. Figures are adapted from Fig. 3b and Fig. S5 in ref. 19

procedure, you may need to increase the number of training sequences to increase statistical strength; (3) if large prediction errors are associated with specific sequence biases (e.g., a particular combination of amino acids at specific sites) consider either adding CFs to describe such scenarios more specifically (e.g., triplet CFs, *see* ref. 16) or eliminate such sequences from estimating error if they are not of interest in the context of the design problem (e.g., score poorly on the target). A more principled statistical approach for choosing appropriate CFs and training set size to maximize the CE accuracy has been described previously [22].

3.7 Design Optimal Peptide Sequences

With an established CE model, evaluating the affinity of a PRD-peptide interaction typically takes less than 1 μs—many orders of magnitude faster than a typical underlying SSM approach. Thus, if the sequence space to consider is only moderately large (i.e., $\leq 10^{10}$ sequences), it is straightforward to simply enumerate all possible peptides against all PRDs of interest (T, U_1 ... Un), deriving predicted target affinity and selectivity scores for each. On the other hand, for larger sequence spaces, CE enables the application of Integer Linear Programming (ILP) for rapid optimization of affinity for T, with constraints on specificity against U_1 ... Un [27]. To this end, a CE model can be represented as a graph, where point clusters are vertices and pair clusters are edges. Specifically, the vertex set V is a union of subsets $V_1 \cup ... \cup V_p$, where set V_i contains vertices associated with amino-acid choices at position i. Each vertex u is assigned a weight E_{uu}, which is the value of the corresponding point CF (or zero if no corresponding point CF exists). Similarly, pair CFs are represented by the edges set D, where an edge exists between two vertices u and v if $u \in V_i$ and $v \in V_j, i \neq j$ and there is a pair CF associated with the amino-acid choices implied by u and v. The edge is assigned the weight E_{uv} according to the value of the corresponding pair CF. Then the binding energy of a designed sequence toward the targeted PRD is expressed as:

$$\varepsilon^{T} = \sum_{u \in V} E_{uu}^{T} x_{uu} + \sum_{u,v \in D} E_{uv}^{T} x_{uv}$$

where x_{uu} and x_{uv} are binary variables (0, 1) determined by vertices and edges (point and pair CFs) involved in the specific binding sequence. The goal of design can then be expressed as optimizing the affinity for the target domain T, under the constraint of selectivity against all undesired competitors $U_1, ... , U_k$, with the following simple ILP:

$$\sum_{u \in V_j} x_{uu} = 1 \text{ for } j = 1,...,p$$

$$\sum_{u \in V_j} x_{uv} = x_{vv}, \text{ for } j = 1,...,p \text{ and } v \in V \setminus V_j$$

$$\varepsilon^{U_1} - \varepsilon^{T} > \Delta, \text{ where } \varepsilon^{U_1} = \sum_{u \in V} E_{uu}^{U_1} x_{uu} + \sum_{u,v \in D} E_{uv}^{U_1} x_{uv}$$

...

$$\varepsilon^{U_k} - \varepsilon^{T} > \Delta, \text{ where } \varepsilon^{U_k} = \sum_{u \in V} E_{uu}^{U_k} x_{uu} + \sum_{u,v \in D} E_{uv}^{U_k} x_{uv}$$

$$x_{uu}, x_{uv} \in \{0,1\}$$

Here Δ indicates the required energy gap between the target domain and any of the competitors (here lower scores are assumed to be more favorable, so a positive Δ means that the design peptide is required to bind T better than any of $U_1 \dots U_n$ by at least Δ score units). We have previously described this framework, called CLASSY, having applied it do design selective coiled-coil inhibitors [18, 21].

Regardless of whether the design sequence space is enumerated explicitly, or via CLASSY, the goal is to identify sequences on the pareto-optimal frontier of the affinity/selectivity space—i.e., all those sequences that cannot be simultaneously improved in both affinity and selectivity. This is done either by solving the above ILP for progressively increasing Δ (as in [18]) or by explicitly identifying the frontier having enumerated all sequences (as in [19]). Conveniently, all other sequences can be considered inferior with respect to the pareto-optimal ones, as the former can be improved in both affinity and selectivity simultaneously and thus need not be considered. Therefore, the entire sequence landscape is reduced to (usually) a handful of sequences than can be manually inspected. Further, pareto-optimal sequences outline the minimal cost in affinity that is required for any incremental gain in specificity—i.e., they make optimal affinity/specificity tradeoffs.

3.8 Select Final Design Sequences for Experimental Characterization

Run the original SSM protocol to rescore each of the pareto-optimal sequences. First, this verifies that CE error is not abnormally high around the pareto frontier. Second, this generates structural models of each candidate design, which should be manually inspected to verify that the predicted affinity and selectivity meet with biophysical/structural intuition. If CE error from training is high, it is advisable to rescore by SSM all/some sequences within a certain distance threshold from the pareto-optimal frontier. For further confirmation, one may apply more detailed simulations (e.g., explicit-solvent molecular dynamics) to generate further insight in support or against specific designs. The relevant timescales of MD may vary depending on systems, but 10–100 ns simulation will provide some information of the local stability of the peptide in the binding site. Final sequences for experimental characterization should be chosen by considering all of the above evidence. Some additional issues are discussed in **Note 2**.

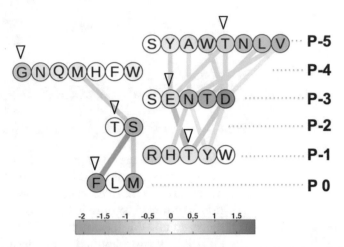

Fig. 2 Graph representation of a CE model. See the detailed description in **Note 1**. The *arrows* indicate a choice of a high affinity peptide alternative to using the left-most amino acids on every position (i.e., the one most favorable by point CF). The figure is adapted from Fig. 3A in ref. 19, demonstrating the model for NHERF-2 PDZ2

4   Notes

1. Graph representation of CE models can help interpret optimal design selections. An example from our PDZ study [19] is shown in Fig. 2, where candidate amino acids at each peptide position (P_{-5}–P_0) are shown as vertices ordered (from left to right) by their increasing point CF contribution, and edges represent pair CFs. The colors of the vertices and edges indicate the sign and the magnitude of each CF value. Although the graph is complicated by the choice of reference amino acids (i.e., some physicochemical interactions are not evident as edges, such as those involving the reference amino acids), one can still infer that the favorable binding to the targeted domain can be achieved either by selecting amino acids with most favorable point CFs, or by balancing self and pair contributions where strongly favorable edges exist. This enables the "leeway" needed to design for selectivity. In our example, two of the sequences emergent from our specificity design framework (SGSTRF and TGETTF) corresponded to these two strategies, respectively, with both exhibiting low-micromolar affinities to the targeted domain [19].

2. The number of candidate sequences on the pareto-optimal front of the affinity-specificity space depends on the specific problem and the size of design alphabet. When it is difficult to experimentally characterize all candidates, carefully selecting the sequences to test can maximize the interpretability of results. We recommend choosing sequences with a range of

predicted specificities (i.e., points spanning the pareto-optimal frontier), as this may provide insight into the accuracy of the predicted affinity/specificity tradeoffs. Also, when possible, choosing sequences representing different structural strategies toward achieving affinity or specificity is advisable.

References

1. Neduva V, Linding R, Su-Angrand I, Stark A, de Masi F, Gibson TJ, Lewis J, Serrano L, Russell RB (2005) Systematic discovery of new recognition peptides mediating protein interaction networks. PLoS Biol 3(12):2090–2099

2. Pawson T, Nash P (2003) Assembly of cell regulatory systems through protein interaction domains. Science 300(5618):445–452

3. Kuriyan J, Cowburn D (1997) Modular peptide recognition domains in eukaryotic signaling. Annu Rev Biophys Biomol Struct 26:259–288

4. Vanhee P, van der Sloot AM, Verschueren E, Serrano L, Rousseau F, Schymkowitz J (2011) Computational design of peptide ligands. Trends Biotechnol 29(5):231–239

5. Chen JR, Chang BH, Allen JE, Stiffler MA, MacBeath G (2008) Predicting PDZ domain-peptide interactions from primary sequences. Nat Biotechnol 26(9):1041–1045

6. Kamisetty H, Ghosh B, Langmead CJ, Bailey-Kellogg C (2014) Learning sequence determinants of protein:protein interaction specificity with sparse graphical models. Res Comput Mol Biol 8394:129–143

7. Gan W, Roux B (2009) Binding specificity of SH2 domains: insight from free energy simulations. Proteins 74(4):996–1007

8. Smith CA, Kortemme T (2010) Structure-based prediction of the peptide sequence space recognized by natural and synthetic PDZ domains. J Mol Biol 402(2):460–474

9. King CA, Bradley P (2010) Structure-based prediction of protein-peptide specificity in Rosetta. Proteins 78(16):3437–3449

10. London N, Lamphear CL, Hougland JL, Fierke CA, Schueler-Furman O (2011) Identification of a novel class of farnesylation targets by structure-based modeling of binding specificity. PLoS Comput Biol 7(10):e1002170

11. London N, Gulla S, Keating AE, Schueler-Furman O (2012) In silico and in vitro elucidation of BH3 binding specificity toward Bcl-2. Biochemistry 51(29):5841–5850

12. Yanover C, Bradley P (2011) Large-scale characterization of peptide-MHC binding landscapes with structural simulations. Proc Natl Acad Sci U S A 108(17):6981–6986

13. Roberts KE, Cushing PR, Boisguerin P, Madden DR, Donald BR (2012) Computational design of a PDZ domain peptide inhibitor that rescues CFTR activity. PLoS Comput Biol 8(4):e1002477

14. DeBartolo J, Dutta S, Reich L, Keating AE (2012) Predictive Bcl-2 family binding models rooted in experiment or structure. J Mol Biol 422(1):124–144

15. DeBartolo J, Taipale M, Keating AE (2014) Genome-wide prediction and validation of peptides that bind human prosurvival Bcl-2 proteins. PLoS Comput Biol 10(6):e1003693

16. Grigoryan G, Zhou F, Lustig SR, Ceder G, Morgan D, Keating AE (2006) Ultra-fast evaluation of protein energies directly from sequence. PLoS Comput Biol 2(6):551–563

17. Zhou F, Grigoryan G, Lustig SR, Keating AE, Ceder G, Morgan D (2005) Coarse-graining protein energetics in sequence variables. Phys Rev Lett 95(14):148103

18. Grigoryan G, Reinke AW, Keating AE (2009) Design of protein-interaction specificity gives selective bZIP-binding peptides. Nature 458(7240):859–U852

19. Zheng F, Jewell H, Fitzpatrick J, Zhang J, Mierke DF, Grigoryan G (2015) Computational design of selective peptides to discriminate between similar PDZ domains in an oncogenic pathway. J Mol Biol 427(2):491–510

20. Negron C, Keating AE (2014) A set of computationally designed orthogonal antiparallel homodimers that expands the synthetic coiled-coil toolkit. J Am Chem Soc 136(47):16544–16556

21. Negron C, Keating AE (2013) Multistate protein design using CLEVER and CLASSY. Methods Enzymol 523:171–190

22. Hahn S, Ashenberg O, Grigoryan G, Keating AE (2010) Identifying and reducing error in cluster-expansion approximations of protein energies. J Comput Chem 31(16):2900–2914

23. Leaver-Fay A, Tyka M, Lewis SM, Lange OF, Thompson J, Jacak R, Kaufman K, Renfrew

PD, Smith CA, Sheffler W, Davis IW, Cooper S, Treuille A, Mandell DJ, Richter F, Ban YEA, Fleishman SJ, Corn JE, Kim DE, Lyskov S, Berrondo M, Mentzer S, Popovic Z, Havranek JJ, Karanicolas J, Das R, Meiler J, Kortemme T, Gray JJ, Kuhlman B, Baker D, Bradley P (2011) Rosetta3: an object-oriented software suite for the simulation and design of macromolecules. Methods Enzymol 487:545–574

24. Raveh B, London N, Zimmerman L, Schueler-Furman O (2011) Rosetta FlexPepDock ab-initio: simultaneous folding, docking and refinement of peptides onto their receptors. PLoS One 6(4):e18934

25. Zheng F, Grigoryan G (2016) Design of specific peptide-protein recognition. Methods Mol Biol 1414:249–263

26. Tonikian R, Zhang YN, Sazinsky SL, Currell B, Yeh JH, Reva B, Held HA, Appleton BA, Evangelista M, Wu Y, Xin XF, Chan AC, Seshagiri S, Lasky LA, Sander C, Boone C, Bader GD, Sidhu SS (2008) A specificity map for the PDZ domain family. PLoS Biol 6(9):2043–2059

27. Kingsford CL, Chazelle B, Singh M (2005) Solving and analyzing side-chain positioning problems using linear and integer programming. Bioinformatics 21(7):1028–1036

Chapter 12

Binding Specificity Profiles from Computational Peptide Screening

Stefan Wallin

Abstract

The computational peptide screening method is a Monte Carlo-based procedure to systematically characterize the specificity of a peptide-binding site. The method is based on a generalized-ensemble algorithm in which the peptide sequence has become a dynamic variable, i.e., molecular simulations with ordinary conformational moves are enhanced with a type of "mutational" move such that proper statistics are achieved for multiple sequences in a single run. The peptide screening method has two main steps. In the first, reference simulations of the unbound state are performed and used to parametrize a linear model of the unbound state free energy, determined by requiring that the marginal distribution of peptide sequences is approximately flat. In the second step, simulations of the bound state are performed. By using the linear model as a free energy reference point, the marginal distribution of peptide sequences becomes skewed towards sequences with higher binding free energies. From analyses of the sequences generated in the second step and their conformational ensembles, information on peptide binding specificity, relative binding affinities, and the molecular basis of specificity can be achieved. Here we demonstrate how the algorithm can be implemented and applied to determine the peptide binding specificity of a PDZ domain from the protein GRIP1.

Key words Protein–peptide interaction, Binding free energy, Affinity, PDZ domains, Monte Carlo simulations, Generalized-ensemble techniques

1 Introduction

The interactions between relatively short polypeptide segments, often found within longer regions of structural disorder or in loops in proteins, and domains with well-ordered structures, are widespread in cellular processes [1] and in signaling and regulation in particular [2]. Examples of protein domains that specialize on recognizing linear peptide sequences include well-characterized and prevalent domains such as SH2, WW, and PDZ. Peptide–protein interactions have been estimated to mediate a large fraction (up to around 40%) [3] of all protein–protein interactions in many genomes, and additional domains with peptide-binding functions are likely to be discovered [4].

Ora Schueler-Furman and Nir London (eds.), *Modeling Peptide-Protein Interactions: Methods and Protocols*, Methods in Molecular Biology, vol. 1561, DOI 10.1007/978-1-4939-6798-8_12, © Springer Science+Business Media LLC 2017

There are several reasons to explore the molecular basis of protein–peptide interactions. Because short peptide segments usually cannot assume a specific structure on their own, it is typically assumed that they undergo a disorder-order process as they bind their targets [5]. Interestingly, this coupled folding and binding process provides these protein interactions with biophysical properties that can be functionally advantageous [6, 7]. Two examples with particular relevance for regulatory processes are the ability of a single flexible peptide to bind structurally different targets, [8] and the ability of peptides to achieve highly specific but short-lived interactions [9]. Even after binding, some peptides may retain some degree of conformational heterogeneity. For example, a peptide from the C terminal region of the tumor suppressor p53 remains partially dynamic even after binding to its target [7, 10]. Conformational heterogeneity in protein–peptide interactions may be more common than previously thought and it is unclear how it impacts specificity and affinity [11, 12].

The computational peptide screening procedure [13] is an algorithm that in a single enhanced simulation generates a large number of sequences biased according to their binding free energies. More precisely, if $N(\bar{\sigma})$ denotes the number of generated sequences, then $N(\bar{\sigma}) \propto e^{-\Delta F(\bar{\sigma})/k_B T}$, where $\Delta F(\bar{\sigma})$ is the binding free energy of sequence $\bar{\sigma}$ at temperature T and k_B is Boltzmann's constant. This property of the algorithm makes it useful for a few different purposes. First, to estimate relative binding free energies for a large number of sequences, by using the fact that $\Delta F(\bar{\sigma}') - \Delta F(\bar{\sigma}) = k_B T \ln[N(\bar{\sigma}')/N(\bar{\sigma})]$. This usage requires a good coverage of sequence space, i.e., $N(\bar{\sigma}) \gg 1$ for the $\bar{\sigma}$'s of interest. Second, to determine the specificity profiles of a given peptide-binding site, i.e., the preference for different amino acid types at different positions on the peptide chain. Third, to provide insight into the structural basis of specificity, including the role of conformational heterogeneity. This is possible because the algorithm generates not only a large number of binding competent $\bar{\sigma}$'s, but also their bound state equilibrium conformational ensembles. Finally, the algorithm may be used as a tool to discover unknown binding sites on a given protein surface [13], although the feasibility of this usage has not yet been fully explored.

In this chapter, we demonstrate how the computational peptide screening can be implemented and applied to determine the specificity profile of the 6th PDZ domain of the protein GRIP1 [14]. Like most other PDZ domains [15], this GRIP1 PDZ domain consists of a conserved fold with 6 β-strands and two α-helices, as shown in Fig. 1. Peptides bind by β-sheet augmentation [16], i.e., the β-sheet of the PDZ domain is extended by a further strand from the tail of the peptide. The focus of this chapter is, in particular, on the technical challenges that come with performing simulations with updates that can alter the amino acid type at dynamic positions on the peptide, while still generating conformational ensembles with proper statistics.

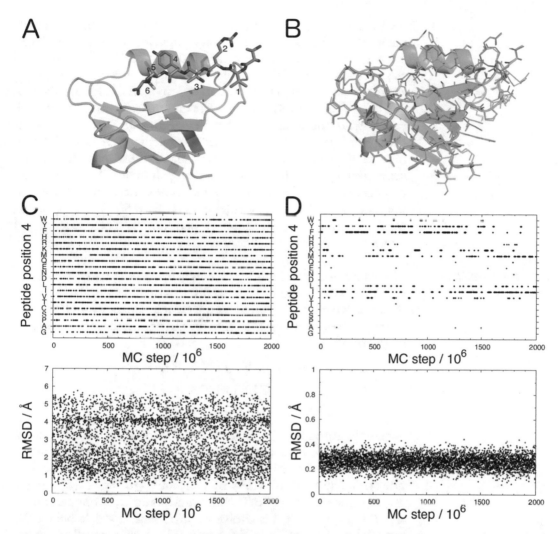

Fig. 1 (a) X-ray structure of the 6th PDZ domain of GRIP1 in complex with a C-terminal peptide from human liprin-α1 (PDB ID 1N7F) [14]. The peptide (amino acid sequence TVRTYSC) is shown in stick representation and the PDZ domain is shown in *green cartoon*. In this work, the peptide screening method is applied to the 6 most C-terminal peptide positions, numbered in the figure such that position 6 is the C-terminal amino acid. **(b)** Comparison between the experimental structure 1N7F (*green*) and the initial model start conformation (*salmon*) of the PDZ–peptide complex. **(c)** Example of part of a multisequence Monte Carlo simulation trajectory of the isolated peptide showing the evolution of the amino acid type in position 4 and the root-mean-square deviation (RMSD) from the liprin-α1 peptide structure in (**a**), calculated over $C\alpha$ coordinates. **(d)** Example of a multisequence Monte Carlo simulation trajectory of the bound state. RMSD is measured in the same way as in (**c**)

---

2 Materials

2.1 Computational Model

A prerequisite of the computational peptide screening method, as currently formulated, is that the underlying physical protein model describes (1) the solvent in an implicit manner and (2) the protein chain using only torsional angles, i.e., bond lengths and angles are held fixed at standard values. To this end, our method is implemented together with the all-atom, implicit solvent model developed in refs. [17, 18]. In this model, the potential energy function can be written as $E(\bar{\sigma},\bar{r})$, where $\bar{\sigma} = \{\sigma_i\}_{i=1}^{N}$ and $\bar{r} = \{\phi_i,\psi_i,\bar{\chi}_i\}_{i=1}^{N}$ are the sequence and chain conformation of an N-amino acid protein chain, respectively, and σ_i is the amino acid type at position i, ϕ_i and ψ_i are the backbone dihedral angles, and $\bar{\chi}_i$ is a set of side-chain torsional angles. The model was initially developed for protein folding [19] and thereafter adapted for protein–peptide interactions [17, 18]. Its potential energy function $E(\bar{\sigma},\bar{r})$ is based on effective hydrophobic attractions, context-dependent hydrogen bonding, and electrostatic interactions. In refs. [17, 18], it was tested on the blind docking of peptides to both bound and unbound PDZ domain structures.

2.2 Multisequence Monte Carlo

The computational peptide screening method is based on multi-sequence Monte Carlo, a generalized-ensemble simulation technique in which $\bar{\sigma}$ is treated as a dynamical parameter [20, 21]. More precisely, our method relies on simulating the joint probability distribution

$$P(\bar{r},\bar{\sigma}) = Z^{-1}e^{-\beta E(\bar{r},\bar{\sigma})+g(\bar{\sigma})}, \tag{1}$$

$$Z = \sum_{\bar{\sigma}}\int_{\bar{r}}d\bar{r}e^{-\beta E(\bar{r},\bar{\sigma})+g(\bar{\sigma})},$$

where the sum should be taken over all allowed $\bar{\sigma}$ and the integral over all possible $\bar{r}$. Importantly, the model parameters $g(\bar{\sigma})$ control the marginal distribution $P(\bar{\sigma})$ and must be chosen. The peptide screening method works by choosing $g(\bar{\sigma}) = \beta F_U(\bar{\sigma})$, where $F_U(\bar{\sigma})$ is the free energy of the unbound state, U, at temperature T and $\beta = 1/k_B T$. With this choice of $g(\bar{\sigma})$, and upon sampling the probability distribution in Eq. 1 over the restricted part of conformational space corresponding to the bound state, B, sequences will be generated according to

$$P(\bar{\sigma}) \propto e^{-\beta(F_B(\bar{\sigma})-F_U(\bar{\sigma}))} = e^{-\beta\Delta F(\bar{\sigma})}, \tag{2}$$

where $F_B(\bar{\sigma})$ is the free energy of the bound state. This is the desired distribution.

2.3 Linear Approximation of the Unbound State Free Energy $F_U(\bar{\sigma})$

If there are M amino acid positions on the peptide that are allowed to change during the multisequence simulations, the number of parameters $g(\sigma)$ to be determined is, in principle, 20^M. To reduce this number, construct the linear approximation

$$g(\bar{\sigma}) = \sum_i h_i(\sigma_i), \tag{3}$$

where the sum goes over the M dynamic amino acid positions on the peptide, and the $h_i(\sigma_i)$'s are discrete functions of the amino acid type at position i, σ_i. Because each $h_i(\sigma_i)$ can be tabulated with 20 values, the total number of parameters to be determined is reduced to $20M$ from 20^M (*see* **Note 1**).

2.4 Monte Carlo Updates

To simulate the probability distribution in Eq. 1 using Monte Carlo methods, two different types of updates are necessary: (1) conformational updates, $\bar{r} \to \bar{r}'$, and (2) mutational updates, $\bar{\sigma} \to \bar{\sigma}'$. Perform relatively frequent mutational updates, around 1 attempt every $10^3 - 10^4$ MC step. Below, R is a uniformly distributed random number between 0 and 1.

Conformational updates:

1. Use two types of conformational moves: (1) Biased Gaussian Step (BGS; an approximately local chain update) with bias parameter $b = 300$ [22], and (2) sidechain rotations (pick a random χ angle and assign a new value between $-\pi$ and π).

2. Select a random move type and perform the move, i.e., calculate the updated atomic coordinates.

3. Calculate the change in energy $\Delta E = E(\bar{r}', \bar{\sigma}) - E(\bar{r}, \bar{\sigma})$.

4. Accept new state if

$$\ln R < -\beta \Delta E, \tag{4}$$

which is the ordinary Metropolis condition [23].

5. If rejected, restore the conformational state $\bar{r}$.

Mutational updates:

1. Randomly select a dynamic amino acid position i and a new amino acid type σ'.

2. Build the new sidechain σ' at position i, i.e., calculate the sidechain atom positions from the torsional angles $\bar{\chi}_i$ (*see* **Note 2**).

3. Calculate the change in energy $\Delta E = E(\bar{r}, \bar{\sigma}') - E(\bar{r}, \bar{\sigma})$, where $\bar{\sigma}'$ is the new sequence.

4. Accept move if

$$\ln R < -\beta \Delta E + \Delta h, \tag{5}$$

where $\Delta h = h_i(\sigma') - h_i(\sigma_i)$.

5. If move is rejected, restore the state by performing the reverse mutation $\bar{\sigma}' \to \bar{\sigma}$.

2.5 Bound State Constraints

In the second step of the peptide screening method, constraints are needed to keep the peptide chain in place in its bound state. This is necessary because even though sequences are generated predominantly with good binding properties, sequences with poor binding properties will also be sampled although they are statistically suppressed. This can cause the peptide to unbind and diffuse away from the binding pocket.

Bound state constraints:

1. To the potential energy function of the model, $E(\bar{r}, \sigma)$, add the term $E_{\text{constr}} = 10 \sum_k f(|\bar{x}_k - \bar{x}_k^{1\text{N7F}}| - 5\,\text{Å})$, where the sum runs over all amino acids k in the protein and peptide chains, $\bar{x}_k$ and $\bar{x}_k^{1\text{N7F}}$ are the $C\alpha$ atom positions in the current conformation $\bar{r}$ and the 1N7F structure, respectively, and $f(a) = \max(0, a)$ (*see* **Note 3**).

3 Methods

Unless otherwise indicated, all simulations refer to multisequence Monte Carlo simulations carried out at $kT = 0.45$ (model units).

3.1 Step 1: Reference Simulations of the Isolated Peptide

3.1.1 Preparation

1. Select the dynamic positions on the peptide. Here, we choose to designate the 6 most C terminal positions on the liprin-α peptide as dynamic, i.e., simulations will be carried out on the peptide TX_6, where T is threonine and X denotes a dynamic amino acid position.

2. Choose a random initial peptide sequence and conformation, i.e., set each torsional angle to a random number between $-\pi$ to π and each dynamic σ_i to a random amino acid type.

3. Set $h_i(\sigma) = 0$ for all positions i, all amino acid types σ.

4. Thermalize the system by carrying out 10^6 elementary MC steps, with mutational updates turned off.

3.1.2 Simulate the Unbound State

1. In an iterative procedure, perform increasingly longer multi-sequence simulations of the isolated peptide, up to runs of around 10^{10} elementary MC steps. Continue until the probability distributions $p_i(\sigma)$ are approximately flat for all positions i, where $p_i(\sigma)$ is the probability of observing amino acid σ at position i.

2. In between each run, update the h parameters according to $h_i(\sigma) \rightarrow h_i(\sigma) - \ln p_i(\sigma)$ for all i and σ. Around 5–10 iterations may be necessary (*see* **Note 4**).

3. Carry out a final simulation of the unbound peptide with 10^{10} elementary steps using the final choice of $h_i(\sigma)$ parameters. This will take around 4 days on a standard desktop computer. The deviation from the ideal $p_i(\sigma) = 0.05$ is at most a few percent, as can be seen in Fig. 2.

3.2 Step 2: Simulation of Protein-Peptide Bound State

3.2.1 Prepare an Initial Structure of the Complex

1. Having obtained appropriate h-parameters, prepare an initial model conformation of the protein–peptide complex (1N7F [14]). In a protein model with only torsional degrees of freedom, the coordinates of an experimental structure cannot be used as a starting conformation. Instead, construct a model conformation that as closely as possible resembles the experimental

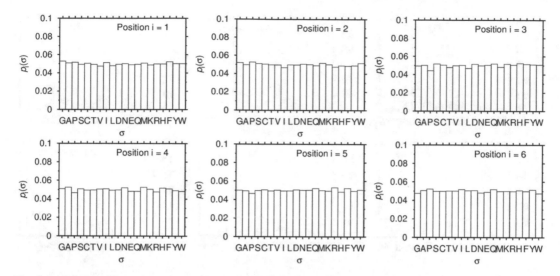

Fig. 2 Probability distribution of amino acid types, $p_i(\sigma)$, at peptide positions 1 through 6 as obtained from simulations of the peptide in isolation with the final choice of method parameters $g(\sigma)$

coordinates, while still not exhibiting any major clashes between atoms. For each chain, this can be done by following the steps 2–5 below, which involves building a model conformation in stages from the N terminus, maximizing the similarity to the experimental structure in each step.

2. Use the atom coordinates of the experimental structure (1N7F) to determine the torsional angles $\{\phi_i, \psi_i, \bar{\chi}_i\}_{i=1}^{N}$.

3. Calculate the cartesian coordinates of the model conformation with these angles. This structure will relatively accurately reproduce the local chain features of the experimental structure, but will deviate significantly globally and may contain severe atomic clashes.

4. In an iterative process, minimize the root-mean-square deviation, RMSD(n), between the model conformation and the experimental coordinates, 1N7F, calculated over the first n amino acids. Perform the minimization using 10,000 BGS Monte Carlo updates at $kT = 0$ (accept moves that decrease RMSD, reject otherwise). Apply only conformational updates that affect the torsional angles of the first n amino acids.

5. Start with $n = 4$ and, following each minimization, increase n by two units (repeat from step 4) until the entire chain is included.

6. To remove potential clashes in the structure, perform a short (fixed-$\bar{\sigma}$) simulation with 10,000 BGS Monte Carlo steps at $kT = 0.45$, using the ordinary energy function E. To keep structural changes very small, accept new states only if $\text{RMSD}_{new} \leq 1.01 \times \text{RMSD}_{old}$.

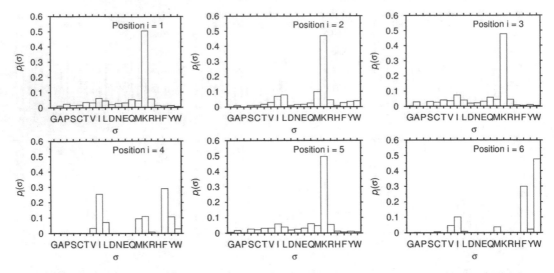

Fig. 3 Probability distributions of amino acid types, $p_i(\sigma)$, at the peptide positions 1 through 6 as obtained from simulations of the bound state. The distributions have become heavily skewed compared to those obtained in the unbound state (cf. Fig. 2)

7. The initial model conformation obtained is shown in Fig. 1. It has RMSD = 0.28 Å calculated with respect to the 1N7F structure, taken over all non-H atoms of the protein and peptide.

3.2.2 Simulate the Bound State

1. Add structural constraints to keep conformations restricted to the bound state (*see* Subheading 2.5).

2. Thermalize the system by performing 10^6 MC steps, with sequence updates turned off.

3. Perform a set of 5 independent simulations, each of 2×10^9 MC steps. Each simulation takes around 9 days on a standard desktop computer.

4. Save snapshots of the system state, i.e. conformation $\bar{r}$ and sequence $\bar{\sigma}$, every 10^5 MC steps, for analysis.

3.3 Data Analysis

1. Using the saved sequences $\bar{\sigma}$, calculate $p_i(\sigma)$, i.e., the probabilities of different amino acids types at the dynamic positions i. Figure 3 shows that $p_i(\sigma)$ have become heavily skewed as compared to the unbound state. At positions 4 and 6, there is a strong preference for hydrophobic amino acids, in line with the known specificity profile of this protein domain [4]. The other positions, 1–3, and 5, reveal a predicted preference for lysine. Although more work is needed to explore this result, we note that several polar and negatively charged amino acids are lining the peptide-binding pocket on the GRIP1 PDZ domain [14]. Flexible lysine sidechains on the peptide might therefore be able to provide favorable interactions with the protein through a combination of sidechain–sidechain hydrogen bonding and electrostatic attraction.

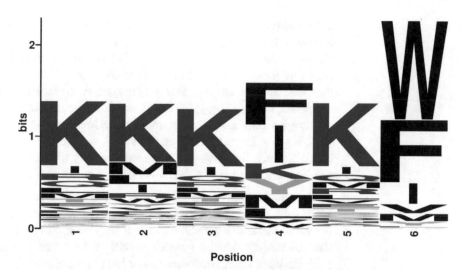

Fig. 4 The specificity profile of the GRIP1 PDZ6 domain, as predicted by the peptide screening method. The size of each letter is proportional to the probability of the corresponding amino acid type at the respective positions, $p_i(\sigma)$, as found from the second step of the peptide screening procedure, and the overall size of each letter column is determined by the sequence conservation expressed in bits (the maximum possible height is $\log_2 20 \approx 4.32$ bits). The figure is produced using the freely available Weblogo tool [24]

2. To visualize the probabilities $p_i(\sigma)$, pick a representative sample of around 1000 generated sequences (the total number of generated snapshots is 10^5 in the 5 simulations). Use the freely available Weblogo tool (http://weblogo.berkeley.edu/logo.cgi) to construct a sequence profile of $p_i(\sigma)$, as shown in Fig. 4. The column height indicates the sequence conservation at each position. Accordingly, Fig. 4 shows that the strongest binding preference of the PDZ domain is at positions 4 and 6.

4   Notes

1. For simulations with very few dynamic positions on the peptide ($M = 1$ or 2), it is easier to skip the construction of a linear model and instead work directly with the model parameters $g(\bar{\sigma})$.

2. There are two ways to deal with the fact that the number of sidechain torsional angles, q, is different for different amino acid types (from $q = 0$ for glycine to $q = 5$ for lysine). The most rigorous way is to explicitly introduce 5 χ_i angles for all dynamic amino acid positions on the peptide. These 5 angles should be updated with sidechain rotational moves as any other degree of freedom in the model. However, when an amino acid type σ with $q < 5$ is assigned to position i, one or more of the 5 χ_i angles will behave as "ghost" angles because their values do not affect the potential energy E. Therefore, these angles will always accept sidechain rotations and thus tend towards a uniform

probability distribution. An alternative way that avoids the introduction of extra χ-angles is the following: For a mutational update with $q_{new} \leq q_{old}$, i.e., when the new amino acid type has fewer or the same number of χ angles, simply discard any unmatched angles. For a mutational update in which q increases, i.e. $q_{new} > q_{old}$, let the new amino acid inherit the first q_{old} angles and assign new, random values to the remaining $(q_{new} - q_{old} + 1)$ χ-angles. Both ways will realize the distribution in Eq. 2, and thus give the same results, but will lead to different numerical values of the $g(\bar{\sigma})$ model parameters.

3. This constraint is designed to keep the peptide (and protein) $C\alpha$ atoms close to the native coordinates while still allowing for some flexibility. In some cases, it may be important to make the constraints on the peptide looser, e.g., when the peptide exhibits significant conformational heterogeneity in the bound state. A looser constraint can be implemented, e.g., by letting the summation over k in Eq. 6 include only 1 or a few of the peptide $C\alpha$ atoms.

4. In the iterative process, some probabilities $p_i(\sigma)$ may be extremely small or identically zero (especially in the early steps) due to a lack of sampling in sequence space, and the correction of the corresponding $h_i(\sigma)$ may become too big or diverge. To get around this issue, use instead the update rule $h_i(\sigma) \to h_i(\sigma) - \ln[\max(p_i(\sigma), 0.01/20)]$, i.e., assume there is always a small but finite minimal probability for every σ.

References

1. Pawson T, Nash P (2003) Assembly of cell regulatory systems through protein interaction domains. Science 300:445–452

2. Wright PE, Dyson HJ (2015) Intrinsically disordered proteins in cellular signalling and regulation. Nat Rev Mol Cell Biol 16:18–29

3. Neduva V, Linding R, Su-Angrand I, Stark A, de Masi F, et al (2005) Systematic discovery of new recognition peptides mediating protein interaction networks. PLoS Biol 3:e405

4. Teyra J, Sidhu SS, Kim PM (2012) Elucidation of the binding preferences of peptide recognition modules: SH3 and PDZ domains. FEBS Lett 586:2631–2637

5. London N, Movshovitz-Attias D, Schueler-Furman O (2010) The structural basis of peptide-protein binding strategies. Structure 18:188–199

6. Wright PE, Dyson HJ (2009) Linking folding and binding. Curr Opin Struct Biol 19:31–38

7. Chen J (2012) Towards the physical basis of how intrinsic disorder mediates protein function. Arch Biochem Biophys 524:123–131

8. Hsu WL, Oldfield CJ, Xue B, Meng J, Huang F, et al (2013) Exploring the binding diversity of intrinsically disordered proteins involved in one-to-many binding. Protein Sci 22:258–273

9. Zhou HX (2012) Intrinsic disorder: signaling via highly specific but short-lived association. Trends Biochem Sci 37:43–48

10. Staneva I, Huang Y, Liu Z, Wallin S (2012) Binding of two intrinsically disordered peptides to a multi-specific protein: a combined Monte Carlo and molecular dynamics study. PLoS Comput Biol 8:e1002682

11. Sharma R, Raduly Z, Miskei M, Fuxreiter M (2015) Fuzzy complexes: specific binding without complete folding. FEBS Lett 589:2533–2542

12. Uversky VN, Dunker AK (2015) The case for intrinsically disordered proteins playing contributory roles in molecular recognition without a stable 3D structure. F1000 Biol Rep 5:1

13. Bhattacherjee A, Wallin S (2013) Exploring protein-peptide binding specificity through computational peptide screening. PLoS Comput Biol 9:e1003277

14. Im YJ, Park SH, Rho SH, Lee JH, Kang GB, et al (2003) Crystal structure of GRIP1 PDZ6-peptide complex reveals the structural basis for class II PDZ target recognition and PDZ domain-mediated multimerization. J Biol Chem 278:8501–8507

15. Lee HJ, Zheng JJ (2010) PDZ domains and their binding partners: structure, specificity, and modification. Cell Commun Signal 8:8

16. Remaut H, Waksman G (2006) Protein-protein interaction through beta-strand addition. Trends Biochem Sci 31:436–444

17. Staneva I, Wallin S (2009) All-atom Monte Carlo approach to protein-peptide binding. J Mol Biol 393:1118–1128

18. Staneva I, Wallin S (2011) Binding free energy landscape of domain-peptide interactions. PLoS Comput Biol 7:e1002131

19. Irbäck A, Samuelsson B, Sjunnesson F, Wallin S (2003) Thermodynamics of alpha- and beta-structure formation in proteins. Biophys J 85:1466–1473

20. Irbäck A, Peterson C, Potthast F, Sandelin E (1998) Monte Carlo procedure for protein design. Phys Rev E 58:5249–5252

21. Irbäck A, Peterson C, Potthast F, Sandelin E (1999) Design of sequences with good folding properties in coarse-grained protein models. Structure 7:347–360

22. Favrin G, Irbäck A, Sjunnesson F (2003) Monte Carlo update for chain molecules: biased Gaussian steps in torsional space. J Chem Phys 114:8154–8158

23. Metropolis N, Rosenbluth AW, Rosenbluth MN, Teller AH, Teller E (1953) Equations of state calculations by fast computing machines. J Chem Phys 21:1087–1092

24. Crooks GE, Hon G, Chandonia JM, Brenner SE (2004) WebLogo: a sequence logo generator. Genome Res 14:1188–1190

Chapter 13

Enriching Peptide Libraries for Binding Affinity and Specificity Through Computationally Directed Library Design

Glenna Wink Foight, T. Scott Chen, Daniel Richman, and Amy E. Keating

Abstract

Peptide reagents with high affinity or specificity for their target protein interaction partner are of utility for many important applications. Optimization of peptide binding by screening large libraries is a proven and powerful approach. Libraries designed to be enriched in peptide sequences that are predicted to have desired affinity or specificity characteristics are more likely to yield success than random mutagenesis. We present a library optimization method in which the choice of amino acids to encode at each peptide position can be guided by available experimental data or structure-based predictions. We discuss how to use analysis of predicted library performance to inform rounds of library design. Finally, we include protocols for more complex library design procedures that consider the chemical diversity of the amino acids at each peptide position and optimize a library score based on a user-specified input model.

Key words Library design, Integer linear programming, Peptide engineering

1 Introduction

The increasing use of peptides as reagents for diagnostic, therapeutic, and basic research purposes highlights the need for engineered molecules with particular affinity and specificity profiles. Natural peptide interaction partners often do not have the high affinity or specificity required for such applications. Two main approaches exist for developing peptide reagents with desired binding characteristics: screening of large peptide libraries and computational design of peptide sequences [1].

Library screens that use cell surface display, phage display, mRNA or ribosome display, and smaller, synthetic peptide libraries screened on beads are powerful techniques for discovering peptide reagents [2–5]. High-throughput screening technologies can routinely survey 10^8 (yeast display)–10^{15} (mRNA display) DNA sequences, but the enormous theoretical sequence spaces of peptides greater than ~6–8 residues exceeds even those large numbers.

Ora Schueler-Furman and Nir London (eds.), *Modeling Peptide-Protein Interactions: Methods and Protocols*, Methods in Molecular Biology, vol. 1561, DOI 10.1007/978-1-4939-6798-8_13, © Springer Science+Business Media LLC 2017

Thus, randomly generated sequences may not sample the best molecules. A common alternative approach to identifying optimized peptide binders is to mutagenize a known interaction partner. When using random mutagenesis, whether by error-prone PCR or the use of NNK or other degenerate codons, the mutational load per sequence may be difficult to tune because different protein positions have different sensitivities to mutation. Too high of a mutagenesis load will yield many nonbinders, whereas low mutagenesis may not achieve sequences sufficiently diverged from the original sequence to meet challenging affinity or specificity goals [6].

Meanwhile, computational modeling of protein–peptide interactions is advancing. Several methods have used structural information to successfully predict interacting peptide sequences in the proteome [7–11]. Physical detail in the models used ranges from high, in methods based on molecular mechanics calculations, to low in methods that use simple distance tabulations. However, relatively few examples of purely computational design of novel peptide binding partners have been reported [12–15]. This is in contrast to the field of protein–protein interaction design, in which computational design of novel interaction partners is becoming increasingly common [16–18]. It remains difficult to achieve adequate conformational sampling of peptide conformations, and inaccuracies in standard energy functions limit the accuracy of scoring complexes that involve only a small number of residue contacts [19].

In recent years, the strengths of computational design and library screening technologies have been combined in methods that utilize computational algorithms to design libraries that reflect predictions about stability or binding made by a computational or data-based model [1]. Designing a large library that can be screened by high-throughput technologies, rather than just a small handful of sequences, overcomes the requirement for detailed and accurate information on all peptide positions. Researchers can make the best use of their screening capabilities by limiting variation to a productive sequence space, e.g., to peptide positions at which models predict affinity or specificity-enhancing mutations.

In the laboratory, libraries can be made using degenerate codons to include variation at different protein positions, can incorporate mixtures of defined codons at different positions, or can be composed of members of defined protein sequence. Libraries made with degenerate codons are the most economical and are the focus of this methods paper; such libraries are widely used. Degenerate codons include mixtures of nucleotides at each codon position that, collectively, encode a set of amino acid residues. There are conventions for naming such codons, e.g. NNK stands for a codon with a mixture of A, C, G, T at the first and second positions, and a mixture of G and T at the last position. The NNK degenerate codon can code for any amino acid or a single stop codon. See the standard IUPAC nomenclature for definitions of other degenerate codons (http://www.bioinformatics.org/sms/iupac.html).

Much of the published work on computationally directed library design has been done in the context of enriching libraries for functional sequences of single-domain proteins such as green fluorescent protein (GFP), cytochrome P450, and β-lactamase [20–24]. A common general approach is to use a list of sequences (e.g., a multiple sequence alignment (MSA) or the output of protein design calculations) thought to be enriched in functional proteins and then try to match the amino acid preferences at positions of interest by intelligently choosing degenerate codons. Several different algorithms and methods have been developed to guide the choice of degenerate codons, or, for directly synthesized, defined-sequence libraries, the choice of amino acids. Existing methods for optimization of degenerate codon choice have recently been well summarized by Jacobs et al. [25]. Relatively simple methods have used brute force enumeration of all possible libraries composed of degenerate codons that approximate an amino acid distribution found in an MSA or ranked list of protein designs [26, 27]. This leaves the choice of an individual library design to be made by the user based on library size or score. The OCoM method of Parker et al. also requires an MSA as input, but chooses degenerate codons based on dynamic programming and integer programming. This method considers pairwise frequencies, and also allows design of defined-sequence libraries [20]. The optimization method balances a quality objective (matching the MSA frequencies) with a novelty objective (minimizing sequence identity to individual members of the MSA) to meet the goal of a library enriched in beneficial mutations. In a further advancement, Jacobs et al. used dynamic programming to choose degenerate codons to represent the amino acid distributions in a sequence list, but allowed multiple degenerate codons at each position in the library design phase. This additional flexibility can be used to minimize how much the size of the library in DNA space exceeds the number of protein sequences encoded [25].

Alternative methods used for library design have borrowed from techniques used in protein structure design. An early method by Hayes et al. generated a ranked list of sequences based on a Monte Carlo search around the calculated global minimum energy sequence and conformation [28]. This list of sequences was then converted to an amino acid probability table, and a defined-sequence library was constructed to meet certain size and score cutoffs. Treynor et al. created libraries encoding GFP variants by several different methods including methods based on Hayes et al., error-prone PCR, and a new method, DBIS (diversity benefits applied to interacting sets) [23]. The DBIS method used dead end elimination to optimize degenerate codon choices based on average rotamer interaction energies for the amino acid sets encoded by the degenerate codons. Comparison of the success rate of different library design methods in generating functional, fluorescent GFP variants revealed that structure-based design methods had greater success rates than methods based on an MSA, and all intelligent library design methods performed better than error-prone PCR. Additionally, Treynor et al. showed that the success rate when screening designed libraries increased with the mutational load in the library, but this was not true for the error-prone PCR libraries. Guntas et al. also found that a naïve library that randomly mutated protein interface residues failed to produce binders in the design of a novel protein–protein interaction, while Rosetta-based libraries were successful [29]. These studies underscore the advantages of intelligently designed libraries over random mutagenesis in increasing the probability of success for diverse protein function and binding goals. Another recent example of structure-based library design used cluster expansion to convert structure-based Rosetta energies of variants to sequence-level scores [24]. The authors then used integer linear programming to optimize the library composition, with options to use pairwise energies and output a degenerate codon or defined-sequence library.

For the design of peptide libraries, flexibility in the ability to use many different types of input information is advantageous. Multiple sequence alignments, which are the preferred input for many previous library design methods, may not be useful for all protein–peptide interaction families, due to either too few validated binding partners, or extreme diversity in the binding site sequences. For many systems, there may also exist experimental data of varying types (SPOT arrays, alanine mutagenesis, deep sequencing data from single point mutant libraries, etc.) that a researcher would like to take into account. To incorporate diverse information sources that include both experimental and computationally derived data, we present a method for the computational design of peptide libraries enriched in sequences with a desired affinity or specificity profile. The basic method presented here was first used by Chen et al. to design libraries of Bcl-x_L variants with enhanced specificity for binding to BH3 peptides. Library design in that instance was based on Rosetta energies [30]. The method was then adapted to design BH3 peptide libraries enriched in specific binders of Bfl-1 based on SPOT array data, with a constraint imposed to ensure sampling of chemical diversity [31]. We have further demonstrated the utility of the method by designing two more BH3 peptide libraries with specificity for other Bcl-2 family members using SPOT array data and computational predictions from STATIUM [32]. Diverse, quantifiable experimental data or computational predictions can be used as input to this general framework, which applies integer linear programming to optimize library composition based on an easily modified set of parameters. The output is a degenerate codon library with size and characteristics tuned to the desired experimental screening strategy and end goal defined by the researcher.

2 Materials

The library optimization method presented here can be run with a simple set of scripts in a terminal environment on a Mac or Linux machine. Two Perl scripts, writeCodon.pl and runILP.pl implement the complete design process. Two additional files are needed to run these scripts. File codon_combos.txt includes a database of codons used by writeCodon.pl. File library_design.mod is the file formatted for use with the ILP solver, glpsol, and it is edited and run by the runILP.pl script. These files are included in file LibraryDesignScripts.tar.gz. All scripts and example files are available on the GitHub KeatingLab/LibraryDesign repository.

Additionally, example input and output files are included: file_ pref, file_req, codons_output, library_design_output. The files are referred to by these names throughout Subheading 3, and they are included in the file LibraryDesignExample.tar.gz. The tar files can be opened using the Archive Utility on a Mac or the command "tar −xvzf LibaryDesignExample.tar.gz" in the terminal.

The ILP problem is solved using the glpsol solver in the GNU Linear Programing Kit (GLPK). This is available as a free download from https://www.gnu.org/software/glpk/. Install on a Linux or Mac machine by following the installation instructions included with the software. You may need to use the command "sudo make install" to install in the default location. Make note of the path where the glpsol solver is installed, as you will need to direct the runILP.pl script to its location (default is /usr/local/bin/glpsol).

Finally, for the advanced multi-option method presented in Subheading 3.5, scripts and example input and output files are included in MultiOption.tar.gz. File names are as given in Subheading 3.5.

3 Methods

The method for library design presented here proceeds through the three steps outlined in Fig. 1, followed by optional analysis of predicted library performance and further rounds of design. Because the available input information will vary for every protein–peptide interaction study, we present a general framework for formalizing various input datasets. Likewise, library design objectives will vary. We describe and provide a basic framework for optimizing a library to include peptides with high affinity or specificity for

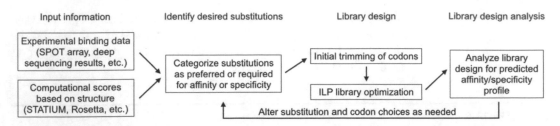

Fig. 1 Flowchart of the library design process. The first two steps of gathering information from binding experiments or structure-based models and prioritizing substitutions will depend on the information available for the protein–peptide interaction of interest, so we present some general guidelines. The optimization of a degenerate codon library to encode the desired substitutions then proceeds in two parts: initial trimming of the codon choices based on codon size and score, followed by ILP library optimization to yield a library of a desired size with an optimal score. Finally, suggestions are given for analysis of the predicted behavior of the library based on input models, which can inform further rounds of library design to improve predicted library characteristics

one target interaction partner. After a library design is output by the ILP code, a researcher can analyze it for its predicted behavior based on any experimental or computational models available. The results of these analyses can then be used to manually alter the allowed substitution and codon choices to improve output for further rounds of design. We have found this iterative process of design and evaluation very useful in exploring the tradeoffs that a protein designer inevitably faces when devising a screen.

3.1 Formalization of Prior Knowledge of Binding Preferences

In deciding which positions to mutate, a researcher should use all information available (**Note 1**). This may include SPOT arrays, in which peptide positions are mutated to all 20 amino acids and binding is semi-quantitatively measured to the peptides synthesized on a membrane. Similarly, alanine scanning or hydrophile scanning can provide information on which positions are most important for binding [33]. Deep sequencing data from random mutagenesis libraries or deep mutational scanning experiments can provide similar positional information [34, 35]. Lacking any experimental binding data, a structure of the peptide bound to the target is a valuable source of information (**Note 2**). Computational methods such as STATIUM or Rosetta can be used to generate scores for all possible peptide point mutants based on a structure of the complex, in effect generating a virtual SPOT array [36, 37]. In this section, we cover how to convert SPOT array intensities or deep sequencing data to a position-specific scoring matrix (PSSM). We also discuss the use of computational tools for mutational scoring [37].

3.1.1 Generation of a PSSM from SPOT Array Data

1. SPOT array intensities can be quantified using imaging software. Several wild-type peptide spots (ideally distributed throughout the array to control for variation in exposure) can

be used for normalization. In our protocols, we average the intensity of all wild-type peptide signals.

2. To compute a PSSM score for a mutation, use the following equation where M is the mutant SPOT array intensity, and W is the average wild-type intensity

$$\text{PSSM score} = \log_{10} \frac{M}{W}$$

Mutants showing weaker binding than wild type will have scores below zero, and mutants with tighter binding will have positive scores.

3. If SPOT arrays are available for the target and a competitor, a difference PSSM score can be calculated for use in a library designed for specificity. Because the range of intensities observed on SPOT arrays is likely to vary for different binding partners, each SPOT array-derived PSSM should be normalized and a Z-score calculated. The Z-score difference can then be used as a metric of specificity for each possible substitution. The equation below can be used to calculate the standardized PSSM difference, where μ is the mean intensity across each array (target, T or competitor, C), and σ is the standard deviation across each array.

$$\Delta\text{PSSM}_{T-C} = \log_{10} \left\{ \left(\frac{M - \mu}{\sigma} - \frac{W - \mu}{\sigma} \right)_T - \left(\frac{M - \mu}{\sigma} - \frac{W - \mu}{\sigma} \right)_C \right\}$$

3.1.2 Generation of a PSSM from Deep Sequencing Data

The type of metric generated from deep sequencing data will depend on the type of experiment that it was produced from. For a single mutant dataset in which both the input and selected libraries were sequenced, variant frequencies can be converted into a PSSM-like matrix via a variety of previously published methods [38–40]. If multiple positions were mutated at once, a PSSM can be generated based on the frequencies of substitutions in unique sequences. Using unique sequences limits the biases that can arise in cell surface display or phage display datasets, e.g. from growth rate differences or background mutations. To further improve the quality of the dataset used for generating a sequencing-based model, one can limit the sequences included to those that had some minimum number of counts in the sequencing. This minimizes noise from sequencing errors. If similarly generated datasets are available for multiple binding partners, a difference PSSM score can be calculated and used as a specificity metric.

3.1.3 Use of Structure-Based Scores

When a structure of the protein-peptide complex of interest is available, or a homology model built on a close homolog, a variety of computational scoring methods can be used to provide predictions of the effect of mutations on binding affinity. Scoring methods that can easily score all possible mutations in a peptide include STATIUM, Rosetta, FoldX, and Discovery Studio [41]. The effect of each point mutation can be calculated as the difference from the wild-type peptide score (e.g., ΔSTATIUM = STATIUM$_{wt}$ – STATIUM$_{mutant}$). If structures are available for the peptide bound to the target protein and competitors, a specificity score can be computed as the difference between these relative scores (e.g., $\Delta\Delta$STATIUM = ΔSTATIUM$_{target}$ – ΔSTATIUM$_{competitor}$). As discussed above for the PSSM specificity scores, the range of scores for different structures may be different. Therefore, it is advisable to compute the positional scores for all 20 amino acids at each peptide position and then normalize these scores for each structure. A Z-score difference can then be calculated as in Subheading 3.1.1. If the wild-type peptide binds to the target and competitors with different affinity, the score difference will be a difference in the effect on binding relative to wild-type.

3.2 Categorization of Mutations

Before designing a library on the DNA level, a researcher must first choose which peptide positions to vary and which amino acid substitutions to favor. These choices will depend on the goal for the library screening experiment, particularly whether the goal is simply to obtain high-affinity peptides for one protein target, or to obtain peptides that show both high affinity for the target and much lower binding to other proteins (competitors), in other words, specificity for the target. The length of the region to mutate can depend on physical considerations, such as how much of the peptide comes in contact with its binding partner, as well as on practical considerations, such as the length of oligonucleotides required for library assembly and the length of sequencing reads if the enriched library pools will be deep sequenced. Given advances in DNA synthesis and sequencing in recent years, most peptides will be well within standard length limits.

In our protocol, two categories of substitutions must be chosen: *required* and *preferred*. Substitutions categorized as required will always be included in the library design. Wild-type residues are generally included as required. Additional required residues may include substitutions for which there is strong evidence (experimental or computational) suggesting that they will have the desired effect on affinity or specificity. Preferred substitutions are included in the library as space and other criteria permit, as determined by the optimization algorithm. For a library designed to optimize binding affinity, preferred residues could include all residues predicted to be nondisruptive for binding to the target (neutral to beneficial). For specificity library design, the preferred set might be further narrowed to require that residues also weaken binding to competitors, according to some metric. Choices of how to define sets of required and preferred residues will depend on the data available for the protein–peptide interaction system and are ultimately made by the user. For example schemes used to designate preferred and required residues, please *see* refs. 30–32. The number of positions input into the design process can exceed the number that will be varied in the output library design. Thus, a designer can be generous at this stage and provide information on more positions than they ultimately want to vary.

The following steps will define the two required sets of residues:

1. Make a plain text file for the required residues (see example file_req). Each line should include the wild-type residue and peptide position number followed by a list of the one-letter amino acid codes of the residues that the designer wants to require at that position. For example, the line I4 IRY, would mean that in place of the wild-type isoleucine at position 4, the designer wants to require sampling of isoleucine, arginine, and tyrosine.

2. Make a plain text file for the preferred residues (see example file_pref). Each line lists the wild-type residue and position number followed by a list of the one-letter amino acid codes of the preferred residues, each followed by the number 1. The preferred residues must include all of the required residues. For example, I4 I 1 K 1 R 1 T 1 A 1 V 1 L 1 M 1 Y 1.

3.3 Library Optimization

The choice of degenerate codons used to encode the library proceeds through two steps. First, for each peptide position, a list of all degenerate codons capable of encoding all of the required residues is output. This list is narrowed to exclude codons that encode fewer preferred residues but more trinucleotides than another codon in the list. Second, the list of possible degenerate codons at each position is fed into an ILP solver and a codon is chosen for each position such that the library score is maximized and the library size restraint is met. The default library score is the number of protein sequences encoded by the library that are composed entirely of preferred residues (i.e., the product of the number of preferred amino acids encoded by the chosen codons at each position). Users can define other scores that are linear functions of the codon choices, *see* Subheading 3.5.

3.3.1 Initial Trimming of Codons

1. Put four files into one directory: the two files specifying the preferred and required residues (e.g. file_pref and file_req), a file containing all codons (codon_combos.txt), and the Perl script that makes the initial codon choices (writeCodon.pl).

2. In a terminal, in the directory with the four files, run the script with the following command:
    perl -w writeCodon.pl file_req file_pref codons_output,
    where "codons_output" is any name the designer chooses for the output file. This file (e.g. codons_output) lists each

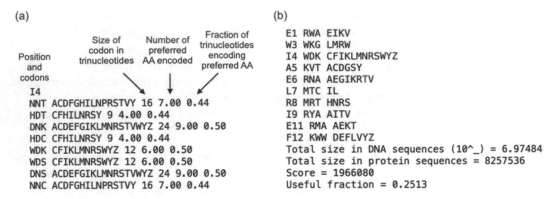

Fig. 2 Output of the initial codon trimming step and the ILP library optimization step. (a) An example of the degenerate codon choices for one position as output by writeCodon.pl. The columns are labeled with their corresponding properties. (b) An example library design output by runILP.pl. The three columns are position, degenerate codon, and amino acids encoded. The total size in DNA sequences is under the limit set by the user (in this case 10⁷). The total size in protein sequences is the product of the number of amino acids encoded by each chosen codon. The score is the optimized value, the number of protein sequences composed entirely of preferred amino acids. The useful fraction is the product of the fraction of trinucleotides encoding preferred amino acids for each chosen codon

peptide position specified in the required file (file_req) followed by a list of the degenerate codon choices for each position (Fig. 2a). Each degenerate codon line includes the following information: degenerate codon, amino acids encoded, the number of trinucleotides encoded (codon size), the number of preferred residues encoded, and the percentage of trinucleotides that encode for preferred amino acids. Standard IUPAC nomenclature is used for the degenerate codons (http://www.bioinformatics.org/sms/iupac.html). Thus, the line: DNK ACDEFGIKLMNRSTVWYZ 24 9.00 0.50, is interpreted as the degenerate codon DNK encodes the amino acids ACDEFGIKLMNRSTVWYZ (Z is a stop codon) using 24 trinucleotides. This codon encodes nine preferred amino acids, with 50 % of the 24 trinucleotides encoding preferred amino acids. Note that some positions may have different codons encoding different amino acid sets that have the same number of preferred residues and same size. The designer should look through the codons_output file for such examples and manually choose one codon to keep based on criteria such as chemical diversity or scores in the input models. If this is not done, the ILP script in the next step will simply use the first codon listed of a given size and score.

3.3.2 ILP Library Optimization

The script runILP.pl reads in the codons_output file created by the step above and writes out a text file with the degenerate codon chosen for each peptide position and information on the library score and size (**Note 3**). To make the codon choices, the ILP solver is instructed by the file library_design.mod to optimize the library score, with the constraint that the library be smaller than a specified size. The library score is the number of sequences that are entirely composed of preferred residues.

1. Put the following three files into your design directory: the codons_output file created in the step above, library_design. mod, and runILP.pl, the Perl script that directs the ILP solver and creates the output file.

2. Edit the following line in runILP.pl to include the correct path to where the ILP solver, glpsol, was installed on your computer (replace path/to):

   ```
   my $glpsol = "/path/to/glpsol";
   ```

   If the GLPK package was installed on your local computer using default installation settings, it will likely be located in /usr/local/bin/glpsol.

3. Edit library_design.mod to set the library size constraint (**Note 4**). This file contains the constraints that go into the ILP solver. Find the line "subject to totalsize: sum {v in V} costVTOT[v] * X[v], <- 7.0;", and change 7.0 to another number (e.g. for a library size constraint of 10^5 change to 5.0). This library size is the size in DNA sequences. A good rule-of-thumb is to set this tenfold lower than the maximum transformation efficiency or screening throughput of the library-screening platform to be used, in order to sample most of the library.

4. Run the ILP optimization:

   perl -w runILP.pl codons_output library_design_output

   The file "library_design_output" is whatever the designer chooses to name their output file, and codons output is the file output by writeCodon.pl.

If a solution is found, the standard output will say "Optimal solution found", and a text file with the library design will be created (e.g. library_design_output_example). An example library design output is shown in Fig. 2b, with an explanation of the outputted metrics given in the figure legend. This is a very fast process, completed in <1 s on a standard laptop. If a solution is not found, there is no solution possible that encodes all of the required residues within the library size constraint. The standard output will say "Problem has no feasible solution". In this situation, the designer will need to go back to the categorization of mutations step and reduce the number of required mutations, or increase the library size constraint and run the codon trimming and ILP steps again.

Once a library design is output, the basic process is complete, and the designer can order oligonucleotides encoding the library from a DNA synthesis company using machine mixing of nucleotides to encode the degenerate codons. Below, we present further steps that a designer can take to analyze the predicted characteristics of the library to inform modifications for further rounds of the library design process. Additionally, we provide a more complex protocol that allows consideration of chemical diversity and optimization based directly on a user-defined scoring system, rather than on the number of preferred residues.

3.4 Analysis of Library Designs

Designed libraries can be evaluated on several levels to get an idea of the predicted performance for the affinity or specificity objective. Adjustments can be made to the inputted lists of preferred and required residues, as well as the initial codon lists, in order to improve predicted performance. Predictors of performance or library quality include the simple statistics output by the library design script, or more in-depth analysis of the scores of all theoretical library sequences based on quantitative models available for the peptide interaction system.

A first-pass library design analysis would look at the statistics output by the library design script including the library score, the number of protein sequences encoded, and the percent of the library that is predicted to be useful (i.e., the percentage of the DNA sequences that encode only preferred residues). To maximize these statistics, you can change the amino acids you list as preferred and required, or alter the codon choices by manually editing the codons_ output file before running the ILP script. Some combinations of required amino acids may necessitate the choice of large codons, which may include many amino acids that could be disruptive for binding, lowering the fraction of the library that is predicted to be useful. Consider whether all of the required residues are necessary, or if you can require a chemically similar residue that allows the choice of a smaller codon. Alternatively, if you are willing to use multiple oligonucleotides to construct your library, you can use more than one codon at a position, as done by Chen et al. [30].

It is also important to consider the mutational load of your library, or how many positions are varied. Previous studies have found that a higher mutational load in intelligently designed libraries correlates with a greater chance of success [23]. If some positions turn out to not contribute as much to affinity or specificity as predicted by an input model, then allowing a few amino acid choices at many positions will provide a better chance of success than allowing a large diversity of amino acids at a few positions. However, it is also important to consider how many potentially disruptive mutations you are including. For example, if a given position includes a choice between just two amino acids, and one of these disrupts binding, then half of the library will not bind. To spread the diversity of your library across many positions, you can adjust the size of codons chosen. Manually edit the writeCodons. pl output (codons_output) to remove large codons (particularly codons that encode stop codons or potentially disruptive amino acids) at positions where a large amount of diversity is not a high priority, and then re-run the ILP optimization on the edited file. Focus diversity on positions that are most likely to impart high affinity or specificity, based on the information available for the peptide system.

If quantitative scoring models are available for your peptide system (e.g., PSSMs based on SPOT arrays, or fast-to-evaluate structure-based scores), these models can be used to score the theoretical library and predict how many sequences are likely to have the desired affinity or specificity characteristics. First, write out the theoretical library by creating sequences for all possible combinations of the amino acids at each position in the library design. Then, compute the score for each sequence as the sum of the scores for each position in the peptide. If you are designing a library for specificity and have models for competitor interaction partners, you can score the theoretical library on those models for comparison and analysis of predicted specificity. The score of the wild-type peptide sequence that the library is based upon can be used as a cutoff to calculate the proportion of the library that has wild-type-like or greater affinity. For analysis of a library designed for affinity, a simple histogram can be used to visualize the distribution of library scores. For specificity library design, we use two-dimensional histograms (density plots) to compare the library sequence scores for the target and competitor interaction partners [31].

We recently designed and enriched a BH3 peptide library for specific binding to the viral Bcl-2 homolog KSBcl-2 over the competitor human Bcl-2 homologs [32]. In evaluating different library designs, we scored the libraries using PSSMs derived from SPOT arrays for KSBcl-2 or human Bcl-2 homologs binding to BH3

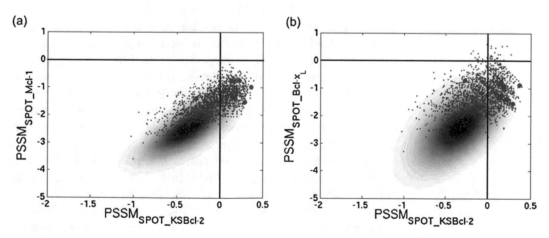

Fig. 3 Analysis of sequence scores for a library designed and then screened for specificity. Predicted KSBcl-2 binding is shown on the x-axis. Predicted binding to competitors Mcl-1 (**a**) or Bcl-x$_L$ (**b**) are shown on the y-axis. A density plot of scores for the theoretical library is shown in *gray scale*. Scores for sequences from a library pool enriched for binding specificity to KSBcl-2 are overlaid in *red*. The *blue points* are for peptides that were tested in solution binding experiments and showed at least some margin of specificity for KSBcl-2 binding

peptide mutants. Figure 3 shows the scores for the theoretical library in gray scale density plots. The wild-type peptide scores are marked with lines, creating quadrants. The proportion of the library that was predicted to have both greater affinity than wild type for KSBcl-2 and weaker binding to one of two human homologs (Mcl-1 in panel 3a and Bcl-x$_L$ in panel 3b) falls in the lower right quadrant. We went through several rounds of refinement of our choice of preferred and required residues, and additionally edited the codons_output file in order to maximize the number of sequences that fell in this quadrant. Sequences from clones that survived experimental enrichment for KSBcl-2-specific binding are overlaid in red, and a small selection of peptides that were directly tested in solution binding assays and shown to bind preferentially to KSBcl-2 are shown in blue. These pools that are enriched in specific sequences cluster near the lower right quadrant lending support for this approach to library design optimization. For more details, see Foight and Keating [32].

3.5 Alternative Library Optimization Protocols

More complex library optimization strategies can be envisioned. In this section we present an additional set of scripts and example files (in MultiOption.tar.gz) that follow the same basic approach as in Subheading 3.3, but allow consideration of chemical diversity and permit optimization based directly on positional scores. Chemical diversity criteria were used to favor codons with more chemically diverse sets of amino acids in the design of a peptide library to bind specifically to Bfl-1 [31]. In analogy to optimizing the number of sequences that contain only preferred residues, optimization can be done using any score that can be converted to a PSSM, i.e., a table with peptide positions as columns and scores for all 20 amino acids as rows. In the example that we present here, we use a table with the frequencies of the 20 amino acids at 21 peptide positions in a multiple sequence alignment (MSA) of BH3 peptides. This library design favors sequences composed of residues with high frequencies, as outlined below, but users can choose their own PSSM. The library design protocol can be run in six different modes: there are three scoring modes, each of which can be used with or without consideration of chemical diversity. The scoring modes are: (1) the count of sequences composed entirely of preferred amino acids ("preferred score", i.e., that used in Subheading 3.3), (2) the MSA frequencies-based score (here referred to as the MSA score), or (3) a sum of the preferred and MSA scores. The scripts included require input of four criteria (chemical diversity classes, preferred and required amino acids, and another score) regardless of which mode is being used.

3.5.1 Initial Trimming of Codons

1. Make input files for writeCodon_MOp.pl, which does the initial trimming of codons based on size and score(s). Make two

files of preferred and required residues as in Subheading 3.2 (example files file_pref_MOp, file_req_MOp). Make a comma-separated value file (csv) of your PSSM-formatted scores (example file BSA_MSA_table.csv). This can be constructed in Excel and saved as a .csv file. The position names in the column header of the .csv file should be of the format: number, lower-case letter (e.g., '2a'). The position names in the required and preferred files should be of the format: capital letter of wild-type amino acid, number, lowercase letter (e.g. 'E2a'). The regular expressions that recognize the position names in both Perl scripts need to be changed if a different position naming convention is used.

2. Edit the chemical diversity classes set in writeCodon_MOp.pl. Open the file in a text editor and go to the section with the header "# settings for chemical diversity classes". Replace the example positions and sets with your own position names and amino acid sets. Create a line for each peptide position, again, using the same position nomenclature as used in **step 1**. When using the chemical diversity criteria, the ILP solver will count the number of "misses" in chemical diversity classes for each codon. The user will set a constraint on the maximum number of misses to allow across all positions, and the solver will fail if it can not find a solution that meets the library size and chemical diversity misses constraints.

3. Put all of the input files and the script into the same directory (file_pref_MOp, file_req_MOp, BSA_MSA_table.csv, codon_combos.txt, and writeCodon_MOp.pl). In a terminal, in that directory, run:

    perl −w writeCodon_MOp.pl file_req_MOp file_pref_MOp BH3_MSA_table.csv codons_output_MOp

    The file codons_output_MOp is the outputted list of positions and codon selections. An example codon line is: NDS CDEFGHIKLMNQRSVWYZ 24 15.00 0.80 0 0.88. The information included is, from left to right: degenerate codon, amino acids encoded, number of trinucleotides, number of preferred amino acids, the sum of the MSA scores for all of the amino acids included, number of chemical diversity class misses, and the fraction of trinucleotides encoding preferred amino acids.

3.5.2 Run the ILP Optimization

The library optimization script runILP_MOp.pl can be run in the six modes described above. As in Subheading 3.3, it outputs library design that maximizes the score while meeting the library size constraint set by the user, with an optional constraint of number of chemical diversity class misses. The "score" output by runILP is the raw score optimized by the ILP, which varies depending on the mode that you are running. For preferred score only, the score is the same as in Subheading 3.3, $\log_{10}$(number of sequences composed of preferred residues). For MSA score only, the score is $\log_{10}$(product of MSA score for each codon). The MSA score for each codon is the sum of the MSA scores for all amino acids included in that codon. When using both scores, "score" is the sum of both, each on a $\log_{10}$ scale. Note that the MSA score and preferred score may be on different scales, which will affect which score dominates the optimization when using both. If possible, rescale so that the magnitudes are similar, or only use one scoring method at a time.

1. Set up the runILP_MOp.pl script according to the mode you want to run it in. Six lines in the script are preceded by the header "###EDIT###".

    (a) The first line includes the name of the library_design.mod file, which contains the library size and chemical diversity misses constraints; this directs the glpsol. Change this file name to library_design_CD_enabled.mod if using chemical diversity, or library_design_CD_disabled.mod if not using chemical diversity.

    (b) The second line to edit tells the script to use chemical diversity or not. Set the variable "$use_chemical_diversity" equal to 1 if using chemical diversity, or 0, if not.

    (c) Edit the third line to contain the correct path to the glpsol on your machine.

    (d) Set the scoring mode. A set of three lines preceded by "###EDIT###" start at line number 120 in the script. Comment out (add a "#" at the beginning of the line) the two scoring modes that you do not want to use. The choices from top to bottom are: preferred score only, MSA score only, or both.

2. If using chemical diversity, edit library_design_CD_enabled. mod to set the constraint on the maximum number of chemical diversity class misses to allow. Go to the line "subject to trs: sum {v in V} costVTRS[v] * X[v], <= 5.0;" and set the number at the end to the number of misses to allow. For example, for a peptide in which ten positions are being varied, 5–10 misses would be a reasonable place to start. The number of misses will depend on how many chemical diversity classes you set in the writeCodon_MOp.pl script.

3. Run the ILP optimization. Include the following files in the directory you are working in: codons_output_MOp, library_design_CD_enabled/disabled.mod, and runILP_MOp.pl. In a terminal enter:
    perl -w runILP_MultiOp.pl codons_output_MOp library_design_output_MOp

    If the ILP solver finds a solution, the library design will be output to library_design_output_MOp (or whatever you decide to name the output file). If no solution is possible, the library size constraint, the chemical diversity misses constraint, or the numbers of positions varied and required residues should be adjusted. The format of the library design output is the same as shown in Fig. 2, with the exception that the "Score" value will correspond to one of the $\log_{10}$ values as described above, depending on which scoring method is used. The number of sequences composed entirely of preferred amino acids (the "Score" for the original optimization method presented in Subheading 3.3) is given as a separate value. Running this protocol without considering chemical diversity and using only the preferred score is equivalent to the protocol described in Subheading 3.3.

4 Notes

1. When the objective is to obtain a peptide with binding specificity for a target protein over competitor proteins, careful consideration of affinity and specificity trade-offs must be made at both the library design and experimental screening stages. To obtain specificity between very similar target and competitor proteins, you may need to include residues that impart specificity but are predicted to be somewhat disruptive for binding the target. However, at the screening stage, if the stringency for binding to the target is too great, these mutations may not make it through the screen.

2. Computational models can provide hypotheses about binding at positions for which there is no experimental data. Modeling can be especially valuable for specificity predictions, because experimental data may be limited to the peptide positions that are most important for affinity. However, positions other than such conserved "motif" residues, including residues near peptide termini, are often important for specificity [32, 42].

3. The runILP.pl script creates temporary input and output files that are deleted at the end of the script. The temporary input file contains the codon scores and is used by the glpsol. The temporary output file generated by glpsol contains the information that is processed by runILP.pl to create the library design output. If you would like to see these files, comment out the line in runILP.pl below "#Include to delete temporary input and output files".

4. The library_design.mod script is based on a script for integer linear programming optimization of rotamer choice by Kingsford et al. [43].

Acknowledgements

This work was supported by the National Institutes of General Medical Sciences award R01 GM110048 to A.E.K.

References

1. Chen TS, Keating AE (2012) Designing specific protein-protein interactions using computation, experimental library screening, or integrated methods. Protein Sci 21:949–963. doi:10.1002/pro.2096

2. Liu BA, Engelmann BW, Nash PD (2012) High-throughput analysis of peptide-binding modules. Proteomics 12:1527–1546. doi:10.1002/pmic.201100599

3. Levin AM, Weiss GA (2006) Optimizing the affinity and specificity of proteins with molecular display. Mol Biosyst 2:49–57. doi:10.1039/b511782h

4. Olivos HJ, Bachhawat Sikder K, Kodadek T (2003) Quantum dots as a visual aid for screening bead-bound combinatorial libraries. Chembiochem 4:1242–1245. doi:10.1002/cbic.200300712

5. Rezaei Araghi R, Ryan JA, Letai A, Keating AE (2016) Rapid optimization of Mcl-1 inhibitors using stapled peptide libraries including nonnatural side chains. ACS Chem Biol 11:1238–1244. doi:10.1021/acschembio.5b01002

6. Goldsmith M, Tawfik DS (2013) Enzyme engineering by targeted libraries. Methods Enzymol 523:257–283. doi:10.1016/B978-0-12-394292-0.00012-6

7. Hou T, Li N, Li Y, Wang W (2012) Characterization of domain-peptide interaction interface: prediction of SH3 domain-mediated protein-protein interaction network in yeast by generic structure-based models. J Proteome Res 11:2982–2995. doi:10.1021/pr3000688

8. London N, Lamphear CL, Hougland JL, Fierke CA, Schueler-Furman O (2011) Identification of a novel class of farnesylation targets by structure-based modeling of binding specificity. PLoS Comput Biol 7:e1002170. doi:10.1371/journal.pcbi.1002170

9. DeBartolo J, Taipale M, Keating AE (2014) Genome-wide prediction and validation of peptides that bind human prosurvival Bcl-2 proteins. PLoS Comput Biol 10:e1003693. doi:10.1371/journal.pcbi.1003693

10. Sánchez IE, Beltrao P, Stricher F, Schymkowitz J, Ferkinghoff-Borg J, Rousseau F, Serrano L (2008) Genome-wide prediction of SH2 domain targets using structural information and the FoldX algorithm. PLoS Comput Biol 4:e1000052. doi:10.1371/journal.pcbi.1000052

11. Fernandez-Ballester G, Beltrao P, Gonzalez JM, Song Y-H, Wilmanns M, Valencia A, Serrano L (2009) Structure-based prediction of the Saccharomyces cerevisiae SH3-ligand interactions. J Mol Biol 388:902–916. doi:10.1016/j.jmb.2009.03.038

12. Fu X, Apgar JR, Keating AE (2007) Modeling backbone flexibility to achieve sequence diversity: the design of novel alpha-helical ligands for Bcl-xL. J Mol Biol 371:1099–1117. doi:10.1016/j.jmb.2007.04.069

13. Roberts KE, Cushing PR, Boisguerin P, Madden DR, Donald BR (2012) Computational design of a PDZ domain peptide inhibitor that rescues CFTR activity. PLoS Comput Biol 8:e1002477. doi:10.1371/journal.pcbi.1002477

14. Zheng F, Jewell H, Fitzpatrick J, Zhang J, Mierke DF, Grigoryan G (2015) Computational design of selective peptides to discriminate between similar PDZ domains in an oncogenic pathway. J Mol Biol 427:491–510. doi:10.1016/j.jmb.2014.10.014

15. Sammond DW, Bosch DE, Butterfoss GL, Purbeck C, Machius M, Siderovski DP, Kuhlman B (2011) Computational design of the sequence and structure of a protein-binding peptide. J Am Chem Soc 133:4190–4192. doi:10.1021/ja110296z

16. Fleishman SJ, Whitehead TA, Ekiert DC, Dreyfus C, Corn JE, Strauch E-M, Wilson IA, Baker D (2011) Computational design of proteins targeting the conserved stem region of influenza hemagglutinin. Science 332:816–821. doi:10.1126/science.1202617

17. Procko E, Hedman R, Hamilton K, Seetharaman J, Fleishman SJ, Su M, Aramini J, Kornhaber G, Hunt JF, Tong L, Montelione GT, Baker D (2013) Computational design of a protein-based enzyme inhibitor. J Mol Biol 425:3563–3575. doi:10.1016/j.jmb.2013.06.035

18. Strauch E-M, Fleishman SJ, Baker D (2014) Computational design of a pH-sensitive IgG

binding protein. Proc Natl Acad Sci U S A 111:675–680. doi:10.1073/pnas.1313605111

19. London N, Raveh B, Schueler-Furman O (2013) Peptide docking and structure-based characterization of peptide binding: from knowledge to know-how. Curr Opin Struct Biol 23:894–902. doi:10.1016/j.sbi.2013.07.006

20. Parker AS, Griswold KE, Bailey-Kellogg C (2011) Optimization of combinatorial mutagenesis. J Comput Biol 18:1743–1756. doi:10.1089/cmb.2011.0152

21. Pantazes RJ, Saraf MC, Maranas CD (2007) Optimal protein library design using recombination or point mutations based on sequence-based scoring functions. Protein Eng Des Sel 20:361–373. doi:10.1093/protein/gzm030

22. Chen MMY, Snow CD, Vizcarra CL, Mayo SL, Arnold FH (2012) Comparison of random mutagenesis and semi-rational designed libraries for improved cytochrome P450 BM3-catalyzed hydroxylation of small alkanes. Protein Eng Des Sel 25:171–178. doi:10.1093/protein/gzs004

23. Treynor TP, Vizcarra CL, Nedelcu D, Mayo SL (2007) Computationally designed libraries of fluorescent proteins evaluated by preservation and diversity of function. Proc Natl Acad Sci 104:48–53. doi:10.1073/pnas.0609647103

24. Verma D, Grigoryan G, Bailey-Kellogg C (2015) Structure-based design of combinatorial mutagenesis libraries. Protein Sci 24:895–908. doi:10.1002/pro.2642

25. Jacobs TM, Yumerefendi H, Kuhlman B, Leaver-Fay A (2015) SwiftLib: rapid degenerate-codon-library optimization through dynamic programming. Nucleic Acids Res 43:e34–e34. doi:10.1093/nar/gku1323

26. Mena MA, Daugherty PS (2005) Automated design of degenerate codon libraries. Protein Eng Des Sel 18:559–561. doi:10.1093/protein/gzi061

27. Allen BD, Nisthal A, Mayo SL (2010) Experimental library screening demonstrates the successful application of computational protein design to large structural ensembles. Proc Natl Acad Sci 107:19838–19843. doi:10.1073/pnas.1012985107

28. Hayes RJ, Bentzien J, Ary ML, Hwang MY, Jacinto JM, Vielmetter J, Kundu A, Dahiyat BI (2002) Combining computational and experimental screening for rapid optimization of protein properties. Proc Natl Acad Sci U S A 99:15926–15931. doi:10.1073/pnas.212627499

29. Guntas G, Purbeck C, Kuhlman B (2010) Engineering a protein-protein interface using a computationally designed library. Proc Natl

Acad Sci U S A 107:19296–19301. doi:10.1073/pnas.1006528107

30. Chen TS, Palacios H, Keating AE (2013) Structure-based redesign of the binding specificity of anti-apoptotic Bcl-x(L). J Mol Biol 425:171–185. doi:10.1016/j.jmb.2012.11.009

31. Dutta S, Chen TS, Keating AE (2013) Peptide ligands for pro-survival protein Bfl-1 from computationally guided library screening. ACS Chem Biol 8:778–788. doi:10.1021/cb300679a

32. Foight GW, Keating AE (2015) Locating herpesvirus Bcl-2 homologs in the specificity landscape of anti-apoptotic Bcl-2 proteins. J Mol Biol. doi:10.1016/j.jmb.2015.05.015

33. Boersma M, Sadowsky J, Tomita Y (2008) Hydrophile scanning as a complement to alanine scanning for exploring and manipulating protein-protein recognition: application to the Bim BH3 domain. Protein Sci 17:1232–1240. doi:10.1110/ps.032896.107

34. Fowler DM, Araya CL, Fleishman SJ, Kellogg EH, Stephany JJ, Baker D, Fields S (2010) High-resolution mapping of protein sequence-function relationships. Nat Methods 7:741–746. doi:10.1038/nmeth.1492

35. Fowler DM, Fields S (2014) Deep mutational scanning: a new style of protein science. Nat Methods 11:801–807. doi:10.1038/nmeth.3027

36. London N, Gullá S, Keating AE, Schueler-Furman O (2012) In silico and in vitro elucidation of BH3 binding specificity toward Bcl-2. Biochemistry 51:5841–5850. doi:10.1021/bi3003567

37. DeBartolo J, Dutta S, Reich L, Keating AE (2012) Predictive Bcl-2 family binding models rooted in experiment or structure. J Mol Biol 422:124–144. doi:10.1016/j.jmb.2012.05.022

38. Araya CL, Fowler DM, Chen W, Muniez I, Kelly JW, Fields S (2012) A fundamental protein property, thermodynamic stability, revealed solely from large-scale measurements of protein function. Proc Natl Acad Sci U S A 109:16858–16863. doi:10.1073/pnas.1209751109

39. Fowler DM, Araya CL, Gerard W, Fields S (2011) Enrich: software for analysis of protein function by enrichment and depletion of variants. Bioinformatics 27:3430–3431. doi:10.1093/bioinformatics/btr577

40. Starita LM, Young DL, Islam M, Kitzman JO, Gullingsrud J, Hause RJ, Fowler DM, Parvin JD, Shendure J, Fields S (2015) Massively parallel functional analysis of BRCA1 RING domain variants. Genetics. doi:10.1534/genetics.115.175802

41. Sirin S, Apgar JR, Bennett EM, Keating AE (2015) AB-Bind: antibody binding mutational database for computational affinity predictions. Protein Sci. doi:10.1002/pro.2829

42. Stein A, Aloy P (2008) Contextual specificity in peptide-mediated protein interactions. PLoS One 3:e2524. doi:10.1371/journal.pone.0002524

43. Kingsford CL, Chazelle B, Singh M (2005) Solving and analyzing side-chain positioning problems using linear and integer programming. Bioinformatics 21:1028–1036. doi:10.1093/bioinformatics/bti144

Part IV

Design of Inhibitory Peptides

Chapter 14

Investigating Protein–Peptide Interactions Using the Schrödinger Computational Suite

Jas Bhachoo and Thijs Beuming

Abstract

The Schrödinger software suite contains a broad array of computational chemistry and molecular modeling tools that can be used to study the interaction of peptides with proteins. These include molecular docking using Glide and Piper, relative binding free energy predictions with FEP+, conformational searches using MacroModel and Desmond, and structural refinement using Prime and PrimeX. In this review we provide a comprehensive overview of these tools and describe their potential application in the identification and optimization of peptide ligands for proteins.

Key words Peptides, Docking, Glide, Prime, Piper, Free Energy, Conformational search, Molecular dynamics, Protein refinement

1 Introduction

The interest in oligo- and polypeptides as therapeutics has reemerged in recent years, as a result of the rising need to inhibit protein–protein interactions and other difficult targets, coupled with advances in screening and synthesis technologies that have made peptides increasingly viable as drug candidates. Over recent years, the discovery of small-molecule therapeutics has benefited from the availability of accurate computational methods that predict the binding mode and affinity of ligands to their protein targets. With increasing focus on the development of peptide-based therapeutics, including peptidomimetics and macrocycles, there is a growing need to extend computational technologies to include these types of molecules as well, which tend to be larger and more conformationally flexible than traditional drug-like molecules. The Schrödinger software platform provides a plethora of methods for the identification and optimization of small molecules against protein targets, and many of these can be applied and/or adapted for use in peptide drug discovery. In this review we will describe

Ora Schueler-Furman and Nir London (eds.), *Modeling Peptide-Protein Interactions: Methods and Protocols*, Methods in Molecular Biology, vol. 1561, DOI 10.1007/978-1-4939-6798-8_14, © Springer Science+Business Media LLC 2017

several of those tools, including molecular docking, conformational analysis, and QSAR. The tools described here are summarized in Table 1.

2 Methods

2.1 Docking

Structure-based modeling of novel ligands requires the prediction of binding modes for these ligands against their putative targets using molecular docking, which can provide information about key interactions and targets for optimization. In addition, docking calculations can be used to separate actives from inactives in virtual screening campaigns. The program Glide is among the most widely used and validated small-molecule docking algorithms [1]. In a recent paper, we described how peptide-specific modifications of the docking algorithm can expand the applicability of Glide to oligo-peptides, and demonstrated that this modified approach performs promisingly when applied to a peptide docking benchmark set [2]. Below, we provide a step-by-step description of how the

Table 1
Compendium of tools in the Schrödinger suite that can be used for investigating protein–peptide interactions

Category	Name	Function
Data preparation	Protein Preparation Wizard	Assignment of protein and peptide bond orders, tautomeric and ionization states
Docking	Glide	Rigid receptor docking
	Induced Fit Docking	Flexible receptor docking
	Piper	Protein–Protein Docking
Scoring	GlideScore	Docking and virtual Screening
	MM/GBSA	Implicit solvation model for accurate scoring of congeneric and diverse peptides
	WaterMap	Scoring using explicit water, comparing congeneric peptides
	WM/MM	Hybrid of MM/GBSA and WaterMap
	FEP+	Free Energy Perturbation
Conformational searches	MacroModel	Molecular Mechanics based peptide conformations
	Desmond	Explicit solvent Molecular Dynamics simulations
	Prime	Rapid prediction of tethered peptides
Statistical methods	Peptide QSAR/Canvas	Descriptor-based modeling of peptide activity

Glide docking process is applied to peptides, including protein and peptide preparation steps, and docking postprocessing calculations that can be used to refine the results.

2.1.1 Docking Using Glide

A typical peptide docking workflow using Glide consists of the following sequential steps: (1) protein and peptide preparation, (2) grid setup, (3) peptide docking, and (optionally) (4) docking postprocessing.

1. *Protein and peptide preparation*: Accurate prediction of peptide-binding modes requires accurate representation of both the peptide and the protein, which is handled using the Protein Preparation Wizard in Maestro [3, 4]. A crucial step in this preparation is the assignment of rotameric states for peptide residues with polar hydrogen atoms, such as Ser, Thr, Tyr, Asn, and Gln, as well as the orientations of water molecules, to optimize the hydrogen bonding network in the protein structure. In addition, tautomeric/ionization states of His residues are evaluated using the same criteria, and the protonation states of ionizable residues are predicted using the PROPKA algorithm [5]. Extensive benchmarking studies have shown the importance of adequate preparation of the protein structure for accurate small molecule virtual screening results [4]. In addition to preparation of the protein, the tautomeric and ionization states of the peptide needs to be enumerated, and in the case of peptides containing His residues it may be advisable to dock multiple versions of the same peptide with different tautomeric states. Finally, the user needs to decide whether to treat the peptide termini as zwitter-ionic, capped, or neutral. Preparation is required even for proteins where a crystal structure is available, since the positions of the hydrogen atoms are rarely observable in such structures.

2. *Grid setup*: Glide docking is a two-step process, with the initial step the setup of a protein grid, which represents pre-calculated properties of the protein, primarily those related to electrostatics and the van der Waals nonbonded interactions. This grid representation of the protein significantly speeds up the docking calculations, but reflects a mostly rigid model of the protein—hence the importance of properly predicting the hydrogen bonding network in the protein preparation step.

   The grid is generated to enclose the expected binding region (typically inferred from the location of a known ligand in a crystallographic structure). There are numerous advanced options that can be applied during the process of grid generation. These include: scaling back of the van der Waals radii in order to approximate the induced fit effect by allowing a small amount of overlap between the docked peptide and protein; the definition of multiple and diverse receptor-constraints; the flexible treatment of hydroxyls; the control of parameters in the Glide filtering funnel; and, finally, control over the output, in terms of the number and quality of final poses delivered. For an average sized protein (5000–6000 atoms) a grid generation step typically takes 2–3 min and produces a .zip file that contains all of the relevant receptor information for docking. For docking of larger molecules (e.g. peptides) using Glide, it is recommended one prepare a grid file specifically generated for docking of peptides, which increases the dimensions of the grid box to be more appropriate for larger molecules.

3. *Glide Sampling and Scoring Precision*: The underlying docking methodology in Glide has been described in several papers [6–8]. In short, running Glide involves initial generation of a conformational ensemble for the ligand followed by filtering of millions of possible poses using crude geometric and energetic criteria. Next, the evaluation of successively smaller sets of poses is performed using scoring functions of increasing complexity and accuracy. The protein is kept fixed during the entire protocol and only peptide orientational and torsional spaces are sampled.

   Input for a typical peptide docking calculation requires the following components: the grid file prepared in the previous stage and one or more prepared peptides to dock. Once again, numerous advanced options can be specified, such as the choice of scoring function (HTVS, SP, XP, or the peptide-specific SP-Peptide mode—described below), optional protein–ligand constraints, reducing the van der Waals radii specific to the ligands, etc. The time required for docking each ligand is dependent on the chosen scoring function, as well as the size and flexibility of ligands being studied.

   For the traditional application of Glide to small molecules, there are three main levels of scoring precision in Glide that can be deployed depending upon the numbers of compounds to screen, the available computational resources, and the desired turnaround time. Glide is highly parallelized, to the point where each ligand can be run on a separate processor. Glide has been successfully run on more than 1000 processors simultaneously using the Amazon Cloud. The High Throughput Virtual Screening (HTVS) mode is designed to screen the most compounds (typically in the millions for in-house pharmaceutical databases or commercially available compounds), while the Standard Precision (SP) mode is about ten times slower than HTVS and therefore suitable for fewer compounds while Extra Precision (XP) is even more precise and more computationally intensive and so is suitable for more narrow screening, typically near the end of a virtual screening campaign. The SP scoring function (called GlideScore) has been optimized using virtual screening data and is also used in the HTVS mode of Glide, while the XP GlideScore includes a number of penalties and rewards to capture specific protein–ligand recognition motifs. The inclusion of penalty terms in the XP scoring function, which is important for eliminating false positives (inactives), requires additional sampling, thereby increasing the computational costs. Considering these factors, the HTVS mode typically takes 1–2 s per ligand, SP takes 10–20 s per ligand and XP takes 5–10 min per ligand, with the range depending primarily on the number of rotatable bonds in the ligand. Finally, it is worth noting that Glide uses the so-called "Emodel" scoring function to compare poses for the same ligand, while the GlideScore is used to compare different compounds based on a rough predicted affinity.

For the docking of peptides, a version of the SP scoring function was developed that significantly increases the nature and extent of sampling. Several internal parameters that control the Glide funnel width (specifically the number of conformations that enter and exit the rough-scoring stage of Glide) were systematically changed and optimized to maximize docking accuracy for a set of 19 non-α helical peptides [2]. Those optimized parameters are exposed to the end-user in the SP-peptide mode of Glide. Examples of such peptides are shown in Fig. 1, and show varying degrees of accuracy with which they have been docked.

Docking multiple input conformations. Glide results depend on the bond length and bond angles of the input conformation of the ligand [9], since these degrees of freedom are not sampled with the docking algorithm (torsions are the only internal degree of freedom sampled). Hence, it is possible to increase the extent of sampling by running Glide multiple times using

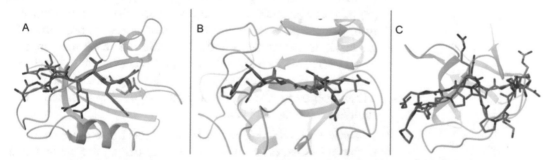

Fig. 1 Three examples of peptide docked using Glide, as described previously [2]. The protein structure is shown in *cyan ribbons*, peptide structures in *orange*, and predicted binding modes in *green*. (**a**) Typical example of a well-docked peptide using the Glide SP-Peptide mode. The interaction between the C-terminal peptide of Liprin with the sixth PDZ domain of GRIP1 (PDB 1N7F) is predicted with high accuracy. (**b**) The central portion of the predicted HIV-1 capsid protein binding to cyclophilin A (PDB 1AWR) is predicted accurately, while there are differences for the termini. This is partially due to the intrinsic flexibility of these peptides as revealed by their high B-factors. (**c**) Peptides containing poly-proline motifs are difficult to dock due to a high degree of internal symmetry

different input conformations and aggregating the results. If the user wants to remove the input-dependence, it is possible to run Glide with a canonicalization option to regularize the input geometry.

4. *Docking postprocessing*: In the case of docking peptides, it has been demonstrated that one can significantly improve docking results by postprocessing each pose using a Molecular Mechanics Generalized Born Surface Area (MM-GSBA) approach [2]. A more detailed description of the MM-method is given below in Subheading 2.2.1. In the Schrödinger suite this functionality can be accessed from the command line or via the MM-GBSA GUI.

5. *A convenient workflow for docking of peptides in the BioLuminate module*: The BioLuminate Suite provides a dedicated set of tools for the application of computational chemistry to biologics systems, including peptides, antibodies, and others. The peptide docking GUI in BioLuminate can be used to carry out the various steps of the docking workflow outlined above, including (1) grid generation, (2) peptide preparation (including directly from sequence data), (3) generation of multiple conformers, and (4) postprocessing using MM-GBSA.

6. *Retrospective docking accuracy and regime of applicability for docking of peptides using Glide*: A recent publication described the application of the Glide peptide docking protocol to a benchmark set of peptide data. This study demonstrated that Glide does reasonably well in docking small (up to eight residues), linear, and relatively nonpolar peptides [2]. Glide produces docked poses of significantly lower accuracy when docking peptides that are larger, nonextended (e.g. peptides with an internal hydrogen bond or significant degree of folding) or highly charged. α-helical peptides can be docked, but require constraints in the conformation generation stage to ensure relevant poses are produced. In general, success for peptide docking still has to be measured in terms of its ability to reproduce the correct pose in an ensemble of solutions (i.e. the best pose appears among the top ten results) rather than by examining the accuracy of the top pose only, as is standard for small molecules.

7. *Induced Fit Docking*: The Glide-based methodologies outlined below are based on sampling of the peptide in the context of a rigid receptor structure. Treatment of receptor flexibility can be performed using the Induced-Fit Docking (IFD) workflow [10]. This protocol consists of an initial docking stage with a softened van der Waals potential to generate an ensemble of poses, followed by receptor optimization around each of the initial ligand/protein complexes using the Prime protein structure prediction program [11], and finally another round of docking using standard Glide settings into these optimized

protein-binding sites. Select side chains can also be removed in the initial docking stage to allow for greater exploration of poses within the receptor-binding site. The IFD method has been shown to correctly model induced-fit effects in a large number of small molecule docking cases where rigid receptor docking fails [10]. The IFD approach is significantly slower than rigid receptor Glide docking, and is typically used for pose prediction and generating induced-fit receptor structures, and not for high-throughput applications such as virtual screening. The SP-Peptide scoring function in Glide is compatible with the IFD workflow.

2.1.2 Docking Using Piper

Background: Glide imposes an upper limit in terms of number of atoms and rotatable bonds that can be handled with conformational flexibility, and as such is appropriate only for docking of small peptides up to a length of around 12 residues (depending on the number of rotatable bonds in the side chains). For larger peptides or even protein domains, the rigid protein–protein docking algorithm Piper should be employed [12]. For peptides for which an accurate structure is not available, conformational variability can be introduced into the process by docking a peptide multiple times with different pre-generated conformations. The conformational ensemble can be generated by various means, such as with MacroModel [13].

Piper is a highly regarded program for the docking of peptides and proteins, with a reputation built on blinded-competitions such as CAPRI [14, 15], where the automated version of Piper (ClusPro) consistently outperforms all other fully automated approaches to protein docking. Within the Schrödinger Suite, Piper powers the Protein–Protein Docking panel of the BioLuminate program and can be used to perform rigid docking of two proteins, or of a peptide to a protein. Because Piper docks the proteins as rigid bodies, it is capable of docking peptides or proteins of significant size, and unlike Glide, is not limited to relatively modest oligo-peptide chains. However, since Piper only samples the relative orientation of the input peptide (in a fixed conformation) with respect to the protein it requires either (1) knowledge of the peptide structure from experimental techniques such as NMR or X-ray crystallography (and then only if the peptide can be reasonably described by a single conformer) or (2) docking of multiple pre-calculated peptide conformations and aggregating the results. While the Piper methodology has not been benchmarked for accuracy in retrospective docking calculations of peptides, it currently provides the only possible method for docking large peptides (i.e. with more then 500 atoms) using the Schrödinger suite, and as such a brief description is included here.

Methodology: Piper is carried out in two steps: conformational sampling, followed by structural clustering to identify and rank likely docked protein poses. For the first step, Piper employs an efficient FFT (Fast Fourier Transform) approach that makes it possible to evaluate a large number of poses in a computationally efficient manner—reducing the required compute time (on a single processor) from days to a few hours. Relative poses are generated and scored using a simple atomistic energy function that can efficiently separate potentially acceptable poses from those that are very unlikely. Typically, 70,000 poses are evaluated, and from these, the 1000 best scoring poses are kept for the second step. In the second step, the 1000 poses are clustered on the basis of structure. Then, from each cluster, the member of the cluster with the most near neighbors is taken as representative of that cluster. Finally, the selected poses from the clusters are presented to the user, and rank ordered based on the size of the cluster from where they were obtained (so the highest ranked pose comes from the largest cluster). Piper also allows for the addition of a biasing term to reflect any knowledge that the user may have regarding residues that, either, are, or, are not believed to be region at which the proteins interact. This biasing term is applied when scoring the generated poses, so it will influence the reduced set of 1000 structures which are ultimately clustered, but it does not affect the efficiency of the sampling itself.

2.1.3 Structural Refinement Using Prime and PrimeX

The Prime program contains a number of tools for sampling and scoring of peptide-protein complexes [11]. Indeed, the problem of predicting the interactions of peptides with proteins can be thought of to be similar to prediction of loop structures, a demonstrated strength of Prime [16–18]. In this context, it is possible to perform an ab initio loop prediction on a portion of peptide, keeping part of the structure fixed (e.g. the N- or C-terminal residue) while allowing sampling of the remainder. This can be useful in the case of peptides interacting with protein domains where parts of the peptide interact with the protein in a highly conserved manner, i.e. the coordination of the C-terminus of peptides by the GLGF loop in PDZ domains [19]. In that case one would orient the C-terminal residue by superposition on a suitable complex template, and sample all other residues using Prime loop prediction. Protein flexibility, including binding-site loop movements, can also be easily incorporated into any Prime sampling strategy. In cases where a relatively good model for the starting conformation is available, Prime offers facilities for focused local optimization starting with that model—including hybrid Monte Carlo and rigid body optimizations.

An interesting extension of Prime (PrimeX) allows incorporation of X-ray crystallographic density during structure determination. PrimeX [20] is an all-atom (including hydrogens) crystal structure refinement package. The PrimeX objective function for optimization combines a weighted sum of the force field energy and the experimental density, with the weight varying depending on the quality of the electron density. All-atom refinement with PrimeX has been shown to produce improved crystal structure models with respect to nonbonded contacts and yields changes in structural details that can dramatically impact the interpretation of some protein–ligand interactions [20]. PrimeX features maximum-likelihood reciprocal-space minimization, simulated-annealing refinement, ligand placement, loop building, and side-chain placement. PrimeX is integrated with the Maestro molecular graphics program, which provides an easy-to-use graphical interface. In addition, command-line access to PrimeX tools provides for their scripting into complex workflows.

2.2 Affinity Prediction

2.2.1 Molecular Mechanics-Generalized Born Solvation Approximation Method (MM-GBSA)

MM-GBSA is a popular scoring approach that can be used to predict relative binding free energies of small molecules to proteins, providing in many cases an acceptable level of accuracy for ranking ligands to a protein target [21, 22]. Unlike more rigorous and time-consuming methods such as molecular dynamics or free-energy perturbation (see below), MM-GBSA allows one to profile potential molecules for future rounds of synthesis with limited computational resources. Data sets that can be evaluated by MM-GBSA can range from tens to thousands of ligands in a lead optimization campaign. Since MM-GBSA energies are evaluated for a static model, each ligand must be pre-oriented prior to running MM-GBSA. Typically, molecular docking using Glide is sufficient to obtain reasonable poses for MM-GBSA postprocessing.

The basic form of the MM-GBSA calculation is described below: the free energy of binding is related to the difference in the free energy of the complex, minus that of the free receptor, minus that of the free ligand.

$$\Delta G(\text{binding}) = \Delta G(\text{complex}) - \Delta G(\text{free receptor}) - \Delta G(\text{free ligand})$$

These energetic terms are approximated by a summation over E_{MM}, E_{GB}, and E_{SA}, where E_{MM} is the sum over the bonded (internal) and nonbonded electrostatic energies, and $E_{GB} + E_{SA}$ represent the polar and nonpolar components of the solvation free energy, respectively. E_{GB} is evaluated using the generalized Born model [23, 24], while E_{SA} is determined from the solvent accessible surface area.

MM-GBSA incorporates an advanced solvation model, the variable dielectric model VSGB [23, 24], which reflects heterogeneous polarization in the protein environment. Rather than treating the protein with a single internal dielectric constant (traditionally set to 1, 2, or 4), it auto-assigns distinct dielectric contributions to different polar residues. While in practice MM-GBSA energies are typically evaluated using the default implicit water solvent model, additional models for solvents such as chloroform are also available. Over time, additional terms have been included to the basic form of the equation in order to augment it with the finer details of protein–ligand interactions, including physics-based corrections for hydrogen bonding, π–π interactions, self-contact interactions and hydrophobic interactions [25].

MM-GBSA can be easily setup from the GUI in Maestro. The prerequisite is a set of pre-docked ligands and a single receptor structure. The protocol can be run in a minimization mode or a conformational searching mode, and one can include the ligands alone and or parts of the receptor as part of the search. The user has the option to adjust the level of protein flexibility desired and to select the sampling method deemed necessary.

As described above, rescoring docked poses using MM-GBSA has been shown to improve pose prediction accuracy in a recent comprehensive peptide docking benchmark [2]. In addition, MM-GBSA can be used to predict the effect of amino acid substitutions on the affinity of a peptide for a protein [26]. For that purpose, the Residue Scanning functionality in the BioLuminate modeling platform allows for the rapid set up and analysis of many simultaneous mutations [27].

2.2.2 Free Energy Perturbation Using FEP+

The most rigorous approaches to the prospective estimation of protein–ligand affinity are free energy simulations, including free energy perturbations (FEP). In drug discovery lead optimization projects, the calculation of binding affinities relative to a lead or reference ligand with experimental data is sufficient (i.e. absolute energies are not needed once one reference activity is known), and affords significant reduction in computational effort as compared to absolute binding free energy calculations. We recently reported a FEP protocol called FEP+, that achieves an unprecedented level of accuracy across a broad range of target classes, with validation results encompassing more than 200 ligands and a variety of chemical perturbations, many of which are large by the standards of prior relative binding free energy calculations and involve major changes in ligand chemical structures [28, 29]. Furthermore, in this same study, the method was applied in a prospective fashion in two drug discovery projects; the results are consistent with those obtained from the retrospective studies and show the ability of this approach to drive decisions in lead optimization. While FEP has been applied to molecular systems for nearly three decades, it is only recently that this approach has become both reliable and fast enough to be generally useful for drug discovery. Recent improvements have included faster computers, massive parallelization, improved force fields, and algorithmic improvements that improve sampling efficiency. While the traditional application of FEP has been for predicting the relative free energy of binding of small molecules to proteins, the FEP method is equally applicable to peptide-protein complexes.

2.2.3 WaterMap

A major contribution to the free energy of binding comes from the displacement of water from the surfaces of the protein and the ligand. The magnitude of this effect depends highly on the thermodynamic properties of the water molecules prior to displacement. These properties in turn depend on the environment of the bound water molecules. For example, water molecules evacuated from a largely hydrophobic environment have an enthalpically favorable effect on binding, while water molecules tightly constrained by a number of hydrogen bonds can produce a strongly favorable entropic effect. The WaterMap algorithm [30] has been developed to quantify this effect using explicit molecular dynamics simulations (MD). In brief, water molecules from an MD trajectory are spatially clustered into hydration sites and their interaction free energies are determined using inhomogeneous solvation theory [31, 32].

WaterMap has been shown to accurately predict the affinities of peptides for several PDZ domains [33]. For example, the high affinity of a peptide with Trp at the P_{-1} position in Erb2 for the Erbin PDZ domain was shown to be due to an effective displacement of a hydration site from the surface of the peptide-binding sub-pocket (Fig. 2a). Peptides with a smaller residue at P_{-1} were not able to displace this water from the sub-pocket. A quantitative analysis showed that the energetics of hydration sites in the peptide-binding pocket were highly correlated with peptide affinity (Fig. 2b). Finally, the equivalent hydration site in the HTRA family of PDZ domains was shown to be a mediator of peptides selectivity (Fig. 2c).

2.3 Peptide Conformational Searches

2.3.1 Sampling with MacroModel

Conformational searching is a key method for generating ensembles of 3D structures and obtaining energies associated with structures in that ensemble. The usual aim is to attain the low energy minimum equivalent to the bioactive conformation, or to attain a collection of conformations that provide information on the degree of flexibility/motion of a molecule. In the context of predicting peptide- or peptidomimetic-binding modes using docking, it is often necessary to perform a conformational search using MacroModel [13] prior to docking. For instance in the case of large, flexible complex cyclic peptides or synthetic macrocycles there are high energy barriers that cannot be sampled adequately within Glide itself, and require additional sampling outside of the docking workflow.

Depending on the application, there are a wide variety of different searching modes available within MacroModel, which can be applied to many types of molecules: small drug-like molecules; peptides of varying length; macrocycles of varying complexity; protein active sites, entire protein domains, or a combination of the above in the case of protein/ligand complexes. Search methods fall into two categories, the first a search in torsional space, randomly or systematically varying torsions in the molecule of

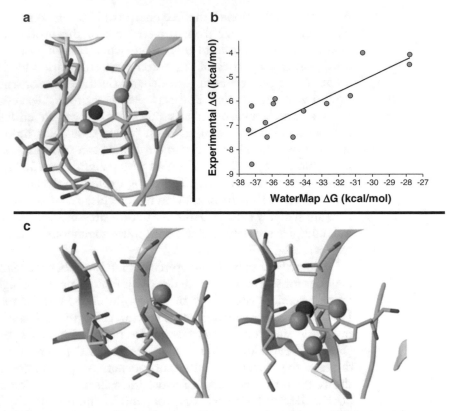

Fig. 2 WaterMap explains peptide affinity and selectivity for PDZ domains [33]. *Red* and *orange spheres* correspond to high-energy hydration sites on the protein surface, leading to high-affinity binding when displaced by a ligand. (**a**) In the Erbin PDZ domain, several high-energy hydration sites in the P-1 pocket are effectively displaced by a Trp in the phage-display optimized peptide [34]. (**b**) WaterMap scores correlate well with measured peptide affinities for a series of Erbin-binding peptides. (**c**) The presence or absence of these sites in HTRA proteins explains relative preference for Trp residue in HTRA2 (*right*) but not HTRA1 (*left*)

interest, and the second a low-mode-based search, where the intrinsic vibrational modes of the molecule are used to generate alternative conformations. Importantly, a combination of the two methods (alternating between torsional sampling and low-mode sampling) is a powerful combination for jumping on the potential energy surface, and has been shown to be extremely effective in generating a representative ensemble of structures for many types of ligands [35]. MacroModel conformational searches are typically performed using an implicit representation of the solvation using generalized Born methods. In addition, a wide range of force fields are supported, including the recent OPLS3 force field [36], which has improved parameters for proteins relative to previous versions of the OPLS force field (see below). Finally, the ConfGen algorithm [37, 38] is a type of MacroModel conformational search that has been developed to generate drug-like (i.e. extended) conformations of small molecules. If one is reasonably certain that the bioactive conformation of a peptide is extended (i.e. in the case of PDZ domains), the ConfGen algorithm provides a convenient and rapid conformational search solution.

Within MacroModel, various types of calculations are available, ranging from simple tasks like "current energy" and "coordinate scan" to more complex procedures such as conformational sampling and dynamics. Options for the Potential present a choice of force fields, solvents, and atomic charge model, while constraint options allow users to easily place atom, distance, angle, and torsional force constants as needed. MacroModel also allows force constants to be applied over selected parts of a protein–ligand complex in order to build up shells of varying flexibility. The substructure facility in MacroModel can be used to easily define regions as flexible, fixed, and frozen atoms for use in various molecular mechanics simulations.

Sampling Macrocycles. Of late, considerable effort has been expended to determine how to better explore the conformational spaces of macrocycles. There is growing commercial interest in macrocycles as potential drugs, particularly as inhibitors of protein–protein interactions. Macrocycles are both diverse and highly flexible, and benefit from "cyclizing" to restrict conformational space. A dedicated search protocol for these complex systems has been developed based on a unique combination of large-scale low-mode search steps with simulation cycles composed of alternating stochastic dynamics and minimization calculations. All of these options are available separately in MacroModel, but for ease of use have been combined into one convenient script (macrocycle_conformational_sampling.py). In addition to being the method of choice for sampling macrocycles, this process can also be used very effectively for sampling peptide structures. Thorough conformational exploration of a macrocycle can be useful for characterizing the internal hydrogen bond networking for the macrocycle, and this can be tied to such phenomena as cell permeability [39]. Extensive conformational exploration is also relevant to docking macrocycles, because Glide uses a look-up table for ring conformations, rather than performing an *ab initio* search as it does for rotatable torsions, and as such docking macrocycles into protein structures requires ring conformations to be precomputed, ideally with the custom protocol described above.

To illustrate the complexity of these molecules, Fig. 3a shows an example of a macrocycle found to bind to BACE-1 (PDB code 2QZK), with the computed conformation (purple) superimposed onto the crystal structure (green). The native ligand buries the macrocycle ring into a mostly enclosed pocket, which in turn requires a particular ring conformation that forms key interactions to ligand carbonyl and protonated amine moieties. With MacroModel derived ring templates, a suitable macrocycle conformation is found and the docked pose has a low RMSD of 0.22 Å. Without the ring templates, even with a visually similar ring conformation (of RMSD = 0.71 Å for macrocycle ring atoms), the old protocol positions the ring outside the pocket leading to a very large RMSD of 10.23 Å (data not shown). Another example of the power of the augmented macrocycle ring templates is the correct conformational prediction (Fig. 3b, left) and subsequent docking of a macrocycle bound to α-β tubulin (PDB code 1JFF) (Fig. 3b, right).

2.3.2 Analyzing Peptide Secondary Structure Using Molecular Dynamics

The Desmond program [40] is a state-of-the-art explicit solvent Molecular Dynamics (MD) program that can be used for a variety of different simulations, in addition to being the driver for MD-based solutions such as FEP+ and WaterMap. In the case of peptides, MD can be used to effectively study conformational properties when implicit solvent search techniques, as implemented in MacroModel, are insufficient.

A recently described application for MD is the prediction of α-helical content of so called "stapled" peptides [41, 42]. These synthetically rigidified α-helical peptides have promising pharmacokinetic, metabolic stability, and cell-penetrating properties, which has sparked some interest in their development for a plethora of therapeutic targets that have otherwise been deemed "undruggable" by more conventional small-molecule strategies. Determining the stability of these α helical peptides is important for their use as therapeutic agents.

Desmond replica exchange simulations [43] were employed to study a series of stabilized stapled α-helical peptides over a range of temperatures in solution. In such a calculation, multiple simultaneous simulations of a single system are performed, using small variations in simulation conditions (e.g. temperature). Swapping conformations among the different replicas allows for very effective crossing of high-energy barriers.

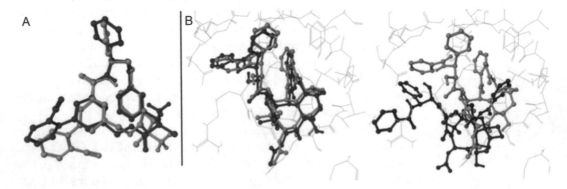

Fig. 3 Conformation and docking of macrocycles. (**a**) 2QZK. A computed macrocycle conformation (*purple*) of 2QZK overlaid with the crystal structure (*green*) showing a low RMSD of 0.22 Å. (**b**) Comparison of the 1JFF protein with the crystallographic ligand conformation (*green*) overlaid with Glide's top-ranked docked conformation (*blue*). On the *right* is the top-ranked pose obtained without a suitable template while on the *left* is the much more accurate top-ranked pose with a template

The predicted α-helical propensities derived from the simulations were in good agreement with the experimentally observed circular dichroism (CD) melting curves. The local flexibility of key residues could be related to differences in relative affinities of the stapled peptides for MDM2. These simulations provided new insights into the design of α-helical stapled peptides and the development of potent inhibitors of α-helical protein–protein interfaces.

Another enhanced sampling technique to model experimental melting temperatures is Simulated Annealing (SA) [44]. The "Alpha Helical Stability Tool" as implemented in BioLuminate allows for the automatic setup of such SA simulations, as well as the analysis of the resulting secondary structure of peptide over the course of a simulation.

2.4 Force Fields in the Schrodinger Suite

Good quality force fields are imperative to the accurate modeling of molecules. The goal of a force field is to accurately represent the systems' chemistry, with distinct force fields suitable for different systems. For example, the well-known force fields Amber [45] and CHARMM [46] have been popular choices to model large molecules such as peptide and proteins in long time scale MD simulations. Force fields in the OPLS and MMFF families have been traditionally used to model small drug-like or fragment molecules.

There is continued development of force fields as part of Schrödinger research, with the OPLS3 force field the most recent version [36]. While earlier version of the algorithm (e.g. OPLS2) have focused on fully covering chemical space for small molecules, current modifications have been made to improve accurate behavior of proteins in MD simulations, resulting in improved predictions of protein–ligand binding.

A number of studies have been done to demonstrate the ability of OPLS3 to preserve the stability secondary structure elements in proteins. For example, the CLN025 β-hairpin and the K19 and $(AAQAA)^3$ α-helical peptides, all maintained correct second structure over the course of several microseconds of MD simulations (*see* Table 2). This correlation with experimental observations is in line with results from CHARMM and Amber. In addition, a number of globular proteins were simulated for several hundred nanoseconds, with average RMSD values remaining within an acceptable level of 1.5 Å in all cases, a significant improvement over OPLS_2005 and OPLS2.1 force fields, and comparable with the stability observed with CHARMM and Amber force fields (*see* Table 3). The predominant factor attributed to this improvement over earlier OPLS force fields is the better representation of the Chi1 rotamer angle in Asp, Glu, and Asn side-chains.

2.5 Peptide QSAR

Quantitative Structure Activity Relationships (QSAR) is typically used to quantify the relationship between the structures of small

Table 2
Secondary structure propensities of peptides in long microsecond MD simulations

Protein	Experiment (%)	OPLS3 (%)	CHARMM (%)	Amber (%)
K19	50	74	62	21
(AAQAA)3	43	51	30	9
CLN025	91	91	90	80

See ref. 36 for details

Table 3
Protein structural stability: Average RMSD values of the simulated protein with respect to the X-ray or NMR Structure

Protein name	PDB code	OPLS3	CHARMM	Amber
Trpcage	1L2Y	1.2	1.7	1.7
GB3	1P7E	0.8	0.7	0.8
Ubiquitin	1UBQ	0.9	0.8	0.9
SUMO2	1WM3	1.3	1.3	1.4
BPT1	1BPI	1.2	1.1	1.1
Crambin	1CRN	0.8	0.7	0.8
Lysozome	6LYT	1.3	1.1	1.1
Average		1.1	1.1	1.1

See ref. 36 for details

drug-like molecules with experimental observables such as activity or cytotoxicity. If a predictive relationship can be determined, this can be used to design new molecules within the same series with better optimized physical properties. Usually a Multiple Linear Regression or similar statistical method is used, where the experimental measurements represent the Y response variable, and these are modeled as a function of the independent predictor X-variables, typically represented by physicochemical descriptors such as simple counts of structural elements (the number of donors or acceptors for example) or a measure of properties, such as log P or cell permeability.

A very similar approach can be applied to peptides—in this case one attempts to predict the property of a peptide based on its sequence of amino acids alone. This, similar to the application of QSAR in small molecules, can be useful when designing and selecting peptides for application as therapeutics. In peptide QSAR, the Y response variable can be properties of the peptide such as binding affinity or solubility, and this is modeled as a function of the X-variables, which include measured and observable properties of the amino acids that form the peptide sequence.

Peptide QSAR can be run in the Schrödinger Suite using the Peptide QSAR panel in the BioLuminate module. It requires a sequence file with multiple peptides with an experimental measurement provided for each of them. A CSV or FASTA file format is acceptable but data can also be loaded directly from the Maestro project table as well. The panel allows one of two options for the statistical method that is used to generate the QSAR model, namely, Partial Least Squares (PLS), or Kernel-based PLS, with the number of factors user-defined depending on the size of the data. Typically, in order to avoid over-fitting, the maximum number of factors should to be no more than the number of sequences divided by 10.

The most important option in the panel is the "peptide descriptor type" and offers a choice of the type of X-variables that can be used to describe the properties of the amino acids. Three sets of descriptors described in the literature have been included (see Table 4). These include the Z-value [47]; EZ-value [48]; and DPPS [49] sets of descriptors in the model. In addition, it is possible to use a combination of all the three sets using the "All" option.

One example in the literature of successful application of such an approach is the development of a QSAR model for ACE-inhibitory peptides, using an Artificial Neural Networks statistical method [50]. In this study it was demonstrated how good activity is governed by hydrophobic amino acids in the C-terminal tail of the peptide. A second example is a PLS model of the MHC protein and its peptide ligands [49]. Here it was demonstrated how antigen recognition is governed by hydrophobic interactions and hydrogen bonds, especially exerting effects on anchor residues of

Table 4
Peptide descriptor sets used in the Peptide QSAR approach

Peptide descriptor set	Described by	Dataset	Descriptors
Z value [47]	Three Z variables	29 physicochemical variables for the 20 coded amino acids	MW, pKa, pI, side-chain vdW volumes, NMR shifts, retention times, partition coefficients, solvent exposure
EZ value [48]	Five extended Z variables	26 physicochemical descriptors for 87 amino acids (including the 20 coded amino acids)	MW, NMR shifts, partition coefficients, side-chain vdW volumes, HOMO and LUMO energies, heats of formation, polarizabilities, surface areas, hardness, TLC retention times, hydrogen-bond donor and acceptor counts, side chain charges
DPPS [49]	Ten divided physicochemical property scores	20 coded amino acids	23 electronic, 54 hydrophobic, 37 steric, and 5 H-bond properties

peptides. Neural Networks, PLS, and several other statistical model building facilities (e.g. Bayesian Models, PCA) are all available through the Canvas program in the Schrödinger Suite [51].

3 Summary

The past decade has seen a tremendous shift in pharmaceutical drug discovery efforts from small molecules toward peptides and proteins. This has lead to increasing need for structure-based computational tools focused on such biopharmaceuticals. Schrödinger has implemented a large number of tools that can be applied in this area, and many of those have been described in this article. Structure-based design of biologics is still rather young compared to small molecule design. Despite this, there is an increasing recognition that structure-based design of biologics can make real, substantive contributions to drug discovery. The field is still rapidly changing, and adoption of tools such as those described here is currently increasing quite rapidly. We expect that in the next few years the majority of new biopharmaceuticals will be designed incorporating structure-based methods, such as those available from Schrödinger, just as has long been the case for small molecules.

References

1. Repasky MP, Murphy RB, Banks JL, Greenwood JR, Tubert-Brohman I, Bhat S, Friesner RA (2012) Docking performance of the glide program as evaluated on the Astex and DUD datasets: a complete set of glide SP results and selected results for a new scoring function integrating WaterMap and glide. J Comput Aided Mol Des 26(6):787–799. doi:10.1007/s10822-012-9575-9

2. Tubert-Brohman I, Sherman W, Repasky M, Beuming T (2013) Improved docking of polypeptides with Glide. J Chem Inf Model 53(7):1689–1699. doi:10.1021/ci400128m

3. Bioluminate 2.1 (2015) Schrödinger, Inc., Portland, OR

4. Sastry GM, Adzhigirey M, Day T, Annabhimoju R, Sherman W (2013) Protein and ligand preparation: parameters, protocols, and influence on virtual screening enrichments. J Comput Aided Mol Des 27(3):221–234. doi:10.1007/s10822-013-9644-8

5. Bas DC, Rogers DM, Jensen JH (2008) Very fast prediction and rationalization of pKa values for protein-ligand complexes. Proteins 73(3):765–783. doi:10.1002/prot.22102

6. Friesner RA, Banks JL, Murphy RB, Halgren TA, Klicic JJ, Mainz DT, Repasky MP, Knoll EH, Shelley M, Perry JK, Shaw DE, Francis P, Shenkin PS (2004) Glide: a new approach for rapid, accurate docking and scoring. 1. Method and assessment of docking accuracy. J Med Chem 47(7):1739–1749

7. Friesner RA, Murphy RB, Repasky MP, Frye LL, Greenwood JR, Halgren TA, Sanschagrin PC, Mainz DT (2006) Extra precision glide: docking and scoring incorporating a model of hydrophobic enclosure for protein-ligand complexes. J Med Chem 49(21):6177–6196

8. Halgren TA, Murphy RB, Friesner RA, Beard HS, Frye LL, Pollard WT, Banks JL (2004) Glide: a new approach for rapid, accurate docking and scoring. 2. Enrichment factors in database screening. J Med Chem 47(7):1750–1759

9. Feher M, Williams CI (2012) Numerical errors and chaotic behavior in docking simulations. J Chem Inf Model 52(3):724–738. doi:10.1021/ci200598m

10. Sherman W, Day T, Jacobson MP, Friesner RA, Farid R (2006) Novel procedure for modeling ligand/receptor induced fit effects. J Med Chem 49(2):534–553

11. Prime 4.2 (2015) Schrödinger, Inc., Portland, OR

12. Kozakov D, Brenke R, Comeau SR, Vajda S (2006) PIPER: an FFT-based protein docking program

with pairwise potentials. Proteins 65(2):392–406. doi:10.1002/prot.21117

13. MacroModel v11.0 (2015) Schrödinger, Inc., Portland, OR

14. Kozakov D, Hall DR, Beglov D, Brenke R, Comeau SR, Shen Y, Li K, Zheng J, Vakili P, Paschalidis I, Vajda S (2010) Achieving reliability and high accuracy in automated protein docking: ClusPro, PIPER, SDU, and stability analysis in CAPRI rounds 13–19. Proteins 78(15):3124–3130. doi:10.1002/prot.22835

15. Shen Y, Brenke R, Kozakov D, Comeau SR, Beglov D, Vajda S (2007) Docking with PIPER and refinement with SDU in rounds 6–11 of CAPRI. Proteins 69(4):734–742. doi:10.1002/prot.21754

16. Miller EB, Murrett CS, Zhu K, Zhao S, Goldfeld DA, Bylund JH, Friesner RA (2013) Prediction of long loops with embedded secondary structure using the protein local optimization program. J Chem Theory Comput 9(3):1846–4864. doi:10.1021/ct301083q

17. Zhao S, Zhu K, Li J, Friesner RA (2011) Progress in super long loop prediction. Proteins 79(10):2920–2935. doi:10.1002/prot.23129

18. Zhu K, Pincus DL, Zhao S, Friesner RA (2006) Long loop prediction using the protein local optimization program. Proteins 65:438–452

19. Nourry C, Grant SG, Borg JP (2003) PDZ domain proteins: plug and play! Sci STKE 2003(179):RE7. doi:10.1126/stke.2003.179.re7

20. Bell JA, Ho KL, Farid R (2012) Significant reduction in errors associated with nonbonded contacts in protein crystal structures: automated all-atom refinement with PrimeX. Acta Crystallogr D Biol Crystallogr 68(Pt 8):935–952. doi:10.1107/S0907444912017453

21. Greenidge PA, Kramer C, Mozziconacci JC, Sherman W (2014) Improving docking results via reranking of ensembles of ligand poses in multiple X-ray protein conformations with MM-GBSA. J Chem Inf Model 54(10):2697–2717. doi:10.1021/ci5003735

22. Guimaraes CR, Cardozo M (2008) MM-GB/SA rescoring of docking poses in structure-based lead optimization. J Chem Inf Model 48(5):958–970. doi:10.1021/ci800004w

23. Zhu K, Shirts MR, Friesner RA (2007) Improved methods for side chain and loop predictions via the protein local optimization program: variable dielectric model for implicitly improving the treatment of polarization effects. J Chem Theory Comput 3(6):2108–2119. doi:10.1021/ct700166f

24. Li J, Abel R, Zhu K, Cao Y, Zhao S, Friesner RA (2011) The VSGB 2.0 model: a next generation energy model for high resolution protein structure modeling. Proteins 79(10):2794–2812. doi:10.1002/prot.23106

25. Young T, Abel R, Kim B, Berne BJ, Friesner RA (2007) Motifs for molecular recognition exploiting hydrophobic enclosure in protein ligand binding. Proc Natl Acad Sci U S A 104:808–813

26. Ylilauri M, Pentikainen OT (2013) MMGBSA as a tool to understand the binding affinities of filamin-peptide interactions. J Chem Inf Model 53(10):2626–2633. doi:10.1021/ci4002475

27. Beard H, Cholleti A, Pearlman D, Sherman W, Loving KA (2013) Applying physics-based scoring to calculate free energies of binding for single amino acid mutations in protein-protein complexes. PLoS One 8(12):e82849. doi:10.1371/journal.pone.0082849

28. Steinbrecher TB, Dahlgren M, Cappel D, Lin T, Wang L, Krilov G, Abel R, Friesner R, Sherman W (2015) Accurate binding free energy predictions in fragment optimization. J Chem Inf Model 55(11):2411–2420. doi:10.1021/acs.jcim.5b00538

29. Wang L, Wu Y, Deng Y, Kim B, Pierce L, Krilov G, Lupyan D, Robinson S, Dahlgren MK, Greenwood J, Romero DL, Masse C, Knight JL, Steinbrecher T, Beuming T, Damm W, Harder E, Sherman W, Brewer M, Wester R, Murcko M, Frye L, Farid R, Lin T, Mobley DL, Jorgensen WL, Berne BJ, Friesner RA, Abel R (2015) Accurate and reliable prediction of relative ligand binding potency in prospective drug discovery by way of a modern free-energy calculation protocol and force field. J Am Chem Soc 137(7):2695–2703. doi:10.1021/ja512751q

30. Abel R, Young T, Farid R, Berne BJ, Friesner RA (2008) Role of the active site solvent in the thermodynamics of factor Xa ligand binding. J Am Chem Soc 130(9):2817–2831

31. Li Z, Lazaridis T (2006) Thermodynamics of buried water clusters at a protein-ligand binding interface. J Phys Chem B 110(3):1464–1475. doi:10.1021/jp056020a

32. Li Z, Lazaridis T (2012) Computing the thermodynamic contributions of interfacial water. Methods Mol Biol 819:393–404. doi:10.1007/978-1-61779-465-0_24

33. Beuming T, Farid R, Sherman W (2009) High-energy water sites determine peptide binding affinity and specificity of PDZ domains. Protein Sci 18(8):1609–1619. doi:10.1002/pro.177

34. Skelton NJ, Koehler MF, Zobel K, Wong WL, Yeh S, Pisabarro MT, Yin JP, Lasky LA, Sidhu SS (2003) Origins of PDZ domain ligand specificity. Structure determination and

mutagenesis of the Erbin PDZ domain. J Biol Chem 278(9):7645–7654. doi:10.1074/jbc. M209751200

35. Kolossváry I, Guida WC (1999) Low-mode conformational search elucidated. Application to C39H80 and flexible docking of 9-deazaguanine inhibitors to PNP. J Comput Chem 20:1671–1684

36. Harder E, Damm W, Maple J, Wu C, Reboul M, Xiang JY, Wang L, Lupyan D, Dahlgren MK, Knight JL, Kaus JW, Cerutti DS, Krilov G, Jorgensen WL, Abel R, Friesner RA (2015) OPLS3: a force field providing broad coverage of drug-like small molecules and proteins. J Chem Theory Comput. doi:10.1021/acs. jctc.5b00864

37. Chen IJ, Foloppe N (2010) Drug-like bioactive structures and conformational coverage with the LigPrep/ConfGen suite: comparison to programs MOE and catalyst. J Chem Inf Model 50(5):822–839. doi:10.1021/ci100026x

38. Watts KS, Dalal P, Murphy RB, Sherman W, Friesner RA, Shelley JC (2010) ConfGen: a conformational search method for efficient generation of bioactive conformers. J Chem Inf Model 50(4):534–546. doi:10.1021/ci100015j

39. Ahlbach CL, Lexa KW, Bockus AT, Chen V, Crews P, Jacobson MP, Lokey RS (2015) Beyond cyclosporine A: conformation-dependent passive membrane permeabilities of cyclic peptide natural products. Future Med Chem 7(16):2121–2130. doi:10.4155/fmc.15.78

40. Desmond v4.4 (2015) Schrödinger, Inc., Portland, OR

41. Guo Z, Mohanty U, Noehre J, Sawyer TK, Sherman W, Krilov G (2010) Probing the alpha-helical structural stability of stapled p53 peptides: molecular dynamics simulations and analysis. Chem Biol Drug Des 75(4):348–359. doi:10.1111/j.1747-0285.2010.00951.x

42. Guo Z, Streu K, Krilov G, Mohanty U (2014) Probing the origin of structural stability of single and double stapled p53 peptide ana-logs bound to MDM2. Chem Biol Drug Des 83(6):631–642. doi:10.1111/cbdd.12284

43. Zhou R (2007) Replica exchange molecular dynamics method for protein folding simulation. Methods Mol Biol 350:205–223

44. Karplus M, McCammon JA (2002) Molecular dynamics simulations of biomolecules. Nat Struct Biol 9(9):646–652. doi:10.1038/ nsb0902-646

45. Wang J, Wolf RM, Caldwell JW, Kollman PA, Case DA (2004) Development and testing of a general amber force field. J Comput Chem 25(9):1157–1174. doi:10.1002/jcc.20035

46. Vanommeslaeghe K, Hatcher E, Acharya C, Kundu S, Zhong S, Shim J, Darian E, Guvench O, Lopes P, Vorobyov I, Mackerell AD Jr (2010) CHARMM general force field: a force field for drug-like molecules compatible with the CHARMM all-atom additive biological force fields. J Comput Chem 31(4):671–690. doi:10.1002/jcc.21367

47. Hellberg S, Sjostrom M, Skagerberg B, Wold S (1987) Peptide quantitative structure-activity relationships, a multivariate approach. J Med Chem 30(7):1126–1135

48. Sandberg M, Eriksson L, Jonsson J, Sjostrom M, Wold S (1998) New chemical descriptors relevant for the design of biologically active peptides. A multivariate characterization of 87 amino acids. J Med Chem 41(14):2481–2491. doi:10.1021/jm9700575

49. Tian F, Lv F, Zhou P, Yang Q, Jalbout AF (2008) Toward prediction of binding affinities between the MHC protein and its peptide ligands using quantitative structure-affinity relationship approach. Protein Pept Lett 15(10):1033–1043

50. He R, Ma H, Zhao W, Qu W, Zhao J, Luo L, Zhu W (2012) Modeling the QSAR of ACE-inhibitory peptides with ANN and its applied illustration. Int J Pept 2012:620609. doi:10.1155/2012/620609

51. Canvas v2.5 (2015) Schrödinger, Inc., Portland, OR

Chapter 15

Identifying Loop-Mediated Protein–Protein Interactions Using LoopFinder

Timothy R. Siegert, Michael Bird, and Joshua A. Kritzer

Abstract

Peptides are an increasingly useful class of molecules, finding unique applications as chemical probes and potential drugs. They are particularly adept at inhibiting protein–protein interactions, which are often difficult to target using small molecules. The identification and rational design of protein-binding epitopes remains a bottleneck in the development of bioactive peptides. One fruitful strategy has been using structured scaffolds to present essential hot spot residues involved in protein–protein recognition, and this process has been greatly advanced by computational tools that can identify hot spot residues. Here we discuss LoopFinder, a program that uses structures from the Protein Data Bank to comprehensively search for protein–protein interactions that are mediated by nonhelical, nonsheet loop structures. We developed LoopFinder to identify these "hot loops" and to assist in the design of cyclic peptides that mimic these important structures. In this article, we provide all key files, outline step-by-step methods for users to conduct independent LoopFinder searches, and provide guidance on additional potential applications for the LoopFinder program.

Key words Protein–protein interactions, Macrocycles, Cyclic peptides, Peptide design, Inhibitors, Chemical biology

1 Introduction

Protein–protein interactions (PPIs) are a quickly expanding class of drug targets [1]. PPIs play essential roles in intercellular and intracellular signaling. PPIs have been dismissed as "undruggable" because they often have lower-affinity interactions, large interaction surface areas (1500–3000 Å²), and a flat surfaces lacking deep binding pockets [2]. All of these features make it difficult to target these interactions with small molecules [1]. With some notable exceptions, traditional approaches to drug discovery have failed to develop potent, selective inhibitors for most PPIs [3]. Thus, new

Electronic supplementary material: The online version of this chapter (doi:10.1007/978-1-4939-6798-8_15) contains supplementary material, which is available to authorized users.

Ora Schueler-Furman and Nir London (eds.), *Modeling Peptide-Protein Interactions: Methods and Protocols*, Methods in Molecular Biology, vol. 1561, DOI 10.1007/978-1-4939-6798-8_15, © Springer Science+Business Media LLC 2017

chemical space must be applied to the problem, specifically molecules capable of binding a large and flat surface area with maximal ligand efficiency. One class of molecules particularly suited for this task is peptides. Many naturally occurring peptides are used by cells to control intracellular and intercellular signaling, including potent hormones, growth factors, and other ligands for cell surface receptors [4]. Like many other biomolecules, they also have the advantages of high target specificity and good safety properties [5].

An initial bottleneck in the design of peptide inhibitors of PPIs is the identification of regions on protein surfaces that are suitable as peptide binding sites. A major breakthrough in this area came with the understanding that PPIs are mediated by hot spot residues, which are amino acids at the interface that are particularly important for the affinity of the PPI. Hot spots were originally defined as residues for which an alanine substitution resulted in a change in binding free energy of ≥ 2.0 kcal/mol [6]. Regions on the protein surfaces containing multiple hot spots are the most critical binding epitopes [7]. To further expedite the process of hot spot identification, several computational methods have been developed to perform alanine scanning in silico. These include the FOLDEF algorithm, which systematically substitutes residues with alanine and calculates the change in binding energy using FoldX complex energy functions that include terms for desolvation, van der Waals forces, hydrogen bonding, Coulombic interactions, entropy change, and dipole interactions [8]. Another computational alanine scanning engine, developed by Kortemme and Baker, uses the Rosetta software package [9, 10]. Rosetta's combination of explicit terms for Lennard-Jones energies, solvation energies, protein repacking, solvation, and hydrogen bonding are used to account for protein repacking while calculating changes in free energy when side chains are systematically substituted by alanine [9, 10]. These computational methods require only the PDB information for each PPI in question, and can accurately identify side chains at the interface that are critical for the binding interaction.

Some additional approaches to uncovering critical binding sites have been developed that examine protein surfaces rather than performing computational alanine scanning. The Camacho lab developed a tool called ANCHOR that identifies "anchor" residues at protein interfaces [11]. These anchor residues are identified by measuring changes in the solvent-accessible surface area (ΔSASA) of side chains upon receptor binding. This tool, available at the PocketQuery website, is effective in identifying small hydrophobic pockets suitable for developing small molecule ligands [12]. Other computational methods probe protein surfaces by introducing small organic probes or molecule fragments such as ammonium cations, carboxyl groups or methane, to identify pockets suitable for direct targeting using small molecule or peptide inhibitors [13, 14].

Beyond identifying potential binding sites for inhibitors, structural information can more directly inform inhibitor design. Recently, multiple techniques have been developed that identify peptide epitopes that can be directly translated into peptide-based inhibitors of PPIs. For example, the Schueler-Furman lab developed a method called Rosetta PeptiDerive to identify "hot segments," defined as stretches of ten amino acids within a protein that contributes most of the binding energy to complex formation [15]. This program breaks protein chains involved in PPIs into all possible ten-amino acid segments, and then uses Rosetta energy functions to calculate the binding energy of each peptide segment to the receptor protein. The total binding energy of each individual peptide segment is then compared to the estimated energy of the entire PPI to identify hot segments. Using a small starting set of 151 PPIs, it was found that about 60% of PPIs possessed a single ten-mer peptide segment that contributed over half of the interaction energy of the whole complex [15]. This method was used to predict a peptide segment from myeloid differentiation factor 2 (MD2) that should bind and activate MD2's binding partner, Toll-like receptor 4 (TLR4) [16]. The linear peptide was synthesized and tested for its ability to activate TLR4 signaling as measured by increased production of nitric oxide in macrophages, but the linear peptide showed no activation. In order for activity to be observed, further computational modeling was required to introduce a disulfide constraint that would better stabilize the native loop structure. This macrocyclic peptide showed some activation of the inflammatory response, inducing nitric oxide production in macrophages [16]. These results reveal a key limitation of ignoring structure when identifying linear segments for the design of PPI inhibitors, because structural stabilization is often required for high-affinity binding by short peptides.

One of the most comprehensive programs that combines hot spot identification with structural analysis is the helix interfaces in protein–protein interactions (HIPP) program developed by Arora and coworkers [17, 18]. HIPP identified alpha-helical regions at protein interfaces that are critical for mediating PPIs. Alpha-helical regions were structurally identified via Rosetta secondary structure calculations of ϕ and φ angles, and computational alanine scanning was used to determine the energy contribution of each residue within each helix. Helices were then analyzed with respect to their binding pockets, with some binding into deeper clefts and others binding over more extended surfaces. Helices were also sorted based on the proportion of hot spots residing on the helix compared to the overall chain [17]. These helical binding epitopes are important because they can be effectively stabilized using a variety of scaffolds and cross-linking strategies [19, 20]. Recently, Arora and coworkers adapted this strategy to beta-strands, identifying strands that are responsible for a large proportion of the binding energy of their associated PPIs [21]. Together, these tools identify

PPI-mediating epitopes that contain common secondary structures. Existing and novel strategies for mimicking these structures can then be used to translate these epitopes into real-world PPI inhibitors [22–24].

While linear segments, alpha-helices, and beta-strands are all effective starting points for inhibitor design, we were inspired by the rich diversity of cyclic peptide natural products to ask whether epitopes that do not fall into these structural categories were being overlooked. Many cyclic peptide natural products, such as cyclosporine A and the amanitins, lack these common secondary structure elements [25, 26]. Instead, these molecules have intramolecular hydrogen bonds that are rare in protein structures, and their unique conformations are critical for their high potency and bioavailability [27, 28]. We reasoned that, to look for epitopes that could be mimicked by similar classes of cyclic peptides, we should examine loop regions at PPI interfaces. We developed LoopFinder with the hypothesis that, if we can identify loops that are critical for PPIs, they will more easily be translated into natural-product-like cyclic peptides.

LoopFinder was heavily inspired by the computational methods discussed above, but was designed to identify loop regions. Fully 50% of PPI interfaces are regions without a common secondary structure [29]. Many of these are nonsheet extended regions that are not optimal starting points for macrocycle design, so we chose to define "loops" using custom-defined size and shape criteria (*see* **Note 3**). In biochemistry, loops are defined as segments that lack secondary structure, and are often found connecting discrete secondary structure elements. For LoopFinder, we were seeking short segments that have their N- and C-termini in close proximity (sometimes called "omega loops"), to allow for their effective cyclization using a variety of different chemistries. Specifically, loops were defined as stretches of four to eight amino acids in which the C_α carbons of the N- and C-terminal residues are within 6.2 Å (though these parameters can be changed; *see* **Note 3**) [30]. Such loops will be the most straightforward epitopes to translate into cyclic peptides or other macrocycles as potential PPI inhibitors.

Our first implementation of LoopFinder was reported in 2014 [30], and since then we have streamlined its use and expanded its capabilities. In this Methods paper, we provide detailed instructions for comprehensive applications of LoopFinder, as well as a link to a website that automates some of the simpler tasks.

2   Materials

To perform comprehensive loop identification and energy calculations on a large set of protein structures, LoopFinder requires the programs and files listed below. These are provided in the online supporting material to this article, in the directory structure indicated in Fig. 1.

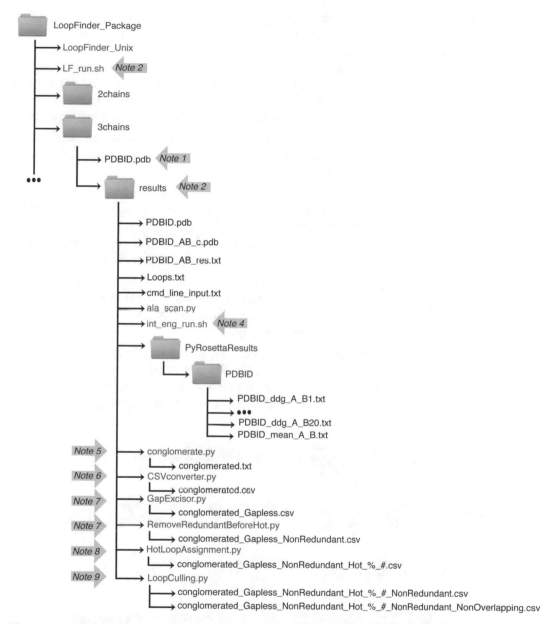

Fig. 1 A specific directory structure, shown here, is required to properly execute LoopFinder and run the computational alanine scan. As provided in online materials, the */LoopFinder_Package* directory contains the Unix-compatible LoopFinder executable file as well as a simple shell script to run LoopFinder with loops defined by the user (*see* **Note 3** for details on custom loop definition). Input PDB files must be sorted into separate directories based on the total number of chains in the assembly. LoopFinder will then search these PDB files for loops and produce a */results* directory containing the identities of the loops and the necessary information to run the computational alanine scan. The loop data and energy data are combined using *conglomerate.py* to produce a conglomerated file containing all loops with their associated computational alanine scan data. Descriptions of each of the executable files and scripts (shown in *green*) and the data files (shown in *black*) are provided in Subheading 2

1. *PyRosetta*: A prerequisite for running LoopFinder is installation of PyRosetta [31]. PyRosetta is a Python-based scripting interface that allows custom molecular modeling using the Rosetta conformational sampling and energy calculation algorithms. PyRosetta is instrumental in the application of Rosetta computational alanine scanning to the identified interface loops. This program is freely available at www.pyrosetta.org

2. *pdb_download.py*: A useful script for downloading large lists of PDB files, freely available from the Harms Lab at *github.com/harmslab/pdbtools*

3. *LoopFinder_Unix*: the Unix-compiled version of LoopFinder, which is provided in the */LoopFinder_Package* directory. This is the program that reads in PDB files, carries out measurements to locate loops at interfaces based on custom parameters, and compiles the resulting list of interface loops.

4. *LF_run.sh*: A simple shell script for running LoopFinder. This file can be edited using a text editing program to define the specific loop parameters of interest to the user. *See* Fig. 2 for an example *LF_run.sh* script.

5. */2chains, /3chains, /4chains, and so on*: These directories contain the input set of PDB files for LoopFinder. The PDBs need to be organized into separate directories, segregated based on the number of chains in the molecular assembly.

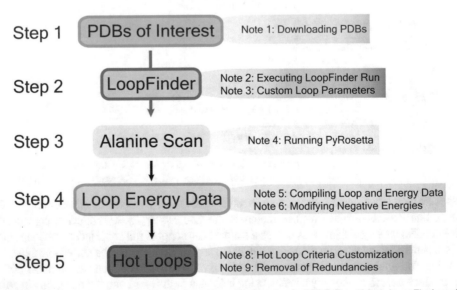

Fig. 2 Workflow for the entire LoopFinder process. From the input set of PDB files (*blue*), LoopFinder will search for loops based on a customizable set of loop size parameters provide by the user (*purple*). Once the loop set is generated, computational alanine scanning is carried out in Rosetta to identify hot spot residues (*yellow*). Finally, the loop data is consolidated with the corresponding energy values for each residue in each loop (*orange*). Using customizable "hot loop" thresholds, the best loops for cyclic peptide inhibitor design (or any other desired application) can be identified (*red*)

6. *Ala_scan.py*: The computational alanine scanning script, run using PyRosetta [9, 32].

7. *int_eng_run.sh*: A simple shell script that loads the PyRosetta module and runs computational alanine scans on each interface for which an interface loop was found by LoopFinder.

8. *conglomerate.py*: A python script that integrates the loop data generated by LoopFinder and the energy data generated by the computational alanine scan. This program compiles overall energy data for each loop (*see* **Note 5**) and adjusts some problematic Rosetta-derived energy values (*see* **Note 6**). This generates a new conglomerated output text file called *conglomerated.txt*.

9. *CSVconverter.py*: A program that converts the output text file to a .csv format that is more readily read into common database programs.

10. *GapExcisor.py*: A program that removes loops with large numbers of unknown residues due to jumps that sometimes occur in the numbering systems of PDB files.

11. *RemoveRedundantBeforeHot.py*: If the loops of the same sequence occur multiple times in a single PDB file due to homomultimerism, this script can be used to remove all but the one with the highest average $\Delta\Delta G$ per residue.

12. *HotLoopAssignment.py*: A script that allows for the identification of hot loops using the user-defined criteria (*see* **Note 8**).

13. *LoopCulling.py*: A simple script that removes redundant loops in the loop set in two ways. First, if one hot loop is contained inside another, only the longer is kept. Second, if two hot loops from the same structure have identical hot spots, only the one with higher average $\Delta\Delta G$ per residue is kept (*see* **Note 9** for details).

During the LoopFinder calculations, several files are created that contain data and/or scripts for later steps in the process. We list these here to help clarify the workflow of LoopFinder and to simplify troubleshooting.

1. *PDBID.pdb*: LoopFinder generates a scrubbed version of each PDB file in question, removing headers and comments and leaving only the atom information. *PDBID* refers to the PDB ID of each individual PDB file.

2. *PDBID_AB_c.pdb*: For each pdb file, LoopFinder creates a new pdb file that contains only two chains that form an interface (in the example filename here, A and B are the chains that form an interface). Multiple of these binary interaction files may be created for each input PDB, depending on how many chains are found to have a binary interface.

3. *PDBID _AB_res.txt*: LoopFinder generates a list of residues found to reside at the interface between the two chains (here, chains A and B).

4. *Loops.txt*: This is the major results file from LoopFinder. It contains information on all the loops found to fit the desired user definitions. This process is described in more detail in **Note 2**.

5. *cmd_line_input.txt*: LoopFinder also generates *cmd_line_input.txt* which contains command line instructions for running the *ala_scan.py* script (for PyRosetta-based computational alanine scanning) on the loops described in *Loops.txt*.

6. */PyRosettaResults*: Once *int_eng_run.sh* is used to run PyRosetta, the output data will be deposited into directories named for each PDB. These will be in a parent directory called */PyRosettaResults*

7. *PDBID_ddg_A_B1.txt*: PyRosetta's alanine scan program creates a new text file containing a list of all residues located at the interface (in this case, between chains A and B) along with their corresponding change in binding free energy value upon mutation to alanine. These are numbered for each trial, with 20 trials recommended for each binary interface.

8. *PDBID_mean_A_B.txt*: All the multiple independent trials are used to calculate a mean energy value for each residue, which are stored in this file.

9. *conglomerated.txt*: The *conglomerate.py* script produces *conglomerated.txt*, which is its main results file (though additional removal of redundant loops can still be performed). It is a conglomeration of the *Loops.txt* data and the energy values associated with each loop residue, as specified by the corresponding *PDBID_mean_A_B.txt* alanine scan energy results file.

10. *conglomerated.csv*: Applying *CSVconverter.py* to *conglomerated.txt* produces this CSV file that can be more easily read by database management software.

11. *conglomerated_Gapless.csv*: Applying *GapExcisor.py* to *conglomerated.csv* produces this file, which has all gap artifacts removed (*see* **Note 7**).

12. *conglomerated_Gapless_NonRedundant.csv*: Applying *RemoveRedundantBeforeHot.py* to *conglomerated_Gapless.csv* produces this file, which effectively eliminates identical loops prior to hot loop identification (*see* **Note 7**).

13. *conglomerated_Gapless_NonRedundant_Hot_%_#.csv*: Running *HotLoopAssignment.py* then produces a new file containing only loops that meet the user specified hot loop criteria. In the filename shown, % is the percent interface energy and # is the minimum value defined as a hot spot (REUs*100) (*see* **Note 8** for details).

14. *conglomerated_Gapless_NonRedundant_Hot_%_#_NonRedundant.csv*: Following hot loop identification, further redundant loops are removed and are deposited in this file by *LoopCulling.py* (*see* **Note 9**).

15. *conglomerated_Gapless_NonRedundant_Hot_%_#_NonRedundant_NonOverlapping.csv: LoopCulling.py* also produces this file, which has nested and overlapping loops removed as well (*see* **Note 9**).

3   Methods

The overall process for identifying hot loops is summarized in Fig. 2. As detailed in Subheading 2, the required resources for a LoopFinder search are: one or more PDB files (the input set), the LoopFinder program, and the Rosetta-enabled alanine scanning engine. The LoopFinder program completes three distinct tasks. The first task is a measurement process that locates loops at protein interfaces based on user-defined loop size and proximity parameters. The second task is computational alanine scanning calculations using PyRosetta [9, 10] on each PPI that contains one or more of the identified loops. Once the computational alanine scan is complete, this data is then merged with the list of identified loops to assign energy values to each residue within each interface loop. The third task is to sort through all interface loops to identify "hot loops" that are particularly promising starting points for inhibitor design.

An overview of the LoopFinder process for identifying hot loops that mediate PPIs is shown in Fig. 2. The */LoopFinder_Package* directory contains all the key elements for successfully running the program, most importantly the *LoopFinder_Unix* executable file. The online materials included with this chapter contain the entire LoopFinder package listed in Subheading 2, except for PyRosetta, which is freely available. For analyzing large input sets of PDB files, it is advisable to install this package on a computer cluster to enable the concurrent running of multiple calculations as each alanine scan run takes about 10–20 min depending on the size of the interface. The process described in this section reflects the use of scripts for cluster submission of large batches of calculations.

Step 1. Download PDB files of interest.

Before the program can be run, the PDB files of interest to the user need to be downloaded. As discussed in **Note 1**, there are many useful tools for downloading large sets of PDB files from the RCSB database, most notable the RCSB bulk download tool at www.rcsb.org/pdb/search/advSearch.do [33]. The input PDBs must be organized into separate directories based on the total number of chains in the molecular assembly, which we accomplished using the simple Python script *pdb_download.py* (*see* **Note 1**).

Step 2. Execute the LoopFinder program.

The *LF_run.sh* shell script should be edited using a simple text editing software to include the directory of the input PDB files, the interface distance definition, the loop size definitions (maximum and minimum loop size), the percent of the loop required to be at the interface, the termini distance measurement, loop size scaling factor, and the number of chains in the PDB files to be considered (as mentioned above, LoopFinder works on sets of PDBs that all have the same number of protein chains). **Note 3** discusses considerations for setting the loop definition parameters, as well as our rationale for the parameters we have used to date. Execution of *LF_run.sh* submits the parameters to the *LoopFinder_Unix* program and generates a */results* directory (Fig. 1). LoopFinder produces several files within */results*, most importantly *Loops.txt*, which contains information on all the loops that fit the given parameters, and *cmd_line_input.txt*, which contains the information required for the computational alanine scan. Examples of each type of output file are shown in **Note 2**. At this step, all loops have been identified and can be analyzed using *Loops.txt*, but their energies have not yet been calculated.

Step 3. Execute the computational alanine scan.

The computational alanine scan is executed using the *int_eng_run.sh* script. This script loads the PyRosetta module and submits each interface listed in *cmd_line_input.txt* for computational alanine scanning. The *int_eng_run.sh* script generates a directory called */PyRosettaResults* (Fig. 1), in which a new directory is generated for each PDB being analyzed with the title being the PDBID. Within the directories named for each PDBID, PyRosetta deposits text files containing energy data from each individual alanine scan (for example, *2flu_ddg_X_P1.txt*), as well as the final file containing the mean energy data among all trials run for that interface (for example, *2flu_mean_X_P.txt*). A more detailed description of this process is outlined in **Note 4**. At this step, computational alanine scanning data has been calculated for all PPIs with loop-mediated interfaces, but the loop data has not yet been combined with the energy data to provide energy values for individual loops.

Step 4. Combine the loop data with the energy data.

Executing the *conglomerate.py* script will combine the LoopFinder loop data generated in **step 2** with the alanine scan energy values generated in **step 3**. A new output file that contains the conglomerated data set for analysis, *conglomerated.txt*, will be generated in the */results* directory (Fig. 1). These files can be converted into a CSV format using *CSVconverter.py* for more fluid compatibility with database management software, producing a file named *conglomerated.csv*.

Step 5. Identify hot loops and remove redundancies.

Before hot loops can be identified, some preliminary cleanup and redundancy removal should be done. First, run *GapExcisor.py* on *conglomerated.csv* to remove any loops that contain amino acids for which energies could not be calculated. These loops are removed and the new list is saved as *conglomerated_Gapless.csv*. These rare gaps occur because of errant numbering in some PDB files (for example, if residue 260 follows immediately after residue 254 with no actual break in the protein backbone). Next, initial redundancy removal using *RemoveRedundantBeforeHot.py*, as described in **Note 7**, yields the file *conglomerated_Gapless_NonRedundant.csv*. This file can be used to identify hot loops using *HotLoopAssignment.py*. As described in **Note 8**, user-defined criteria for hot loop identification are used to produce a list of hot loops called *conglomerated_Gapless_NonRedundant_Hot_%_#.csv*. Note that two of the criteria are denoted in the filename, the percent total interface energy (%) and hot spot cutoff (#, in REUs*100). Following hot loop identification, further redundancies from loop overlap are culled using *LoopCulling.py* (*see* **Note 9**). *LoopCulling.py* first removes redundant loops, and then removes overlapping loops to yield two files, *conglomerated_Gapless_NonRedundant_Hot_%_#_NonRedundant.csv* and *conglomerated_Gapless_NonRedundant_Hot_%_#_NonRedundant_NonOverlapping.csv*. These final lists of hot loops can then be manually sorted, filtered, and organized for specific applications, for instance to identify targets for cyclic peptide design.

4   Notes

1. *Downloading and organizing the PDB input set.*

   The RCSB is a vast resource for protein structure data, and searching for PDB files that fulfill specific criteria is easily done with the advanced search tool www.rcsb.org/pdb/search/advSearch.do [33]. In our implementation of LoopFinder focused on finding loop-mediated interactions that could be inhibited by macrocycles, we used the advanced search tool to find and download all PDB files that had between two and ten chains in the biological assembly. We excluded structures with 90% sequence identity to avoid redundancy. The RCSB search tool can conveniently output the list of PDBIDs matching your criteria into a text file, and the text file can then be used to download PDB files into the proper folder using the *pdb_download.py* script. Using these criteria we downloaded 27,715 PDB files as a comprehensive input set. We chose to start with a very inclusive input set, but it would be simple to conduct a more focused search of the PDB by using input sets based on function or structure. Future implementations could also include structures with high sequence identity, allowing the identification of subtle differences in how highly homologous proteins use loops to mediate PPIs. No matter the input set, it is necessary to place the PDB files in separate folders based on the number of chains in the biological assembly because LoopFinder needs to be run separately for assemblies with different numbers of chains.

2. *Executing LoopFinder to identify interface loops.*

The LoopFinder program can be run using the *LF_run.sh* script (Fig. 3). Several parameters within this script can be edited to customize the loop parameters (*see* **Note 3**). The first parameter is the distance for categorizing which chains within a multichain assembly are in contact, and thus make up a binary interface. In the *LF_run.sh* script, this is the `bdist` parameter. To date, we have defined interfaces using a 6.5 Å distance measurement (`bdist 6.5`), whereby if any atom within a chain is ≤6.5 Å away from any atom within another chain, the chains are characterized as having an interface. This value was chosen based on the similar distance term used by PyRosetta's *ala_scan. py* program to define interface residues [9, 10].

LoopFinder locates all interface loops based on the specified parameters, and then outputs a number of files. First, scrubbed versions of each PDB file that have the header and remarks removed are produced. Next, LoopFinder saves individual PDB-style files for each binary protein–protein interface. For example, the interaction between Nrf2 and Keap1 (PDBID 2FLU) has two chains denoted P and X [34]. LoopFinder finds a loop on chain P that contacts chain X, and makes a new file called *2flu_XP_c.pdb*. If an input PDB file (*PDBID.pdb*) contains chains A, B, and C and each shares an interface with each other chain, LoopFinder outputs separate PDB files that contain only chains A and B, only chains B and C, and only chains A and C (*PDBID_AB_c.pdb, PDBID_ BC_c.pdb* and *PDBID_AC_c.pdb*). This simplifies and accelerates the computational alanine scanning performed in later steps. In addition to the new PDB files, a text file is produced that identifies each residue and assigns a binary value based on

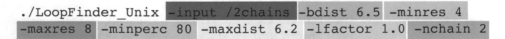

Fig. 3 The *LF_run.sh* script is shown above. Each command-line parameter in the LoopFinder program is customizable. The directory containing the PDB files (*red*) should be organized by the number of chains in the assembly and correspond to the `nchain` value (*gray*). The `bdist` parameter specifies the maximum distance that characterizes an interface (*orange*). The number of residues within the loop is specified by the `minres` and `maxres` parameters (*green* and *purple*, respectively). The `minperc` parameter (*pink*) corresponds to the requirement for how many atoms in the loop within `bdist` from the corresponding partner chain. The maximum termini distance of the loop in Å (*yellow*) is an essential parameter for identifying loops and not extended regions. The final parameter, `lfactor` (*blue*), is used to scale the `maxdist` value for loops of 5 and fewer residues to avoid identifying α-helical turns

whether the residue resides at the interface (*PDBID_AB_res.txt*, *PDBID_BC_res.txt*, *PDBID_AC_res.txt*). The third file produced is a text file called *Loops.txt*, which contains information on all identified loops. The *Loops.txt* file is the primary output of LoopFinder, and contains all interface loops for further analysis. *Loops.txt* includes the following data, in the following order: PDBID, the chain on which the loop resides, the number of residues within the loop, termini distance, the total number of residues on the chain, the identity of each residue in the loop, the identity of the partner chain, and the percent of the loop at the interface. The percent of the loop at the interface is the percentage of the atoms within the loop that are within the specified `bdist` distance to any atom on the partner protein. A representative excerpt from a *Loops.txt* file is shown in Fig. 4. The binary interface PDB files, the lists of residues that indicate which are at the interface, and the *Loops.txt* files are all outputted into directories in a manner that is ready for direct input into PyRosetta for computational alanine scanning. LoopFinder also produces a separate file to use as a command line input file for running PyRosetta, called *cmd_line_input.txt*. This file, an excerpt of which is shown in Fig. 5, contains all information required by PyRosetta for running computational alanine scanning on all loop-containing interfaces, including their PDBIDs, the chains involved in each interaction, the interface cutoff value and the number of trials for each calculation.

3. *Considerations for setting loop definition parameters.*

Loop definitions are provided by the user to the LoopFinder program, allowing for custom definitions for loop size and proximity parameters. In addition to setting the `bdist` parameter described in **Note 2**, users can define three parameters that dictate the size and shape of the loop itself. These three parameters are loop length, which allows the user to specify a minimum and maximum length in residues (for example `minres`

```
2flu P  7 5.9126 num res: 61 LEU 76 ASP 77 GLU 78 GLU 79 THR 80 GLY 81 GLU 82
INTERFACES: XP 100
2flu P  8 4.58   num res: 61 LEU 76 ASP 77 GLU 78 GLU 79 THR 80 GLY 81 GLU 82 PHE 83
INTERFACES: XP 100
2flu P  6 5.3197 num res: 61 ASP 77 GLU 78 GLU 79 THR 80 GLY 81 GLU 82
INTERFACES: XP 100
2flu P  7 6.126  num res: 61 ASP 77 GLU 78 GLU 79 THR 80 GLY 81 GLU 82 PHE 83
INTERFACES: XP 100
```

Fig. 4 The *Loops.txt* file gives data on all of the loops for a given PDB file that fit the loop definition parameters. The file has two lines per loop. The first line contains the PDB ID (*red*), chain on which the loop resides (*orange*), number of residues in the loop (*green*), distance from C_α to C_α of N- and C-terminal residues (*purple*), number of residues at the interface (*pink*), and each residue of the loop itself (*yellow*). The second line has the interface (*blue*) and the percentage of the loop atoms that reside within the given distance parameter of the partner chain (*gray*)

```
--pdb_filename=2fkn_AB_c.pdb --partners=A_B --interface_cutoff=6.5 --trials=20 --trial_output=2fkn_ddg
--pdb_filename=2fl5_LH_c.pdb --partners=L_H --interface_cutoff=6.5 --trials=20 --trial_output=2fl5_ddg
--pdb_filename=2flq_AB_c.pdb --partners=A_B --interface_cutoff=6.5 --trials=20 --trial_output=2flq_ddg
--pdb_filename=2flu_XP_c.pdb --partners=X_P --interface_cutoff=6.5 --trials=20 --trial_output=2flu_ddg
--pdb_filename=2fml_AB_c.pdb --partners=A_B --interface_cutoff=6.5 --trials=20 --trial_output=2fml_ddg
--pdb_filename=2fmy_AB_c.pdb --partners=A_B --interface_cutoff=6.5 --trials=20 --trial_output=2fmy_ddg
--pdb_filename=2fne_AB_c.pdb --partners=A_B --interface_cutoff=6.5 --trials=20 --trial_output=2fne_ddg
--pdb_filename=2fnu_AB_c.pdb --partners=A_B --interface_cutoff=6.5 --trials=20 --trial_output=2fnu_ddg
```

Fig. 5 The LoopFinder program generates a *cmd_line_input.txt* file that can be used for the submission of jobs for PyRosetta alanine scanning mutagenesis. The information contained within this file is the PDB file name (*red*), the partner proteins (*orange*), the interface cutoff distance (*green*), number of trials to run for each interface (*purple*), and the name of the file for data output (*yellow*). The *int_eng_run.sh* script will execute each line of *cmd_line_input.txt* using PyRosetta's *ala_scan.py* program

4 and maxres 8), proportion of the loop at the PPI interface, which allows the user to specify the percentage of the residues with at least one atom within the bdist distance of the binding partner (for example minperc 80), and the distance between the termini of the loop, which allows the user to specify a maximum distance in Å from the C_α of the first residue to the C_α of the last residue (for example maxdist 6.2). A scaling factor is also used for smaller loop sizes (for example lfactor 1) as described below. This Note discusses important factors for setting each of these parameters. All of these parameters are contained in the *LF_run.sh* script for execution of the LoopFinder program as discussed in Subheading 3. LoopFinder searches the input set of PDBs for loops that match the specified parameters, and while our initial parameters (described below) were designed to identify loops suitable as starting points for designing macrocycles that inhibit PPIs, other sets of parameters can be used to apply LoopFinder to other problems in biochemistry and drug design.

Setting the minres and maxres parameters. In our applications to date, we have limited loop lengths to between four and eight amino acids, in order to focus on starting points for the design of small cyclic peptides. Loops with lengths less than four are not particularly meaningful, so setting minres to 4 is effectively setting the minimum loop size as small as possible. Maximum loop length can be defined differently depending on the specific questions being asked. Our upper limit of eight amino acids was informed by our goal to design cyclic peptides of similar size to potent natural products and known peptide inhibitors of PPIs [27]. LoopFinder can certainly be applied to discover larger loops, or even whole protein domains with their termini near each other. Such loop sets would be useful for a variety of applications. For instance, sets of loops ranging from 7 to 20 amino acids would be useful for protein grafting onto loop DARPins [35]. Also, antibodies use their hypervariable regions to facilitate protein binding with loops of 5–17 amino acids, so searching for loops in this size range would produce a set of antibody and antibody-like loops [36]. Setting even larger maximum loop lengths could identify entire protein domains with termini near each other, which could assist in the design of tandem repeat proteins or circularly permuted proteins [37–39]. Users can set `minres` and `maxres` to match their ultimate goals for peptide and protein design.

Setting the minperc parameter. In early development of LoopFinder, we noticed a significant proportion of loops that were not major contributors to the interface, but happened to have one hot spot that was within interaction distance (6.5 Å) of the binding partner. We regarded these as hot spots that happened to be on loops, rather than hot loops suitable for macrocycle design. Thus, to ensure that most loop residues are in close proximity to the binding partner, an interface distance requirement `minperc 80` was introduced as a loop definition parameter. This requires at least 80% of the residues within the loop to have at least one atom residing within 6.5 Å of the binding partner. This parameter can be changed based on the user's goals and expectations. We set the 80% limit empirically, to best identify only those loops that make extended contact with the binding partner. Ultimately, this focused our loop searches on epitopes with high ligand efficiency [2, 40, 41]. Reducing `minperc` below 80 also identifies some internal loops that contribute to the hydrophobic core of their parent proteins. Computational alanine scanning of these internally-facing loops would likely produce residues with high $\Delta\Delta G_{res}$ values, but these would be due to destabilization of overall protein structure and not due to direct contacts with the binding partner. Thus, for cyclic peptide design, we recommend keeping this limit high. For larger loop sizes, however, it might be necessary to reduce `minperc`. For instance, if the ultimate goal is grafting loops onto antibodies or DARPins, setting `minperc` too high may rule out loops that would nonetheless have sufficient surface exposure once grafted onto the scaffold. Thus, for larger loops we anticipate that `minperc` will likely need to be relaxed.

Setting the maxdist parameter. The third criterion for defining a loop in LoopFinder is the termini cutoff distance. While this is not a classic requirement for a "loop" structure, it was critical for the identification of loops that are optimal targets for macrocycle design. For this calculation, we used a maximum termini distance `maxdist 6.2`, denoting that loops must measure ≤6.2 Å from the C_α of the first loop residue to the C_α of the last loop residue. The 6.2 Å value was estimated as the approximate length of a dipeptide-sized linker, which could readily be introduced to cyclize loop epitopes. Importantly, setting `maxdist` to 6.2 also avoids α-helices (α-helices of length 5–8 amino acids have end-to-end distances of 6.3–10.8 Å) and β-strands (β-strands of length 4–8 residues have end-to-end distances of 9.8–22.7 Å). The cutoff of 6.2 Å was applied to loops of length 6, 7, and 8 amino acids, but was too long for shorter loops. For loops of length 4 and 5, we empirically adjusted the termini distance parameter to 3.92 Å and 4.9 Å, respectively. This was done using the lfactor value. The specified value for lfactor is divided 3.5 Å (an estimate for the average length of an amino acid in extended conformation) to yield a fraction. This fraction is the limit for loop length compared to a fully extended structure, which allows for identification of shorter fragments with particularly closed conformations. For example, an lfactor of 1 yields a fraction of 0.28. If the maximum linear length of a five-amino acid peptide would be estimated at 17.5 Å, the lfactor would limit the length to 0.28 times 17.5 Å, or 4.9 Å. An lfactor of 1 was used because it limits the loop length for shorter loops to less than one-third of the total length of the segment if it were in an extended conformation. For less stringent end-to-end distances of small loops, the lfactor value can be increased.

Overall, the end-to-end distance parameter is one of the main attributes that allows LoopFinder to uncover unique PPIs and interaction epitopes, providing a more structure-specific epitope search than the Schueler-Furman's PeptiDerive approach [15, 42] and a completely orthogonal search compared to the Arora lab's search algorithms for α-helix and β-strand PPI epitopes [17, 21, 43].

4. *Details of computational alanine scanning.*

LoopFinder's output files are formatted so they can be directly submitted to PyRosetta v2.012 computational alanine scanning mutagenesis [31]. This program calculates the changes in binding free energy when residues at a protein interface are changed to alanine using the *ala_scan.py* script. Briefly, PyRosetta mutates each residue located at the interface on a given chain to alanine, repacks the side chains of surrounding residues, and calculates the binding free energy of the mutant chain [9]. Binding energy in PyRosetta is calculated using score functions. The contribution of each residue at the interface to the overall PPI energy, denoted $\Delta\Delta G_{res}$, is calculated in Rosetta Energy Units (REUs). The score function used in this project was a modified version of the standard Rosetta score function with modifications that were found to better suit general alanine scanning on a wide range of protein–protein interfaces, but lacks environment-dependent hydrogen-bonding terms [9, 10]. For each interface, PyRosetta runs 20 trials and these results are averaged to calculate the mean energy contribution for each residue. The results of alanine scanning for each PPI of interest are output to separate files, *PDBID_ddg_A_B1.txt* (where A and B are the chains forming this binary interface, for example *2flu_ddg_X_P1.txt*, *2flu_ddg_X_P2.txt* ... *2flu_ddg_X_P20.txt*) and averaged results are output into *PDBID_mean_A_B.txt* (for example, *2flu_mean_X_P.txt*).

To execute the PyRosetta alanine scan on cluster-based computing systems in batch format for large sets of interfaces, the *int_eng_run.sh* script is included. This script loads the PyRosetta module and calls each line from the *cmd_line_input.txt* file to submit individual interfaces for computational alanine scanning using the *ala_scan.py* program. As mentioned above, this simple shell script should be edited to contain the proper directory containing PDB input files, and these should all have the same number of protein chains (Fig. 1). Running *cmd_line_input.txt* on large batches of PDB files can be time intensive, as each interface energy calculation requires approximately 10–15 min.

5. *Combining the loop sets and energy data.*

The *conglomerate.py* script is used to combine the computational alanine scanning mutagenesis data in *PDBID_mean_A_B.txt* files (for example, *2flu_mean_X_P.txt*) with the lists of interface loops in the *Loops.txt* file. This script compiles energy information for each loop in *Loops.txt*, including $\Delta\Delta G_{res}$ for each residue in the loop, the sum of all $\Delta\Delta G_{res}$ for the loop, and the z-score, which is a measurement that assesses the error over the 20 iterations for each interface. It then outputs the data into *conglomerated.txt*, which can be converted to a CSV file using *CSVconverter.py*, producing *conglomerated.csv*.

6. *Modifications to Rosetta-generated alanine scanning energies.*

During LoopFinder development, we examined how raw values from the computational alanine scans affected our overall analysis of loop energies. We noticed two problems that, in a minority of cases, seemed to prevent the accurate interpretation of $\Delta\Delta G_{res}$ values as realistic side chain binding energies. First, some $\Delta\Delta G_{res}$ values were unreasonably large. To prevent extreme effects of such $\Delta\Delta G_{res}$ values on our overall analysis, we reduced all energy values ≥ 4.5 REUs to 4.5 REUs. This "cap" on hot spot energies ensured that the effects of individual hot spots on overall loop energies would not be overestimated. The second issue was that the PyRosetta computational alanine scan sometimes produces negative values for $\Delta\Delta G_{res}$ [9, 10]. Previous applications ignored these negative values [15, 17, 21], but we attempted to minimize their effects on our analysis of overall loop energies, and even to use them to inform hot spot identification. A subset of residues in interface loops that PyRosetta assigned negative $\Delta\Delta G_{res}$ values were manually inspected, and it was observed that highly negative values ($\Delta\Delta G_{res} \leq -2.0$ REUs) nearly always made important hydrogen bonding interactions with the binding partner. This is consistent with prior interpretations of negative $\Delta\Delta G_{res}$ values, which were interpreted as the removal of one partner in a buried hydrogen bond [9]. Large, negative $\Delta\Delta G_{res}$ values were thus corrected such that all $\Delta\Delta G_{res}$ values below -2.0 REUs were set to 1 REU. This ensured that these were counted as hot spot residues in subsequent analyses. All energies that were only slightly negative (-2.0 REUs $\leq \Delta\Delta G_{res} \leq 0.0$ REUs) were set to 0.0 REUs based on our empirical observations that such residues do not typically have obviously important contacts within the interface. This adjustment effectively ignores any contributions by these residues, and prevents the possibility of underestimating total loop energies and total interface energies due to negative $\Delta\Delta G_{res}$ values, or in some cases, having a total interface energy sum with a negative value. These conversion parameters are built into the *conglomerate.py* script.

7. *Removal of artifacts from the loop list.*

   After combining the loop data with the $\Delta\Delta G_{res}$ data and modifying overlarge and negative $\Delta\Delta G_{res}$ values, the interface loops can be processed to check for errors and remove redundancies. *GapExcisor.py* acts on *conglomerated.csv* to remove gaps that arise from nonsequential numbering in some PDB files, and yields the file *conglomerated_Gapless.csv*. *RemoveRedundantBeforeHot.py* takes *conglomerated_Gapless. csv* as an input, locates identical loops, and retains only the loop with the highest average $\Delta\Delta G_{res}$. This predominantly serves to remove redundant loops from homomultimer structures where multiple, identical protein chains are assembled symmetrically. All together, these redundancy checks typically winnow the overall number of interface loops by roughly 60%. The data within the *conglomerated_Gapless_NonRedundant. csv* file is now the input for the next step: identifying the loops of greatest interest based on "hot loop" criteria.

8. *Identifying hot loops using custom criteria.*

   The overall interface loop set can be very large. For example our most recent LoopFinder run identified 83,170 nonredundant interface loops using the parameters described in **Note 3**. Thus, it is critical to rank loops according to criteria important for your specific application, and to set these criteria using the compiled information on loops and their $\Delta\Delta G_{res}$ values. To identify loops of interest, we developed "hot loop" criteria for the identification of loops that are strongly involved in the binding interaction between protein partners. Our initial hot loop criteria, as reported in 2014, were: having an average $\Delta\Delta G_{res}$ (the sum of all $\Delta\Delta G_{res}$ values for the loop divided by total number of residues in the loop) of at least 1.0 REUs, having at least three hot spots (defined as individual residues with $\Delta\Delta G_{res} \geq 1.0$ REUs) within the loop, or having at least two consecutive hot spot residues [30].

Upon further analysis of the hot loops that arise from these criteria, it was apparent that there was a large overlap between hot loops with two consecutive hot spots and the hot loops with three or more hot spots overall. We also concluded that hot loops with two consecutive hot spot residues may be amenable to inhibition by small molecules, since two adjacent hot spots form a contiguous epitope that is unlikely to require loop structure for target binding. Thus, we developed a new set of requirements that more accurately reflect our primary goal of developing cyclic peptides. First, we relaxed the standard for the strength of a hot spot to ≥ 0.6 REUs. This was informed by an analysis of peptide–protein interactions from the PDB, in which we observed that many high-affinity peptides did not have many hot spot residues with $\Delta\Delta G_{res} \geq 1.0$ REUs but did have hot spots with $\Delta\Delta G_{res} \geq 0.6$ REUs. We discarded the consecutive-hot-spot criterion for reasons described above, and replaced it with an estimation of the proportion of the total binding energy attributable to the loop. This was enabled by adding to all LoopFinder runs the calculation of the total interface energy. Since total loop $\Delta\Delta G$ energies were calculated by summing up individual $\Delta\Delta G_{res}$ values, we estimated total interface energy by adding up $\Delta\Delta G_{res}$ values for all residues at the interface. This produced a relevant measure of the percent contribution of the loop to the total interface energy. Using this most recent parameter, we can identify as particularly promising any hot loops that comprise a large proportion of the total interface energy, from 25% contribution to 100% contribution. Different users with different applications can consider these and other criteria. Ultimately, for the development of cyclic peptides and macrocycles, we chose to identify hot loops as those that have either an average $\Delta\Delta G_{res} \geq 0.6$ REUs, those with three or more hot spots, or those that comprise $\geq 50\%$ of the total interface energy. This identified 7225 hot loops from the total interface loop set of 83,170. Figure 6 shows a Venn diagram that illustrates how these criteria identified these loops. As the diagram shows, we recommend using three or more criteria to sort through a large interface loop set, in order to identify those interface loops with multiple desired properties for the desired application. The python script *HotLoopAssignment.py* utilizes the user-defined hot loop criteria of interface energy percentage (%) and hot loop cutoff (#, in terms of REUs*100) to output a file containing loops that meet those criteria (*conglomerated_Gapless_NonRedundant_Hot_%_#.csv*). For example, if the percent total interface energy of the loop is set for 70%, and the hot spot cutoff is set at 60 (≥ 0.6 REUs) a file will be produced named *conglomerated_Gapless_NonRedundant_Hot_70_60.csv*.

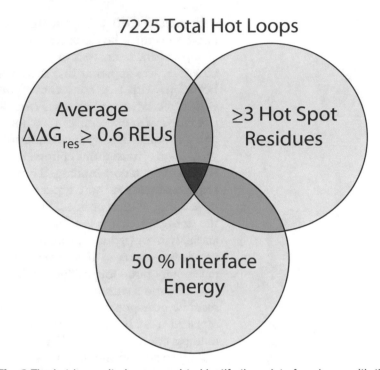

Fig. 6 The hot loop criteria are used to identify those interface loops with the most favorable properties for a specific application. While other criteria are possible, for identifying starting points for macrocycle design we currently use three adjustable criteria: average $\Delta\Delta G_{res}$ of the loop, the number of hot spots contained within the loop (the threshold for defining hot spots can also be customized by the user), and the percent contribution of the loop to total interface energy. In the Venn diagram shown here, we suggest hot loop criteria of average $\Delta\Delta G_{res} \geq 0.6$ REUs, number of hot spots with average $\Delta\Delta G_{res} \geq 1.0$ REUs is at least 3, and percentage of total interface energy contained within the loop is $\geq 50\%$. We expect the central overlapping region to contain hot loops that are the best starting points for cyclic peptide design

9. *Removal of nested and overlapping loops.*

  After hot loop identification, *LoopCulling.py* performs a second redundancy check to remove any hot loop that is entirely contained within another, thereby keeping only the longest possible version of a given hot loop in the output file *conglomerated_Gapless_NonRedundant_Hot_%_#_NonRedundant.csv*. While both versions of the loop may be useful starting points for peptide design, retaining only the longest loop greatly minimizes the numbers of redundant, "nested" loops, while maximizing the useful information provided to the user. *LoopCulling.py* also provides a second output file that removes overlapping loops (*conglomerated_Gapless_NonRedundant_Hot_%_#_NonRedundant_NonOverlapping.csv*). When two loops overlap in sequence but one is not completely nested within the other, the locations of hot spot residues (as defined by the user, as described in **Note 8** we have lowered this threshold to $\Delta\Delta G_{res} \geq 0.6$ REUs) are used to decide which to keep. If two overlapping loops have the same hot spots, *LoopCulling.py* keeps the loop with higher average residue energy. When one loop contains all the hot spots of the other as well as additional hot spots, the loop with more hot spots is kept. In the case that each loop contains at least one hot spot that is not in the other, they are both kept, since these overlapping loops represent potentially independent epitopes for translation to macrocyclic inhibitors.

10. *Using the LoopFinder website for simple tasks.*

In order to make LoopFinder available for wider use, we built loopfinder.tufts.edu, a website where people can submit their own LoopFinder jobs and query the large loop databases we have already produced. LoopFinder's web app allows users to check a single PDB entry of their choice for hot loops according to custom loop definition parameters for `minres`, `maxres`, `maxdist`, and `minperc` (*see* **Note 3**) and our current default hot loop criteria (*see* **Note 8** and Fig. 6). However, users receive the unfiltered list of loops matching their loop definitions and all alanine scanning data, so analysis beyond our criteria for defining hot loops can be performed as desired.

In addition to searching single PDBs for hot loops, all of our own hot loop databases are available through the website. These can be downloaded for analysis using database software, or queried on the website by users for specific proteins, functions, or primary sequences. At present, the hot loop database (7225 hot loops, Fig. 6) is available, and it will be joined soon by databases of hot loops from LoopFinder searches with expanded parameters, including larger loops suitable for protein and antibody grafting (*see* **Note 3**). In the near future, the website will also feature a tool for computational grafting of single hot loops onto a library of scaffold proteins, with a database of scaffolds available for queries and user submission. Finally, data based on clustering analysis of our large loop sets will be made available online. Future publications will describe these grafting and clustering tools in greater detail.

11. *Translating LoopFinder data into cyclic peptide inhibitors of PPIs.*

The careful calculations and judiciously chosen parameters described for our current implementation of LoopFinder has produced a unique set of loop-mediated PPIs. The hot loops within these PPIs, especially those loops that fulfill all three hot loop criteria (Fig. 6, central region), will provide excellent starting points for the design of cyclic peptide inhibitors. In the context of the native PPI, these loops are conformationally stabilized by the global protein structure. To develop the loops into inhibitors, conformational constraints can be applied to synthetic peptides in order to mimic the hot loop. A variety of chemistries are available for this purpose, including head-to-tail cyclization, lactam-bridge cyclization, olefin stapling, disulfide cyclization, and cysteine bis-alkylation [44–46]. We are currently using these synthetic strategies to develop constrained peptides to block loop-mediated PPIs. We also note that these loops can be grafted onto a wide variety of protein scaffolds, with the same goal of stabilizing the hot loop epitope in a manner that allows for high-affinity binding and inhibition of the associated PPI [47].

References

1. Wells JA, McClendon CL (2007) Reaching for high-hanging fruit in drug discovery at protein-protein interfaces. Nature 450(7172):1001–1009

2. Smith MC, Gestwicki JE (2012) Features of protein-protein interactions that translate into potent inhibitors: topology, surface area and affinity. Expert Rev Mol Med 14

3. Arkin MR, Tang YY, Wells JA (2014) Small-molecule inhibitors of protein-protein interactions: progressing toward the reality. Chem Biol 21(9):1102–1114

4. Fosgerau K, Hoffmann T (2015) Peptide therapeutics: current status and future directions. Drug Discov Today 20(1):122–128

5. Craik DJ, Fairlie DP, Liras S, Price D (2013) The future of peptide-based drugs. Chem Biol Drug Des 81(1):136–147

6. Clackson T, Wells JA (1995) A hot-spot of binding-energy in a hormone-receptor interface. Science 267(5196):383–386

7. Cunningham BC, Wells JA (1993) Comparison of a structural and a functional epitope. J Mol Biol 234:554–563

8. Schymkowitz J et al (2005) The FoldX web server: an online force field. Nucleic Acids Res 33:W382–W388

9. Kortemme T, Baker D (2002) A simple physical model for binding energy hot spots in protein-protein complexes. Proc Natl Acad Sci U S A 99(22):14116–14121

10. Kortemme T, Kim D, Baker D (2004) Computational alanine scanning of protein-protein interfaces. Sci STKE 2004:pl2

11. Rajamani D, Thiel S, Vajda S, Camacho CJ (2004) Anchor residues in protein–protein interactions. Proc Natl Acad Sci U S A 101:11287–11292

12. Koes DR, Camacho CJ (2012) PocketQuery: protein–protein interaction inhibitor starting points from protein–protein interaction structure. Nucleic Acids Res 40:W387–W392

13. Gao Y, Wang RX, Lai LH (2004) Structure-based method for analyzing protein-protein interfaces. J Mol Model 10(1):44–54

14. Brenke R et al (2009) Fragment-based identification of druggable 'hot spots' of proteins using Fourier domain correlation techniques. Bioinformatics 25(5):621–627

15. London N, Raveh B, Movshovitz-Attias D, Schueler-Furman O (2010) Can self-inhibitory peptides be derived from the interfaces of globular protein–protein interactions? Proteins 78:3140–3149

16. Gao M et al (2014) Rationally designed macrocyclic peptides as synergistic agonists of LPS-induced inflammatory response. Tetrahedron 70:7664–7668

17. Jochim AL, Arora PS (2010) Systematic analysis of helical protein interfaces reveals targets for synthetic inhibitors. ACS Chem Biol 5(10):919–923

18. Bullock BN, Jochim AL, Arora PS (2011) Assessing helical protein interfaces for inhibitor design. J Am Chem Soc 133(36):14220–14223

19. Patgiri A, Jochim AL, Arora PS (2008) A hydrogen bond surrogate approach for stabilization of short peptide sequences in alpha-helical conformation. Acc Chem Res 41(10):1289–1300

20. Azzarito V, Long K, Murphy NS, Wilson AJ (2013) Inhibition of alpha-helix-mediated protein-protein interactions using designed molecules. Nat Chem 5(3):161–173

21. Watkins AM, Arora PS (2014) Anatomy of beta-strands at protein-protein interfaces. ACS Chem Biol 9(8):1747–1754

22. Jayatunga MKP, Thompson S, Hamilton AD (2014) alpha-Helix mimetics: outwards and upwards. Bioorg Med Chem Lett 24(3):717–724

23. Loughlin WA, Tyndall JDA, Glenn MP, Fairlie DP (2004) Beta-strand mimetics. Chem Rev 104(12):6085–6117

24. Pelay-Gimeno M, Glas A, Koch O, Grossmann TN (2015) Structure-based design of inhibitors of protein-protein interactions: mimicking peptide binding epitopes. Angew Chem Int Ed 54(31):8896–8927

25. Wieland T, Faulstich H (1991) 50 Years of amanitin. Experientia 47(11–12):1186–1193

26. Schreiber SL, Crabtree GR (1992) The mechanism of action of cyclosporine-a and Fk506. Immunol Today 13(4):136–142

27. Bockus AT, McEwen CM, Lokey RS (2013) Form and function in cyclic peptide natural products: a pharmacokinetic perspective. Curr Top Med Chem 13(7):821–836

28. Bock JE, Gavenonis J, Kritzer JA (2013) Getting in shape: controlling peptide bioactivity and bioavailability using conformational constraints. ACS Chem Biol 8(3):488–499

29. Guharoy M, Chakrabarti P (2007) Secondary structure based analysis and classification of biological interfaces: identification of binding motifs in protein–protein interactions. Bioinformatics 23(15):1909–1918

30. Gavenonis J, Sheneman BA, Siegert TR, Eshelman MR, Kritzer JA (2014) Comprehensive analysis of loops at protein-protein interfaces for macrocycle design. Nat Chem Biol 10(9):716–722

31. Chaudhury S, Lyskov S, Gray JJ (2010) PyRosetta: a script-based interface for implementing molecular modeling algorithms using Rosetta. Bioinformatics 26(5):689–691

32. Shulman-Peleg A, Shatsky M, Nussinov R, Wolfson HJ (2007) Spatial chemical conservation of hot spot interactions in protein-protein complexes. BMC Biol 5

33. Berman HM et al (2000) The Protein Data Bank. Nucleic Acids Res 28(1):235–242

34. Lo SC, Li XC, Henzl MT, Beamer LJ, Hannink M (2006) Structure of the Keap1: Nrf2 interface provides mechanistic insight into Nrf2 signaling. EMBO J 25(15):3605–3617

35. Schilling J, Schoppe J, Pluckthun A (2014) From DARPins to LoopDARPins: novel LoopDARPin design allows the selection of low picomolar binders in a single round of ribosome display. J Mol Biol 426(3):691–721

36. North B, Lehmann A, Dunbrack RL (2011) A new clustering of antibody CDR loop conformations. J Mol Biol 406(2):228–256

37. Javadi Y, Itzhaki LS (2013) Tandem-repeat proteins: regularity plus modularity equals designability. Curr Opin Struct Biol 23(4):622–631

38. Yu Y, Lutz S (2011) Circular permutation: a different way to engineer enzyme structure and function. Trends Biotechnol 29(1):18–25

39. Brunette TJ et al (2015) Exploring the repeat protein universe through computational protein design. Nature 528(7583):580–584

40. Villar EA et al (2014) How proteins bind macrocycles. Nat Chem Biol 10(9):723–731

41. Hopkins AL, Groom CR, Alex A (2004) Ligand efficiency: a useful metric for lead selection. Drug Discov Today 9(10):430–431

42. Lavi A et al (2013) Detection of peptide-binding sites on protein surfaces: the first step toward the modeling and targeting of peptide-mediated interactions. Proteins 81:2096–2105

43. Bergey CM, Watkins AM, Arora PS (2013) HippDB: a database of readily targeted helical protein-protein interactions. Bioinformatics 29(21):2806–2807

44. White CJ, Yudin AK (2011) Contemporary strategies for peptide macrocyclization. Nat Chem 3(7):509–524

45. Timmerman P, Beld J, Puijk WC, Meloen RH (2005) Rapid and quantitative cyclization of multiple peptide loops onto synthetic scaffolds for structural mimicry of protein surfaces. Chembiochem 6:821–824

46. Walensky LD, Bird GH (2014) Hydrocarbon-stapled peptides: principles, practice, and progress. J Med Chem 57(15):6275–6288

47. Gould A, Ji YB, Aboye TL, Camarero JA (2011) Cyclotides, a novel ultrastable polypeptide scaffold for drug discovery. Curr Pharm Des 17(38):4294–4307

Chapter 16

Protein-Peptide Interaction Design: PepCrawler and PinaColada

Daniel Zaidman and Haim J. Wolfson

Abstract

In this chapter we present two methods related to rational design of inhibitory peptides:

- **PepCrawler**: A tool to derive binding peptides from protein–protein complexes and the prediction of protein–peptide complexes. Given an initial protein–peptide complex, the method detects improved predicted peptide binding conformations which bind the protein with higher affinity. This program is a robotics motivated algorithm, representing the peptide as a robotic arm moving among obstacles and exploring its conformational space in an efficient way.

- **PinaColada**: A peptide design program for the discovery of novel peptide candidates that inhibit protein–protein interactions. PinaColada uses PepCrawler while introducing sequence mutations, in order to find novel inhibitory peptides for PPIs. It uses the ant colony optimization approach to explore the peptide's sequence space, while using PepCrawler in the refinement stage.

Key words Peptides, Protein–protein interaction inhibitors, RRT, Ant colony optimization, Inhibitor design, Computer aided drug design

1 Introduction

Protein–protein interactions (PPIs) play a crucial role in many cellular processes and pathways. Thus, development of small molecules that affect PPIs has become an intensive research field in Structural Bioinformatics and Computer Aided Drug Design. The task is to discover small molecules that will block a specific PPI, without harming the function (e.g., catalytic activity) of the target proteins [1, 2]. Some of those desired inhibitors include small-molecule inhibitors and especially short peptide inhibitors.

Ora Schueler-Furman and Nir London (eds.), *Modeling Peptide-Protein Interactions: Methods and Protocols*, Methods in Molecular Biology, vol. 1561, DOI 10.1007/978-1-4939-6798-8_16, © Springer Science+Business Media LLC 2017

An interaction between two proteins usually consists of a large interface without a well-characterized binding pocket [3, 4]. Peptides, could be good starting points for new leads in rational design of inhibitory drugs by mimicking part of the interacting surface of one of the proteins [5–9]. Peptides both natural and non-natural, have been used to inhibit PPIs. Some experimental studies have investigated inhibitory peptides as competitive docking partners [10]. Here we will present two methods related to peptide modeling and peptide ligand design.

PepCrawler [11] is a tool for the prediction of protein–peptide complexes at high resolution as well as the derivation of binding peptides out of protein–protein complexes' interface. This method exploits the RRT (Rapidly exploring Random Tree) robotics motivated methodology. It represents the peptide as a robotic arm, with degrees of freedom at the rotational atomic bonds. Thus, the peptide has m+6 degrees of freedom, where m is the number of rotational bonds, and 6 degrees of freedom are for the spatial position of the peptide. The RRT explores this multi-dimensional conformational space, and tries to quickly determine its boundaries and obstacles. Obstacles include clashes between the peptide and the protein, and clashes between the residues of the peptide. In cases of clashes, it tries different side chain conformations for the clashing amino acids, and if it still cannot resolve them, the explored docking solution is considered part of the unfeasible space. The exploration is done in a tree like fashion.

The output is a protein–peptide complex depicting the final predicted peptide conformation (the one with the best computational energy score). It also outputs a funnel score (the slope of the Energy/RMSD plot) which is a measurement of the peptide likelihood to achieve the predicted conformation. It was established on experimental benchmarks that there is a positive correlation between the steepness of the funnel score and the affinity of the peptide binding. In practice, a funnel slope steepness of 5 and above indicated binding of the peptide. For a detailed account of PepCrawler see [11].

PinaColada [12] is a method for computational screening of peptides in order to find novel inhibitors for PPIs. It uses the ant colony optimization method to explore the peptide sequence space. PepCrawler is being used to refine and calculate the energy and funnel score for each tested peptide sequence. The peptide sequence space is the size of 20^n where n is the number of residues (usually 5–15). Thus, there is a need for an efficient way to explore this large space, avoiding getting stuck in local minima. The optimization step of our method is based on the Ant Colony Optimization algorithm, introduced in [13], which has been applied for various tasks including the traveling salesman problem, protein folding [14–16], scheduling, etc. Ant colony optimization is one example of the more general set of swarm algorithms. In nature, ants try to find the shortest path to a food source by spreading pheromones. This is called stigmergic communication. Ants which choose better paths will return faster, and more pheromones will accumulate on their path. The newly arriving ants choose their path based on the pheromone levels on the ground. Thus, they soon converge to the shortest path, or a very good local minimum.

The main idea of ant-colony optimization is having many distributed artificial "ant" processes, which try to complete a given task in a pseudo-random fashion. It also keeps a pheromone network (weighted graph), which the ants update after their life cycle is finished. Ant colony optimization (ACO) fits our problem very well due to its nature of combining good partial solutions to better global solutions as well as due to its inherent balance between exploration (searching the space) and exploitation (drilling down good solutions to minimize them as much as possible). The algorithm is executed in parallel for M ants (in our experiments $M = 40$). For a detailed account of PinaColada, see [12].

2 Materials

PepCrawler has two modes of action. These are the two kinds of input it accepts (all in PDB format):

- A protein–peptide complex.
- A protein–protein complex, from which you would like to extract a strongly binding peptide.

PinaColada has three modes of action. These are the three kinds of input it accepts:

- A protein–peptide complex.
- A protein–protein complex.
- A protein and its binding site (the exact format will be given in Methods).

Note: In both methods, you only need to have one of the above options.

3 Methods

3.1 PepCrawler

PepCrawler can be found at the following URL: <http://bioinfo 3d.cs.tau.ac.il/PepCrawler/>. There is a possibility to download the PepCrawler executable to your computer and apply it locally. We shall focus on the web server application of PepCrawler. The server home is depicted in Fig. 1.

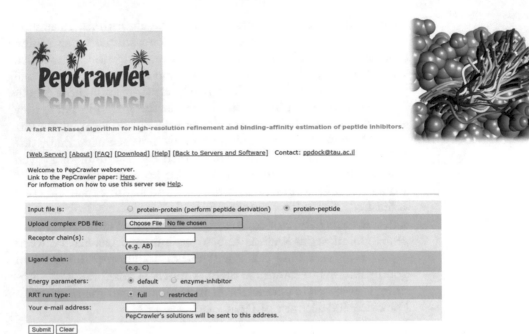

Fig. 1 The web server of the PepCrawler algorithm

Protein-Peptide (default) —the instructions are following the blue lines:

1. Check that the protein–peptide option is chosen in the first line.

2. Upload the protein–peptide complex file (in a PDB format).

3. Write down the receptor chains (the chains must appear in the PDB file), without spaces.

4. Write down the peptide's chain. Usually the peptide is represented as a single chain, and the program requires it. If it is not the case, you could use software like Chimera [17], to save the peptide as a single PDB chain.

5. Determine the program parameters, which govern the energy function. Use the default option, unless your protein is an enzyme, then, the enzyme-inhibitor option is recommended.

6. Determine how wide the RRT exploration is. Use "full" for large exploration area (slower) or "restricted" for a small exploration area (faster).

7. Email address, which the results will be sent to.

Protein-Protein —the instructions are following the blue lines:

1. Check that the protein–protein option is chosen in the first line.

2. Upload the protein–protein complex file (in a PDB format).

3. Write down the receptor chains (the chains must appear in the PDB file), without spaces.

4. Write down the chain of the second protein of which you want the peptide to be extracted from.

5. Determine the program parameters, which govern the energy function. Use the default option, unless your protein is an enzyme, then, the enzyme-inhibitor option is recommended.

6. Determine how wide the RRT exploration is. Use "full" for large exploration area (slower) or "restricted" for a small exploration area (faster).

7. Email address, which the results will be sent to.

3.2 PinaColada

PinaColada can be found at the following URL:: <http://bioin fo3d.cs.tau.ac.il/PinaColada/>. We will go over the steps to use PinaColada server. The server home is depicted in Fig. 2.

- **Protein-Protein (default)** —the instructions are following the blue lines:

  1. Check that the protein–protein option is chosen in the first line.

  2. Normal or extended options.

  3. Upload the protein–protein complex file (in a PDB format).

  4. Write down the receptor chains (the chains must appear in the PDB file), without spaces.

  5. Write down the chain of the second protein of which you want the peptide to be extracted from.

  6. Binding site isn't relevant for this mode.

  7. Specify the peptide's length.

  8. Extended options:

     - Number of rounds (reduce to increase speed).
     - Ants per round (reduce to increase speed).
     - Number of output peptides.
     - Upload biological constraints (format in the help section).

  9. Email address, which the results will be sent to.

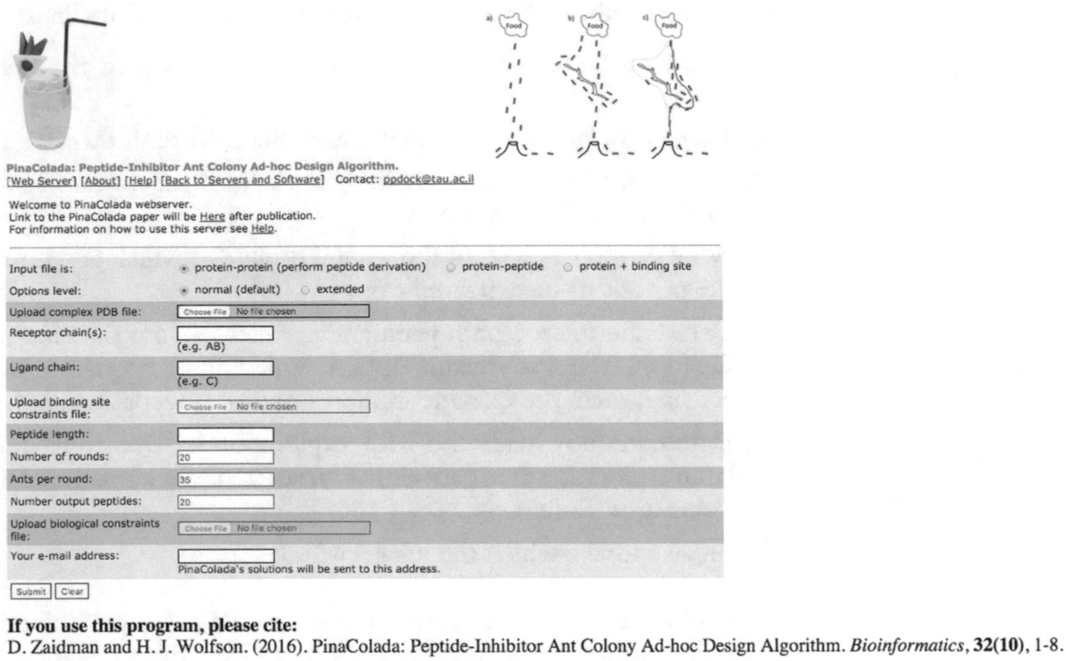

PinaColada: Peptide-Inhibitor Ant Colony Ad-hoc Design Algorithm.
[Web Server] [About] [Help] [Back to Servers and Software]   Contact: ppdock@tau.ac.il

Welcome to PinaColada webserver.
Link to the PinaColada paper will be Here after publication.
For information on how to use this server see Help.

Input file is:	⦿ protein-protein (perform peptide derivation) ◯ protein-peptide ◯ protein + binding site
Options level:	⦿ normal (default) ◯ extended
Upload complex PDB file:	[Choose File] No file chosen
Receptor chain(s):	[ ] (e.g. AB)
Ligand chain:	[ ] (e.g. C)
Upload binding site constraints file:	[Choose File] No file chosen
Peptide length:	[ ]
Number of rounds:	[20]
Ants per round:	[35]
Number output peptides:	[20]
Upload biological constraints file:	[Choose File] No file chosen
Your e-mail address:	[ ] PinaColada's solutions will be sent to this address.

[Submit] [Clear]

If you use this program, please cite:
D. Zaidman and H. J. Wolfson. (2016). PinaColada: Peptide-Inhibitor Ant Colony Ad-hoc Design Algorithm. *Bioinformatics*, **32(10)**, 1-8.

Fig. 2 The web server of the PinaColada algorithm

- **Protein-Peptide** —the instructions are following the blue lines:
  1. Check that the protein-peptide option is chosen in the first line.
  2. Upload the protein–peptide complex file (in a PDB format).
  3. Write down the receptor chains (the chains must appear in the PDB file), without spaces.
  4. Write down the peptide's chain. Usually the peptide is represented as a single chain, and the program requires it. If it is not the case, you could use software like Chimera, [17] to save the peptide as a single PDB chain.
  5. Binding site isn't relevant for this mode.
  6. Specify the peptide's length.
  7. Extended options:
     - Number of rounds (reduce to increase speed).
     - Ants per round (reduce to increase speed).
     - Number of output peptides.
     - Upload biological constraints (format in the help section).
  8. Email address, which the results will be sent to.

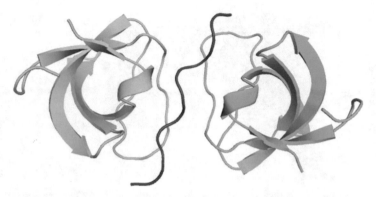

Fig. 3 Case study for PepCrawler. 1ZUK homo dimer of the Yeast BBC1 Sh3 domain complexed with a peptide from Las17. This is the input we used for the example run of PepCrawler

- **Protein-Binding Site**:
    1. Check that the protein + binding site option is chosen in the first line.
    2. Upload the single protein file (in a PDB format).
    3. Write down the receptor chains (the chains must appear in the PDB file), without spaces.
    4. Ligand chain isn't relevant for this mode.
    5. Upload the binding site information, as specified in the help section.
    6. Specify the peptide's length.
    7. Extended options:
        - Number of rounds (reduce to increase speed).
        - Ants per round (reduce to increase speed).
        - Number of output peptides.
        - Upload biological constraints (format in the help section).
    8. Email address, which the results will be sent to.

4   Example of Web Server Applications

4.1   PepCrawler

Input:

entry: 1ZUK (Fig. 3)—a complex of a homo-dimer and a peptide.

Output:

When the algorithm has finished, the user receives a mail with the following output. We'll go over the main files:

1. FunnelScore.txt: Final funnel score is: 17.8798721437.

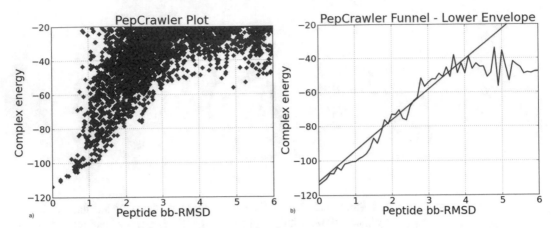

Fig. 4 Funnel plot of PepCrawler conformations. On the *left* (**a**), the graph of computational energy/rmsd of the sampled conformations. This graph is later referred as the funnel graph. On the *right* (**b**) we see the slope of the funnel graph. The slope is later referred to as the funnel score. The funnel score is used to assess the likelihood of the peptide to bind to the receptor. If the funnel score is > 5, we consider the peptide as likely to bind

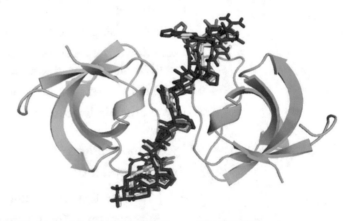

Fig. 5 Five highest scoring conformations of PepCrawler

This is the funnel score. Any number above 5 indicates high chances of binding to the protein.

2. EnergyChartOfMin.png: Fig. 4 shows the funnel shape of the conformations' energy scores. This graph is shown in Fig. 4a

3. LowerEnvOfMin.png: This shows the funnel slope of the energy/RMSD graph. This graph is shown in Fig. 4b

4. StartConfAll, Lig, Prot: These are the starting conformations of both the ligand, the receptor and the whole complex, as shown in Fig. 3.

5. MinEnrgAll, Lig: These are the output peptide in complex with the receptor protein as well as by itself.

6. Sol0, 1, 2, 3, 4-Lig: Fig. 5 shows the best five solutions of peptide conformations, according to the energy score.

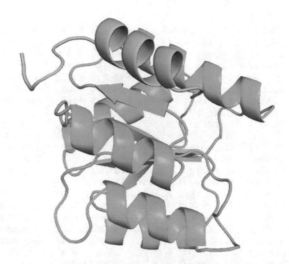

Fig. 6 Case study for PinaColada. 2PL9: Crystal Structure of CheY-Mg(2+)-BeF(3) (-) in Complex with CheZ(C19) Peptide solved from a P2(1)2(1)2 Crystal. For the example run of PinaColada we used only one of the three proteins which comprise the complex

4.2   PinaColada

Input:

PDB entry: 2PL9 (only one of the three units) is depicted in Fig. 6.

Output:

When the algorithm has finished, the user receives a mail with the following output. We'll go over the main files:

1. Best.log: Fig. 7 shows the file containing the peptide suggestions with several parameters including: sequence, energy, funnel score, and interface size. Specifically, energy is the computational energy computed by PepCrawler, which is similar to FireDock energy function, funnel score is the slope of the Energy/RMSD plot, which is used as a measure for affinity, and the interface size is the number of peptide amino acids in the active site of the interaction. One could see a pattern in the results. By running the algorithm a few times (due to its random nature), one could get an idea about the receptor binding specificity profile as done in [12].

2. PinaColada.log: This is the overall log file of the algorithm, including all the ants in each round and their appropriate score. This is more useful in case you want to see what happened at which round. Otherwise, Best.log will be more useful, containing all the best overall suggestions.

3. Best Ants: This is the folder, containing all the predicted PDB of the protein–peptide complexes in the output of PinaColada. You could open them with every software which can handle PDB format like.

```
Best ants list:
Best ant number 1: Energy: -48.51, Funnel: 7.76, Sequence: DSSIWCLDGV, Interface Size: 6
Best ant number 2: Energy: -44.69, Funnel: 7.92, Sequence: EGSIHCLDGV, Interface Size: 6
Best ant number 3: Energy: -42.15, Funnel: 8.57, Sequence: QASIRPLCNV, Interface Size: 6
Best ant number 4: Energy: -43.70, Funnel: 6.83, Sequence: HNSIHLLCGV, Interface Size: 6
Best ant number 5: Energy: -42.99, Funnel: 6.73, Sequence: AGKIMMLGGV, Interface Size: 7
Best ant number 6: Energy: -40.91, Funnel: 7.66, Sequence: NGSITPGYGV, Interface Size: 6
Best ant number 7: Energy: -42.34, Funnel: 6.82, Sequence: TGMIRCLSGV, Interface Size: 6
Best ant number 8: Energy: -40.98, Funnel: 7.01, Sequence: AAKFWCLDGV, Interface Size: 7
Best ant number 9: Energy: -42.19, Funnel: 6.38, Sequence: SGSVRPLYGV, Interface Size: 6
Best ant number 10: Energy: -42.78, Funnel: 5.82, Sequence: AGYVRPIDGV, Interface Size: 6
Best ant number 11: Energy: -40.94, Funnel: 6.65, Sequence: CGYVRPLDGV, Interface Size: 6
Best ant number 12: Energy: -40.44, Funnel: 6.37, Sequence: NGSIRPLDGV, Interface Size: 6
Best ant number 13: Energy: -40.58, Funnel: 6.25, Sequence: ACHIRQLDGV, Interface Size: 6
Best ant number 14: Energy: -37.54, Funnel: 7.64, Sequence: AGSIIGLDVV, Interface Size: 6
Best ant number 15: Energy: -39.63, Funnel: 6.53, Sequence: PVLVEPMDGV, Interface Size: 6
Best ant number 16: Energy: -38.96, Funnel: 6.70, Sequence: DSCHRPLDGV, Interface Size: 6
Best ant number 17: Energy: -40.30, Funnel: 6.00, Sequence: DSTIHCLAVV, Interface Size: 6
Best ant number 18: Energy: -40.24, Funnel: 6.02, Sequence: AGSIRALDGV, Interface Size: 6
Best ant number 19: Energy: -39.75, Funnel: 6.16, Sequence: QVMIRPLCGV, Interface Size: 6
Best ant number 20: Energy: -39.70, Funnel: 6.05, Sequence: GGSIRPLCGV, Interface Size: 6
Best ant number 21: Energy: -37.76, Funnel: 7.05, Sequence: WPSILRLDGY, Interface Size: 2
Best ant number 22: Energy: -36.57, Funnel: 7.54, Sequence: FLTIRPLYSV, Interface Size: 6
Best ant number 23: Energy: -37.49, Funnel: 7.03, Sequence: AGSIQCLVFV, Interface Size: 6
Best ant number 24: Energy: -37.74, Funnel: 6.71, Sequence: AGHFWPLDGV, Interface Size: 4
Best ant number 25: Energy: -38.06, Funnel: 6.33, Sequence: DSTIRPFDGV, Interface Size: 6
Best ant number 26: Energy: -37.62, Funnel: 6.50, Sequence: AGSARPLDVV, Interface Size: 6
Best ant number 27: Energy: -38.95, Funnel: 5.38, Sequence: AGSIRQTYGV, Interface Size: 6
Best ant number 28: Energy: -32.56, Funnel: 9.30, Sequence: AAKFSRLSQV, Interface Size: 7
Best ant number 29: Energy: -39.38, Funnel: 4.79, Sequence: FGAVEKLDGV, Interface Size: 6
Best ant number 30: Energy: -33.05, Funnel: 8.54, Sequence: AGLIGRLDQV, Interface Size: 6
```

Fig. 7 PinaColada's results file. Best.log is the result file, including the best scoring peptide sequences, their score, funnel score, and the surface with the receptor

4. For example, Fig. 8 shows one of the five best overall suggestions.

5. The highest scoring five peptides in PinaColada's output are depicted in Fig. 9.

5   Conclusions and Future Work

We presented two algorithms dealing with peptide design and binding conformation prediction.

PepCrawler receives an initial protein–peptide structure or a protein–protein structure from which to derive a binding peptide. Its output is a refined model of the protein–peptide interaction at high resolution, after searching the vast space of possible conformations. It is very efficient and can predict a peptide conformation in a matter of minutes. PinaColada is a peptide design algorithm. It receives an initial protein–peptide, protein–protein, or protein+ binding site. Its output is a list of suggested peptide sequences for inhibition of a specific protein–protein interaction. This method uses the efficient ant colony optimization scheme, and the fast PepCrawler, in order to explore the computational energy of numerous peptides and suggest those most likely to bind. In the future we plan to explore several ideas for future work, involving

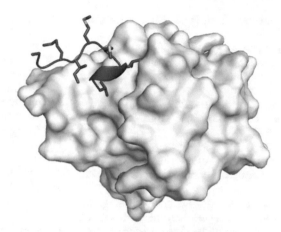

Fig. 8 PinaColada's example result. This is one of the five best scoring peptides from PinaColada's run. We can see that the side chains are designed to match the sub-pockets within the binding site

Fig. 9 PinaColada's five best results. These are the five best scoring peptides from PinaColada. We can see the diversity of choosing the side chains of the peptides

both improving the existing methods and developing new ones. Some of them include improving the ant colony exploration by using several starting points (both with and without an initial peptide, or using several initial peptides at the same time). A promising development could the simultaneous combination of sequence and conformation exploration. It is very challenging because of the exponential nature of both, and the necessity to find a solution in reasonable time.

References

1. Arkin MR, Whitty A (2009) The road less traveled: modulating signal transduction enzymes by inhibiting their protein-protein interactions. Curr Opin Chem 13: 284–290

2. Arkin MR, Wells JA (2004) Small-molecule inhibitors of protein-protein interactions:

progressing towards the dream. Nat Rev Drug Discov 3:301–317

3. Conte L et al (1999) The atomic structure of protein-protein recognition sites. Mol Biol 285:2177–2198

4. Fletcher S, Hamilton AD (2006) Targeting protein-protein interactions by rational design:

mimicry of protein surfaces. J R Soc Interface 3:215–233

5. Nieddu E, Pasa S (2007) Interfering with protein-protein contact: molecular interaction maps and peptide modulators. Curr Top Med Chem 7:21–32

6. Naider F, Anglister J (2009) Peptides in the treatment of AIDS. Curr Opin Struct Biol 19:473–482

7. Mochly-Rosen D, Qvit N (2010) Peptide inhibitors of protein-protein interactions: from rational design to the clinic. Chim Oggi 28:14–16

8. Monfregola L et al (2009) A SPR strategy for high-throughput ligand screenings based on synthetic peptides mimicking a selected subdomain of the target protein: a proof of concept on HER2 receptor. Bioorg Med Chem 17:7015–7020

9. Parthasarathi L et al (2009) Approved drug mimics of short peptide ligands from protein interaction motifs. J Chem Inf Model 48:1943–1948

10. Hancock R et al (2013) Peptide inhibitors of the Keap1-Nrf2 protein-protein interaction with improved binding and cellular activity. Org Biomol Chem 11(21):3553–3557

11. Donsky E, Wolfson HJ (2011) PepCrawler: a fast RRT-based algorithm for high-resolution refinement and binding affinity estimation of peptide inhibitors. Bioinformatics 27(20):2836–2842

12. Zaidman D, Wolfson HJ (2016) PinaColada: peptide-inhibitor ant colony ad-hoc design algorithm. Bioinformatics 32(10):1–8

13. Dorigo M, Gambardella LM (1997) Ant colony system: a cooperative learning approach to the traveling salesman problem. IEEE Trans Evol Comput 1:53–66

14. Hu XM et al (2008) Protein folding in hydrophobic-polar lattice model: a flexible ant-colony optimization approach. Protein Pept Lett 15(5):469–477

15. Shmygelska A et al (2002) An ant colony algorithm for the 2d hp protein folding problem. Lect Notes Comput Sci 2463:40–52

16. Nardelli M et al (2013) Cross-lattice behavior of general aco folding for proteins in the hp model. In: Proceedings of ACM SAC 2013, pp 1320–1327

17. Pettersen EF et al (2004) Ucsf chimera–a visualization system for exploratory research and analysis. J Comput Chem 25(13):1605–1612

Chapter 17

Modeling and Design of Peptidomimetics to Modulate Protein–Protein Interactions

Andrew M. Watkins, Richard Bonneau, and Paramjit S. Arora

Abstract

We describe a modular approach to identify and inhibit protein–protein interactions (PPIs) that are mediated by protein secondary and tertiary structures with rationally designed peptidomimetics. Our analysis begins with entries of high-resolution complexes in the Protein Data Bank and utilizes conformational sampling, scoring, and design capabilities of advanced biomolecular modeling software to develop peptidomimetics.

Key words Peptidomimetics, Protein–protein interactions, Inhibitor design, Computational design

1 Introduction

Recent advances in computational chemistry and structural biology have enabled the systematic targeting of recalcitrant protein–protein interfaces that have traditionally been neglected as inaccessible targets. Protein protein interaction (PPI) inhibitors are sought as probes to reversibly induce a desired cell state, to examine signaling pathways in vitro, and as potential leads to modulate disease states in vivo. Here we describe a successful rational design approach for the development of PPI inhibitors that uses computational methods to identify protein secondary and tertiary structures critical for binding interactions followed by the mimicry of these domains by synthetic scaffolds for the generation of potent inhibitors [1–3].

Several classes of synthetic scaffolds that mimic the backbone or side chain geometries of protein secondary and tertiary structures have been described [4–10]. Elegant strategies for computational evaluation of protein interfaces have also been developed. These include methods that predict functional sites, such as InterProSurf [11], those that construct maps of residue contacts, such as COCOMAPS, and approaches that integrate this

Ora Schueler-Furman and Nir London (eds.), *Modeling Peptide-Protein Interactions: Methods and Protocols*, Methods in Molecular Biology, vol. 1561, DOI 10.1007/978-1-4939-6798-8_17, © Springer Science+Business Media LLC 2017

information with evolutionary conservation data [12, 13]. Analysis of evolutionary conservation is a reliable tool to infer key binding residues and can be especially valuable when high-resolution structures of protein complexes are not available [14]. Public webservers provide useful starting points for such analyses [15–19].

We have used Rosetta for the identification of hot spot residues found on interfacial helices [20–23], β-strands [24], and helix dimers [25]. In combination with the helix [26, 27], β-strand [28–30] and helix dimer [31] scaffolds developed in our group, these computational studies have provided a streamlined approach to discovery of inhibitors for a range of protein–protein interactions (Fig. 1) [31–39]. In this Chapter, we discuss methods that we have found most useful for development of protein-interface disrupting mimetics.

Protein Interfaces

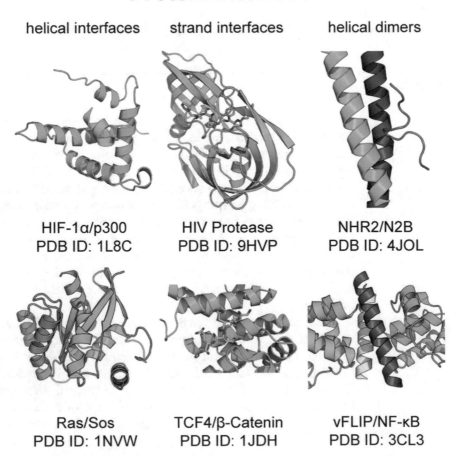

Fig. 1 Protein–protein interactions possess diverse interfacial structures and may thereby be inhibited by a variety of peptidomimetics, which mimic motifs such as the α-helix, β-strand, or helical dimer

2   Theory

We have used the Rosetta software package [40], as well as numerous public servers implementing Rosetta protocols, to analyze protein–protein interfaces and to optimize peptidomimetic compounds [17–19]. Code for analyzing alanine scanning output is available at https://github.com/everyday847/protein-interface-analysis, and Rosetta is available at https://rosettacommons.org/software.

2.1   Alanine Scanning and Solvent-Accessible Surface Area Calculation as Methods to Designate Key Binding Residues

Both alanine scanning and calculation of solvent accessible surface area (SASA) burial upon binding (ΔSASA) are classic methods of interface analysis that produce useful conclusions about the relative importance of different residues. Alanine scanning mutagenesis results may be computed by many methods, from MM/GBSA and MM/PBSA [41, 42] to specially trained energy functions [43]; they may employ only the bound structure or they may explicitly sample bound and unbound state ensembles [44]. The results will vary widely with the choice of force field and sampling level. We perform alanine scanning with Rosetta's talaris2014 scoring function, which combines statistical potentials to describe rotamer and backbone single-body energies with explicit evaluation of multibody Lennard-Jones, electrostatic, hydrogen bonding, and desolvation interactions, and we choose to repack the bound and unbound states. Similarly, multiple algorithms are available for accessible surface area computation [45–47], but the variance between methods is considerably smaller, and the quantity is typically faster to compute. Though SASA is easier to calculate, it is more distantly related to the objective of developing better binders, while an alanine scanning measurement offers an expression of the importance of a particular residue side chain to an interaction's [48]. Alanine scanning can be used directly to estimate a change in binding free energy. SASA and other geometric quantities provide empirical models, along with explicit energetic features [49]. Even though measures like ΔSASA cannot account for the value of polar contacts and hydrogen bonding that contributes to interface stability, they afford an accurate approximation to the overall question of what residues matter most.

2.2   Selecting Appropriate Peptidomimetic Scaffolds

Many peptidic and peptidomimetic scaffolds that possess the approximate backbone and side chain local conformation of particular secondary structure motifs have been described [50–52]. A *peptidic* scaffold stabilizes the conformation of a peptide composed of natural and nonnatural amino acid residues with structural constraints [5, 6, 8, 27, 31, 53–56]. These compounds are useful starting points if many residues, or residues spanning a large range of sequence space or Cartesian space, are necessary for mimicry. We have extensively used hydrogen bond surrogate (HBS) helices to mimic interfacial helical domains and inhibit protein–protein interactions [32, 35, 36, 38, 57]. The development and properties of HBS helices have been previously reviewed [27, 54]. A *nonpeptidic* scaffold or a peptidomimetic is often a small molecule where the backbone features an amide bond isostere in the expected binding mode [26, 30, 52, 58–66]. Peptidomimetics often reproduce the surface of a peptide, i.e. the side chain residues, rather than the backbone conformation [2]. We have developed topographical mimics of strands (triazolamers [28–30]) and helices (oxopiperazine helix mimetics or OHMs [26, 33, 34]) to develop PPI inhibitors mediated by these secondary structures. Peptidomimetics may be more desirable than peptides if the hot spot residues are localized in a small region of protein. A nonpeptidic scaffold might also be preferable if the best binders to an interface are predicted to require highly unusual noncanonical side-chains, for which the conformational repercussions on a peptidic backbone may be hard to predict.

2.3 Conformational Sampling of Peptidomimetic Structures

The systematic variation of free scaffold (i.e. nonsidechain) torsions, followed by energy minimization at a QM level of theory, is a robust but time-consuming method [50]. This method is appropriate for scaffolds one anticipates using repeatedly. For unfamiliar or complex mimetics, software such as OpenEye's OMEGA can rapidly produce libraries of conformers [67]. Care is needed to appropriately filter out conformers that are too similar to efficiently sample conformational space. Of course, whatever in silico sampling method is to be used for subsequent design may be able to produce a set of peptidomimetic conformers in the absence of a target. For example, molecular dynamics codes can simply be run on the solvated inhibitor scaffold in isolation; in Rosetta, Monte Carlo simulations can produce an ensemble of energetically reasonable conformers. This is frequently a reasonable middle ground between the rigor of QM and the rapidity of OMEGA; furthermore, it has the added advantage that the conformers generated are sure to be compatible with the subsequent computational system as well.

2.4 The Target Protein Structure

For the purpose of peptidomimetic design, obtaining an ensemble of native conformations is critical to understand both the error in the protein model and the protein's intrinsic flexibility. When available, both features are partially captured by B-factors or, better, anisotropic B-factors. Thus, protein structural models are contingent on experimental conditions (and, indeed, the experiment in question) and at best represent an estimate given the data—for example, electron density maps from diffraction studies or distance restraints from NMR experiments. High-quality structures contain invaluable information about molecular conformation, but it is essential not to overestimate our certainty in these conformations. Using protein structure prediction and molecular dynamics codes to sample likely near-native conformations can provide us with an ensemble for further work that properly represents the protein and its site-specific uncertainty.

Algorithms can describe whole-protein motions statistically, in terms of normal modes [68] or the RMSF of individual residues [69], which can be treated analogously to a crystallographic B-factor to describe mobility [70]. To provide a set of starting conformations, it is essential for subsequent design and optimization to draw explicit samples from those ensembles. In addition to algorithms like Rosetta's FastRelax and molecular dynamics simulation, the Hilser lab's COREX algorithm provides a conformational ensemble based on local unfolding events that fits well to deuterium exchange data [71].

Particular attention must be given to the nature of the various structures available of a target protein. If *apo* and *holo* structures of the target protein are highly similar, then the structures bound by peptidomimetics are likely similar to both. If they diverge considerably, aggressive sampling in the presence of an example mimetic may be necessary to recover a realistic structure. Appreciable work in multiple computational frameworks attempts to address this issue [72].

3   Methods

Here we describe a rational design process that progresses from target discovery and complex analysis to the design and evaluation of particular peptidomimetic inhibitors (Fig. 2). Typically in our lab, we begin with the classical methodology of interface alanine scanning [73] to discover hot spot residues by the systematic analysis of entries in the Protein Data Bank. These studies provide novel targets for protein–protein interaction inhibitors. Subsequently, we use dock-design algorithms in Rosetta to model diverse peptidomimetic scaffolds and noncanonical functionality [34, 74]. Given its modular design and the demonstration of key components in other settings we expect that this methodology will be general, yielding multiple potential approaches to interface analysis and inhibitor design (Fig. 3).

3.1   Isolating Key Structural Features at the Protein Interface

1. Designate residues at the interface that are critical to the interaction based on the data available.

   (a) If there is limited structural data on the macromolecule to be mimicked, machine learning techniques to infer critical binding residues from sequence [14] or structure features [75] are frequently quite effective.

   (b) The PocketQuery server permits the identification of clusters of anchor residues (high ΔSASA) whose mimicry might be essential to small molecule development (*see* **Note 1**) [76].

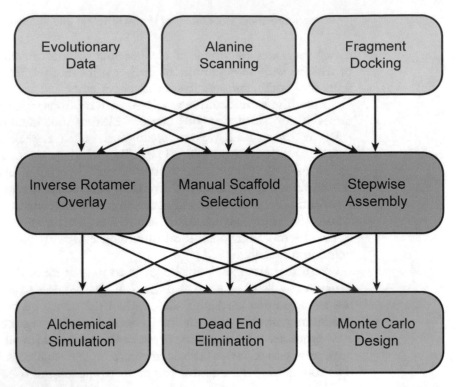

Fig. 2 An example of a peptidomimetic design workflow. We begin with all multichain biological assemblies from the Protein Data Bank with only protein components. Subsequently, we identify all those complexes of good resolution without common defects. We pair all chains from these complexes and perform alanine scanning. Analysis of the resulting data produces high-affinity interfaces with key helices, strands, or helical dimers

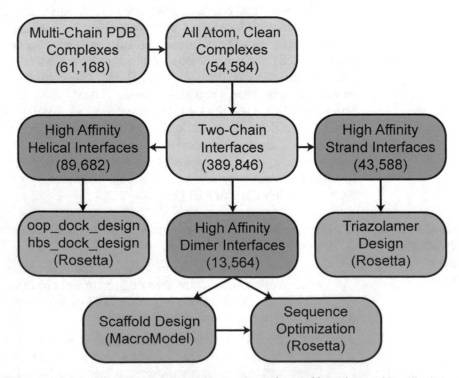

Fig. 3 Different modalities of starting point analysis (*gray*) may be used interchangeably with alternative constructions of an initial pose (*brown*) and, ultimately, inhibitor design (*green*)

(c) The InterProSurf program may be used to perform patch and cluster analyses on interfacial residues to predict functional sites (*see* **Note 2**) [11].

(d) Interface alanine scanning may be performed manually (i.e. by direct truncation of residues to alanine, followed by scoring in some force field) or with specialized tools.

- Alanine scanning can be done with AMBER; within this broad framework for molecular modeling and simulation, MM/PBSA and MM/GBSA are molecular mechanics methods that differ only in solvation model.

- Several Rosetta-based computational protocols can be used: The Robetta webserver, RosettaScripts AlaScan filter, and ddg_monomer application all permit interface alanine scanning analysis (*see* **Note 3**).

(e) If there is no protein–protein complex available, the surface of the protein to be inhibited may be analyzed to find viable binding pockets.

- Docking side chain fragments (e.g. benzyl for phenylalanine, isobutyl for leucine) may produce starting points that could be used, in rather forced analogy, like hot spot or anchor residues (*see* **Note 4**).

- Absent an obvious choice of fragments, methods for interpreting a protein surface as a collection of pockets like Fpocket [77], AlphaSpace [78], and Pocket V.3 [79] provide analogous starting points to identify key regions of the surface with favorable chemical and topological properties.

2. Analyze the protein interface with the secondary structure analysis programs DSSP [47] or STRIDE [80] to determine if multiple critical residues reside on the same secondary structure element: groups of spatially proximal residues, groups of sequence proximal residues, and high-affinity elements are all possible handles for further analysis.

3. Contextualize the energetic role of the high-affinity secondary structure element (or other residue groupings) in question.

(a) The percentage of the protein's total $\Delta\Delta G$ contributed by that element can help ascertain whether it is worth mimicking.

(b) The number of helical faces employed for recognition can affect what inhibitor classes are appropriate choices for its mimicry.

4. Address the possibility that the interface in question has multiple high-affinity secondary structure elements or residue clusters.

(a) In general, only discard an element from consideration if there are more than three in all and if it only contributes 10% or less of the interaction's total $\Delta\Delta G$. Otherwise, eliminating it from consideration will not save enough computation to be worth the risk of prematurely discarding a potential inhibitor binding site.

(b) Key geometric measures can indicate if multiple secondary structure elements may be used for joint mimicry by a complex molecule (*see* **Note 5**).

3.2 Obtaining a Starting Pose with a Peptidomimetic Molecule

1. Select an appropriate set of peptidomimetic scaffolds from those synthetically available and suited to the problem (with conformation matching the identified secondary structure backbone structure).

2. From this set of scaffolds of interest, obtain a conformational ensemble representative of low-energy structures.

3. Map scaffold conformers onto some subset of the array of critical binding residues to obtain a *starting pose*.

    (a) Because a peptidic scaffold typically involves the mimicry of every residue in primary sequence, the problem of scaffold alignment is trivial: every peptidic scaffold may be aligned almost atom-by-atom to the peptide mimicked (*see* **Note 6**).

    (b) In circumstances with a small number of binding residues and a conformationally well-resolved scaffold, simple RMSD minimization of the alpha and beta carbons versus those of the native protein may suffice, and it is certainly the most generally applicable procedure. Other methods may be preferred:

    - If there are many reasonable conformers, many binding residues, and/or many scaffold residue isosteres, the combinatorics of RMSD-aligning all possible conformer alignments explode and manual alignment is unreasonably laborious.

    - If the expected optimal rotamers on a particular scaffold differ from those of a native peptide, the best $C\alpha$-$C\beta$ aligned scaffolds may not be physically realizable with the binding residue rotamers grafted on.

    (c) Even in situations where a complex parameterization of a peptidomimetic is available, it is entirely possible to treat that molecule as a small molecule ligand [81] or as a protein chain and to subject it to global docking [82]. Such methods are not particularly suitable for obtaining a high-resolution orientation of a peptidomimetic in a binding pocket, but they would deliver a critical piece of information: if a peptidomimetic might competitively bind elsewhere on a protein surface and be unsuitable for that reason.

(d) Alternatively, one may take advantage of codes initially built to match catalytic residues to novel protein scaffolds as part of Rosetta's enzyme design framework [83, 84]. The enzyme design code provides constraints that could be used in combination to require that desired peptidomimetic conformations present the same rotamers, in the same location, as the native protein.

(e) Reasonable biomolecular conformations can also be assembled stepwise, either using an enumerative or Monte Carlo approach [85]. The scoring function would be biased using the aforementioned restraints to present key binding residues.

This strategy is preferred if a general conformational ensemble—without a protein target—is difficult or impossible to generate, since the stepwise approach would permit the concurrent creation and evaluation of a compatible conformational ensemble.

3.3 Core Design: Shopping for Sidechains

1. Select an appropriate framework for scoring designs and their conformations (*see* **Note** 7).

(a) Determine the portion of the protein–inhibitor complex to which the scoring function ought to be applied.

- First, attempt to score all residues, but be aware that implicit in this decision is the assumption that the entire complex has been sampled adequately.

- If there is reason to suspect that only the interface has truly converged, then the scores of distant residues will contribute more noise than signal, and these should be discounted. Thus, choose a definition of the protein–protein interface and score only those residues.

- Finally, for an ad hoc definition of what residues are at the interface, explicitly sample both bound and unbound states of the complex in question and take the difference as a putative binding energy. On average, residues that are far from the interface will contribute less to this difference than residues at the interface (although one should check that this is the case).

  - Doing so opens the question of what degree of unbound state sampling is appropriate: it is perhaps far different for an intrinsically disordered protein than for a helix bundle.

  - Since this method requires strictly more sampling, it is only ideal if the above methods do not produce satisfactory results.

(b) Choose a criterion to determine the subset of trajectory results that will be ranked by the selected scoring function and selected for downstream design runs.

- Many structures will have flaws that should disqualify them. Top among these flaws are violation of experimental or biological constraints: for example, perhaps it is known that a particular hydrogen bond is critical. Those decoys lacking it could be excluded.

- Additional filters such as overall score, buried surface area, and so forth might help separate the most physically realistic structures from the chaff (*see* **Note 8**).

- Exclusively analyzing the metrics of single structures is not ideal for assessing their collective physical plausibility. Clustering the structures that result from such a trajectory provides considerable insight into the nature of the conformational minima they might inhabit.

(c) Select a method to proceed from such a set of resulting structures or other simulation results to concrete design recommendations (*see* **Note 9**).

- Typically, exclude any compounds that appear only once or twice in a set of trajectories (*see* **Note 10**).

- Take the best-scoring compounds that also come up frequently (Fig. 4).

- If there is no top-scoring compound that also comes up substantially more than any other, or if the very

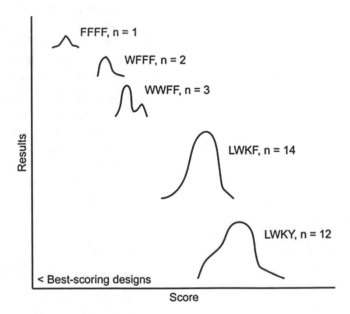

Fig. 4 An example result set from a study of tetrapeptides that might inhibit a target protein. Of 1000 physically reasonable resulting structures, three sequences appear one, two, and three times. Even though they score better on average than LWKF and LWKY, those two designs are likely superior due to their appreciably higher frequency

best few compounds all appear with comparable but unremarkable frequency, it will be necessary to perform parallel simulations, starting with the completed structures of interest, holding the chemical structure of each compound constant (*see* **Note 11**).

2. Obtain and optimize a satisfactory target protein structure (as described in Subheading 2.4). Being able to compare crystal structures of both *apo* and *holo* is best, but a *holo* complex alone can be perfectly suitable, and even an *apo* structure alone can be addressed. An NMR ensemble presents a considerably greater challenge but is by no means prohibitive by careful and extensive sampling.

3. For a subset of starting poses—combinations of protein target structures and scaffold conformers—optimize the peptidomimetic structure.

   (a) To make possible changes in sequence, molecular dynamics permits alchemical simulations where a continuously varied parameter governs the degree to which a given residue's identity of e.g. leucine or isoleucine interacts with its surroundings [86, 87]. Low-energy snapshots from such simulations can constitute an ensemble of peptidomimetic designs.

   (b) Rotameric approximations of side chain conformation permit the inclusion of rotamers of different amino acids in their rotamer sets, and Monte Carlo methods can sample from or anneal these configurations to approximate the minimum energy configuration.

   (c) Alternatively, dead end elimination can prove categorically that particular rotamers cannot be part of the global minimum energy configuration, or GMEC [88, 89].

   (d) Other conformational degrees of freedom must be sampled concurrently.

   • Rosetta, for example, employs a "relax" algorithm that performs repeated sidechain packing and minimization while ramping the repulsive weight of the scoring function.

   • Molecular dynamics simulations can employ simulated annealing for a similar effect [90].

   • The RosettaScripts framework [91] permits the seamless integration of the particular sampling modalities necessary for Monte Carlo simulations of special peptidomimetic scaffolds with those needed for e.g. remodeling a flexible loop (*see* **Note 12**).

4. According to the scoring criteria determined in **step 1**, select a set of compounds to synthesize and evaluate.

5. Employ the results to better inform subsequent rounds of design for this and other targets.

4   Notes

1. PocketQuery (and other methods that find clusters of pertinent residues) might be of particular use if the interface contains many loops distant from each other in primary sequence,
   i.e. in circumstances where few sets of consecutive residues
   would be useful.

2. InterProSurf is particularly useful if the protein–protein interface is large and there exist multiple distinct areas that might
   plausibly be targeted; it permits the elimination of entire irrelevant interface regions.

3. Of these options, submitting a job to the Robetta webserver
   requires none of one's own computing resources and is reasonably fast. The AlaScan filter is reasonably accurate and more
   customizable in its application than Robetta (in terms of scoring function selection and convergence criteria). The ddg_
   monomer application was the result of a systematic study
   comparing different levels of sampling sophistication and scoring functions to optimize the speed versus accuracy tradeoff
   and is the best choice to evaluate a handful of mutations if accurate energies are desired. For merely identifying whether residues are important, the AlaScan filter is more than adequate.

4. Since these moieties are handles for complex analysis, rather
   than literal representations of a bound structure, some of the
   challenges of fragment docking are not pertinent; mostly, the
   aim is finding shape complementarity and considerable contact
   surface area [92, 93].

5. For example, we anticipate that pairs of α-helices with inter-
   helical axes separated by angles of less than 30° and a distance
   of at most 15 Å are likely amenable to inhibition by a designed
   helical dimer mimetic [25]. Similar analysis might govern
   β-hairpins or other arrangements of elements, such as particular arrangements of key loop anchors [94].

6. Aligning a hybrid scaffold (e.g. a mostly peptidic scaffold with
   a set of residue isosteres grafted on in a nonpeptidic sense)
   simply reduces to the challenge of aligning the nonpeptidic
   portion *given* an atom-by-atom peptidic alignment obtained
   from the peptidic section.

7. Even beginning from a particular framework—for example, a
   molecular mechanics force field or Rosetta's all-atom scoring
   function *talaris2013*—there still remain meaningful choices to
   describe interfaces and rank different ligands [95–97]. Be
   comfortable beginning with a well-understood and well validated scoring function, but be equally prepared to modify it to
   suit the system at hand—in doing so, be sure to have a benchmark to quantifiably support that the modified scoring function treats your system more appropriately.

8. It is important to distinguish the notions of an overall score—a combined metric of physical realism—and of an interface or binding score. This way, one may differently judge the physical realism of an *entire structure* versus the expected binding affinity or biological activity of a designed mimetic.

9. The only chemically manipulable choice, following a set of simulations thus described, is the actual structure of the peptidomimetic—not the particular bound structure, to which such a simulation can assign a score. So, given that an identical peptidomimetic structure, in slightly different bound conformations, will be assigned many different scores over a set of simulations, how ought one to decide what compounds to synthesize?

10. Such a result suggests that the bound conformation that passed initial filters and has an excellent interface score is not at a particularly stable equilibrium: if small perturbations from that conformation also scored comparably well, then there would likely be more instances of that compound in the post-simulation data.

11. In doing so, one guarantees the same number of structures will be generated for each compound of interest, and the potential fragility of a given scoring function's preference for one compound or another will be made plain.

12. No particular programming experience is necessary to compose a protocol using RosettaScripts. The previously developed OopDockDesign and HbsDockDesign protocols [74] have been incorporated into the RosettaScripts framework which enables biologists with no programming expertise to compose their own complex protocols. Furthermore, these protocols—which were developed to work on particular scaffolds synthesized in the Arora lab—were generalized into a mover to handle the docking and design of arbitrary noncanonical backbones.

References

1. Watkins AM, Arora PS (2015) Structure-based inhibition of protein–protein interactions. Eur J Med Chem 94:480–488

2. Pelay-Gimeno M, Glas A, Koch O, Grossmann TN (2015) Structure-based design of inhibitors of protein–protein interactions: mimicking peptide binding epitopes. Angew Chem Int Ed 54(31):8896–8927

3. London N, Raveh B, Schueler-Furman O (2013) Druggable protein–protein interactions—from hot spots to hot segments. Curr Opin Chem Biol 17(6):952–959

4. Milroy L-G, Grossmann TN, Hennig S, Brunsveld L et al (2014) Modulators of protein–protein interactions. Chem Rev 114(9):4695–4748

5. Checco JW, Kreitler DF, Thomas NC, Belair DG et al (2015) Targeting diverse protein–protein interaction interfaces with α/β-peptides derived from the Z-domain scaffold. Proc Natl Acad Sci U S A 112(15):4552–4557

6. Boersma MD, Haase HS, Peterson-Kaufman KJ, Lee EF et al (2012) Evaluation of diverse alpha/beta-backbone patterns for functional alpha-helix mimicry: analogues of the Bim BH3 domain. J Am Chem Soc 134(1):315–323

7. Azzarito V, Long K, Murphy NS, Wilson AJ (2013) Inhibition of alpha-helix-mediated

protein-protein interactions using designed molecules. Nat Chem 5(3):161–173

8. Walensky LD, Bird GH (2014) Hydrocarbon-stapled peptides: principles, practice, and progress. J Med Chem 57(15):6275–6288

9. Villar EA, Beglov D, Chennamadhavuni S, Porco JA Jr et al (2014) How proteins bind macrocycles. Nat Chem Biol 10(9):723–731

10. Arkin Michelle R, Tang Y, Wells James A (2014) Small-molecule inhibitors of protein-protein interactions: progressing toward the reality. Chem Biol 21(9):1102–1114

11. Negi SS, Schein CH, Oezguen N, Power TD et al (2007) InterProSurf: a web server for predicting interacting sites on protein surfaces. Bioinformatics 23(24):3397–3399

12. Vangone A, Oliva R, Cavallo L (2012) CONS-COCOMAPS: a novel tool to measure and visualize the conservation of inter-residue contacts in multiple docking solutions. BMC Bioinformatics 13(Suppl 4):S19

13. Vangone A, Spinelli R, Scarano V, Cavallo L et al (2011) COCOMAPS: a web application to analyze and visualize contacts at the interface of biomolecular complexes. Bioinformatics 27(20):2915–2916

14. Shingate P, Manoharan M, Sukhwal A, Sowdhamini R (2014) ECMIS: computational approach for the identification of hotspots at protein-protein interfaces. BMC Bioinformatics 15:303

15. Tuncbag N, Keskin O, Gursoy A (2010) HotPoint: hot spot prediction server for protein interfaces. Nucleic Acids Res 38(Web Server issue):W402–W406

16. Lauck F, Smith CA, Friedland GF, Humphris EL et al (2010) RosettaBackrub—a web server for flexible backbone protein structure modeling and design. Nucleic Acids Res 38(Web Server issue):W569–W575

17. Meireles LM, Domling AS, Camacho CJ (2010) ANCHOR: a web server and database for analysis of protein-protein interaction binding pockets for drug discovery. Nucleic Acids Res 38(Web Server issue):W407–W411

18. Lyskov S, Chou FC, Conchuir SO, Der BS et al (2013) Serverification of molecular modeling applications: the Rosetta Online Server that Includes Everyone (ROSIE). PLoS One 8(5):e63906

19. Chivian D, Kim DE, Malmstrom L, Bradley P et al (2003) Automated prediction of CASP-5 structures using the Robetta server. Proteins 53(Suppl 6):524–533

20. Bergey CM, Watkins AM, Arora PS (2013) HippDB: a database of readily targeted helical protein-protein interactions. Bioinformatics 29(21):2806–2807

21. Bullock BN, Jochim AL, Arora PS (2011) Assessing helical protein interfaces for inhibitor design. J Am Chem Soc 133(36):14220–14223

22. Jochim AL, Arora PS (2010) Systematic analysis of helical protein interfaces reveals targets for synthetic inhibitors. ACS Chem Biol 5(10):919–923

23. Jochim AL, Arora PS (2009) Assessment of helical interfaces in protein-protein interactions. Mol Biosyst 5:924–926

24. Watkins AM, Arora PS (2014) The anatomy of β-strands at protein-protein interfaces. ACS Chem Biol 9(8):1747–1754

25. Watkins AM, Wuo MG, Arora PS (2015) Protein-protein interactions mediated by helical tertiary structure motifs. J Am Chem Soc 137(36):11622–11630

26. Tosovska P, Arora PS (2010) Oligooxopiperazines as nonpeptidic alpha-helix mimetics. Org Lett 12:1588–1591

27. Patgiri A, Jochim AL, Arora PS (2008) A hydrogen bond surrogate approach for stabilization of short peptide sequences in alpha-helical conformation. Acc Chem Res 41(10):1289–1300

28. Jochim AL, Miller SE, Angelo NG, Arora PS (2009) Evaluation of triazolamers as active site inhibitors of HIV-1 protease. Biorg Med Chem Lett 19(21):6023–6026

29. Angelo NG, Arora PS (2007) Solution- and solid-phase synthesis of triazole oligomers that display protein-like functionality. J Org Chem 72(21):7963–7967

30. Angelo NG, Arora PS (2005) Nonpeptidic foldamers from amino acids: synthesis and characterization of 1,3-substituted triazole oligomers. J Am Chem Soc 127:17134–17135

31. Wuo MG, Mahon AB, Arora PS (2015) An effective strategy for stabilizing minimal coiled coil mimetics. J Am Chem Soc 137(36):11618–11621

32. Xie X, Piao L, Bullock BN, Smith A et al (2014) Targeting HPV16 E6-p300 interaction reactivates p53 and inhibits the tumorigenicity of HPV-positive head and neck squamous cell carcinoma. Oncogene 33(8):1037–1046

33. Lao BB, Grishagin I, Mesallati H, Brewer TF et al (2014) In vivo modulation of hypoxia-inducible signaling by topographical helix mimetics. Proc Natl Acad Sci U S A 111(21):7531–7536

34. Lao BB, Drew K, Guarracino DA, Brewer TF et al (2014) Rational design of topographical helix mimics as potent inhibitors of protein-protein interactions. J Am Chem Soc 136(22):7877–7888

35. Kushal S, Lao BB, Henchey LK, Dubey R et al (2013) Protein domain mimetics as

in vivo modulators of hypoxia-inducible factor signaling. Proc Natl Acad Sci U S A 110(39):15602–15607

36. Patgiri A, Yadav KK, Arora PS, Bar-Sagi D (2011) An orthosteric inhibitor of the Ras-Sos interaction. Nat Chem Biol 7(9):585–587

37. Henchey LK, Porter JR, Ghosh I, Arora PS (2010) High specificity in protein recognition by hydrogen-bond-surrogate alpha-helices: selective inhibition of the p53/MDM2 complex. ChemBiochem 11(15):2104–2107

38. Wang D, Lu M, Arora PS (2008) Inhibition of HIV-1 fusion by hydrogen-bond-surrogate-based alpha helices. Angew Chem Int Ed 47(10):1879–1882

39. Wang D, Liao W, Arora PS (2005) Enhanced metabolic stability and protein-binding properties of artificial alpha-helices derived from a hydrogen-bond surrogate: application to Bcl-xL. Angew Chem Int Ed 44:6525–6529

40. Leaver-Fay A, Tyka M, Lewis SM, Lange OF et al (2011) ROSETTA3: an object-oriented software suite for the simulation and design of macromolecules. Methods Enzymol 487:545–574

41. Hou T, Wang J, Li Y, Wang W (2011) Assessing the performance of the MM/PBSA and MM/GBSA methods. 1. The accuracy of binding free energy calculations based on molecular dynamics simulations. J Chem Inf Model 51(1):69–82

42. Genheden S, Ryde U (2015) The MM/PBSA and MM/GBSA methods to estimate ligand-binding affinities. Expert Opin Drug Discov 10(5):449–461

43. Guerois R, Nielsen JE, Serrano L (2002) Predicting changes in the stability of proteins and protein complexes: a study of more than 1000 mutations. J Mol Biol 320(2):369–387

44. Kellogg EH, Leaver-Fay A, Baker D (2011) Role of conformational sampling in computing mutation-induced changes in protein structure and stability. Proteins 79(3):830–838

45. Eisenhaber F, Argos P (1993) Improved strategy in analytic surface calculation for molecular systems: handling of singularities and computational efficiency. J Comput Chem 14(11):1272–1280

46. Fraczkiewicz R, Braun W (1998) Exact and efficient analytical calculation of the accessible surface areas and their gradients for macromolecules. J Comput Chem 19(3):319–333

47. Kabsch W, Sander C (1983) Dictionary of protein secondary structure: pattern recognition of hydrogen-bonded and geometrical features. Biopolymers 22(12):2577–2637

48. Li Z, Wong L, Li J (2011) DBAC: a simple prediction method for protein binding hot spots based on burial levels and deeply buried atomic contacts. BMC Syst Biol 5(Suppl 1):S5

49. Petukh M, Li M, Alexov E (2015) Predicting binding free energy change caused by point mutations with knowledge-modified MM/PBSA method. PLoS Comput Biol 11(7):e1004276

50. Xin D, Ko E, Perez LM, Ioerger TR et al (2013) Evaluating minimalist mimics by exploring key orientations on secondary structures (EKOS). Org Biomol Chem 11(44):7789–7801

51. Ko E, Liu J, Burgess K (2011) Minimalist and universal peptidomimetics. Chem Soc Rev 40:4411–4421

52. Ko E, Liu J, Perez LM, Lu G et al (2010) Universal peptidomimetics. J Am Chem Soc 133(3):462–477

53. Mahon AB, Miller SE, Joy ST, Arora PS (2012) Rational design strategies for developing synthetic inhibitors of helical protein interfaces protein-protein interactions, vol 8. Springer, New York. doi:10.1007/978-3-642-28965-1_6

54. Henchey LK, Jochim AL, Arora PS (2008) Contemporary strategies for the stabilization of peptides in the alpha-helical conformation. Curr Opin Chem Biol 12(6):692–697

55. Freire F, Gellman SH (2011) Macrocyclic design strategies for small, stable parallel beta-sheet scaffolds. J Am Chem Soc 133(31):12318

56. Harrison RS, Shepherd NE, Hoang HN, Ruiz-Gomez G et al (2010) Downsizing human, bacterial, and viral proteins to short water-stable alpha helices that maintain biological potency. Proc Natl Acad Sci U S A 107(26):11686–11691

57. Henchey LK, Kushal S, Dubey R, Chapman RN et al (2010) Inhibition of hypoxia inducible factor 1–transcription coactivator interaction by a hydrogen bond surrogate alpha-helix. J Am Chem Soc 132(3):941–943

58. Lingard H, Han JT, Thompson AL, Leung IKH et al (2014) Diphenylacetylene-linked peptide strands induce bidirectional β-sheet formation. Angew Chem Int Ed 53(14):3650–3653

59. Sutherell CL, Thompson S, Scott RTW, Hamilton AD (2012) Aryl-linked imidazolidin-2-ones as non-peptidic [small beta]-strand mimetics. Chem Commun 48(79):9834–9836

60. Kang CW, Sun Y, Del Valle JR (2012) Substituted imidazo[1,2-a]pyridines as β-strand peptidomimetics. Org Lett 14(24):6162–6165

61. Khasanova TV, Khakshoor O, Nowick JS (2008) Functionalized analogues of an

unnatural amino acid that mimics a tripeptide Œ ≤ -strand. Org Lett 10(22):5293–5296

62. Hammond MC, Harris BZ, Lim WA, Bartlett PA (2006) Beta strand peptidomimetics as potent PDZ domain ligands. Chem Biol 13(12):1247–1251

63. Phillips ST, Rezac M, Abel U, Kossenjans M et al (2002) "@-tides": the 1,2-dihydro-3(6H)-pyridinone unit as a beta-strand mimic. J Am Chem Soc 124(1):58–66

64. Tsai JH, Waldman AS, Nowick JS (1999) Two new beta-strand mimics. Bioorg Med Chem 7(1):29–38

65. Smith AB, Keenan TP, Holcomb RC, Sprengeler PA et al (1992) Design, synthesis, and crystal-structure of a pyrrolinone-based peptidomimetic possessing the conformation of a beta-strand—potential application to the design of novel inhibitors of proteolytic-enzymes. J Am Chem Soc 114(26):10672–10674

66. Loughlin WA, Tyndall JDA, Glenn MP, Fairlie DP (2004) Beta-strand mimetics. Chem Rev 104(12):6085–6117

67. Hawkins PC, Skillman AG, Warren GL, Ellingson BA et al (2010) Conformer generation with OMEGA: algorithm and validation using high quality structures from the Protein Databank and Cambridge Structural Database. J Chem Inf Model 50(4):572–584

68. Bahar I, Lezon TR, Bakan A, Shrivastava IH (2010) Normal mode analysis of biomolecular structures: functional mechanisms of membrane proteins. Chem Rev 110(3):1463–1497

69. Fuglebakk E, Echave J, Reuter N (2012) Measuring and comparing structural fluctuation patterns in large protein datasets. Bioinformatics 28(19):2431–2440

70. Bornot A, Etchebest C, de Brevern AG (2011) Predicting protein flexibility through the prediction of local structures. Proteins 79(3):839–852

71. Hilser VJ, Whitten ST (2014) Using the COREX/BEST server to model the native-state ensemble. Methods Mol Biol 1084:255–269

72. Seeliger D, de Groot BL (2010) Conformational transitions upon ligand binding: holo-structure prediction from apo conformations. PLoS Comput Biol 6(1):e1000634

73. Kortemme T, Kim DE, Baker D (2004) Computational alanine scanning of protein-protein interfaces. Sci STKE 2004(219):pl2

74. Drew K, Renfrew PD, Craven TW, Butterfoss GL et al (2013) Adding diverse noncanonical backbones to rosetta: enabling peptidomimetic design. PLoS One 8(7):e67051

75. Darnell SJ, LeGault L, Mitchell JC (2008) KFC Server: interactive forecasting of protein interaction hot spots. Nucleic Acids Res 36(Web Server issue):W265–W269

76. Koes DR, Camacho CJ (2012) PocketQuery: protein-protein interaction inhibitor starting points from protein-protein interaction structure. Nucleic Acids Res 40(Web Server issue):W387–W392

77. Schmidtke P, Le Guilloux V, Maupetit J, Tuffery P (2010) fpocket: online tools for protein ensemble pocket detection and tracking. Nucleic Acids Res 38(Web Server issue):W582–W589

78. Rooklin D, Wang C, Katigbak J, Arora PS et al (2015) AlphaSpace: fragment-centric topographical mapping to target protein-protein interaction interfaces. J Chem Inf Model 55(8):1585–1599

79. Chen J, Ma X, Yuan Y, Pei J et al (2014) Protein-protein interface analysis and hot spots identification for chemical ligand design. Curr Pharm Des 20(8):1192–1200

80. Frishman D, Argos P (1995) Knowledge-based protein secondary structure assignment. Proteins 23(4):566–579

81. Lemmon G, Meiler J (2012) Rosetta ligand docking with flexible XML protocols. Methods Mol Biol 819:143–155

82. Gray JJ, Moughon S, Wang C, Schueler-Furman O et al (2003) Protein-protein docking with simultaneous optimization of rigid-body displacement and side-chain conformations. J Mol Biol 331(1):281–299

83. Richter F, Leaver-Fay A, Khare SD, Bjelic S et al (2011) De novo enzyme design using Rosetta3. PLoS One 6(5):e19230

84. Rothlisberger D, Khersonsky O, Wollacott AM, Jiang L et al (2008) Kemp elimination catalysts by computational enzyme design. Nature 453(7192):190–U194

85. Sripakdeevong P, Kladwang W, Das R (2011) An enumerative stepwise ansatz enables atomic-accuracy RNA loop modeling. Proc Natl Acad Sci U S A 108(51):20573–20578

86. Procacci P, Cardelli C (2014) Fast switching alchemical transformations in molecular dynamics simulations. J Chem Theory Comput 10(7):2813–2823

87. Meng Y, Dashti DS, Roitberg AE (2011) Computing alchemical free energy differences with Hamiltonian replica exchange molecular dynamics (H-REMD) simulations. J Chem Theory Comput 7(9):2721–2727

88. Hallen MA, Keedy DA, Donald BR (2013) Dead-end elimination with perturbations

(DEEPer): a provable protein design algorithm with continuous sidechain and backbone flexibility. Proteins 81(1):18–39

89. Georgiev I, Donald BR (2007) Dead-end elimination with backbone flexibility. Bioinformatics 23(13):i185–i194

90. Tidor B (1993) Simulated annealing on free energy surfaces by a combined molecular dynamics and Monte Carlo approach. J Phys Chem 97(5):1069–1073

91. Fleishman SJ, Leaver-Fay A, Corn JE, Strauch EM et al (2011) RosettaScripts: a scripting language interface to the Rosetta macromolecular modeling suite. PLoS One 6(6): e20161

92. Kawatkar S, Wang H, Czerminski R, Joseph-McCarthy D (2009) Virtual fragment screening: an exploration of various docking and scoring protocols for fragments using Glide. J Comput Aided Mol Des 23(8):527–539

93. Sandor M, Kiss R, Keseru GM (2010) Virtual fragment docking by Glide: a validation study on 190 protein-fragment complexes. J Chem Inf Model 50(6):1165–1172

94. Lewis SM, Kuhlman BA (2011) Anchored design of protein-protein interfaces. PLoS One 6(6):e20872

95. Leaver-Fay A, O'Meara MJ, Tyka M, Jacak R et al (2013) Scientific benchmarks for guiding macromolecular energy function improvement. Methods Enzymol 523:109–143

96. Vanommeslaeghe K, Hatcher E, Acharya C, Kundu S et al (2010) CHARMM general force field: a force field for drug-like molecules compatible with the CHARMM all-atom additive biological force fields. J Comput Chem 31(4):671–690

97. Hynninen AP, Crowley MF (2014) New faster CHARMM molecular dynamics engine. J Comput Chem 35(5):406–413

INDEX

A

ACCLUSTER 3–8
Affinity prediction 243–245
Alanine scanning 218, 256, 257, 260–264, 266–271, 275, 293, 295–297
AlphaSpace 297
AMBER 56–58, 61, 99, 114, 115, 249, 250, 297
AnchorDock 95–106
Ant colony optimization 280, 281, 288
Apo structure 50, 301
ATTRACT 27–29, 38, 49–65, 96

B

Benchmark 39, 50, 54, 56, 57, 64, 70, 76, 80, 111, 135, 145, 146, 165, 184, 236, 240, 244, 280, 302
Binding
    affinity 76, 139, 142–143, 149, 154, 157, 181, 183, 190, 213–229, 244, 250, 303
    free energy 202, 244, 256, 262, 270, 293
    specificity 142, 143, 154–157, 180, 226, 229
    specificity profiles 201–210, 287
Binding site prediction
    ACCLUSTE 3–8
    Fpocket...297
    PEP-sitefinder 26–32, 38, 45, 145
    PeptiMap 11–19, 106, 145
    pocketquery 258
Blind peptide-protein docking 50, 51, 54, 58, 85–106, 241
Bound 4–8, 14–17, 23, 26, 50, 54, 61, 62, 80, 96, 100, 110, 112, 135, 140, 141, 144, 148, 157, 158, 161, 164, 176, 179–183, 195, 202–208, 210, 218, 220, 245, 248, 293, 295, 299, 302, 303

C

CABS-dock... 38, 69–93, 96
Calibration.......................... 144, 148, 149, 156–159, 161, 165
CAPRI. *See* Critical assessment of prediction of interactions (CAPRI)
Case study 6–7, 12, 14–17, 30–32, 43–45, 72–76, 114, 135–137, 284, 287
Cluster expansion 190, 216
Coarse-grained 27, 28, 32, 49–65, 70, 71, 80, 141, 145

Computational design 213, 214, 216
Computational peptide screening 201–210
Conformational changes...........................22–24, 54, 58, 63, 70, 110, 135, 136, 140, 174, 177
Conformational sampling.........................140, 155, 214, 241, 247, 260, 294
Conformational search 96, 236, 244–249
Constraints 50, 96, 103, 147–149, 154, 155, 157–159, 161, 163, 196, 205–206, 208, 210, 216, 223, 227–229, 237, 238, 240, 247, 257, 276, 283–285, 293, 299, 300
Critical assessment of prediction of interactions (CAPRI) 39, 43, 44, 61, 64, 96–97, 135, 136, 241
Cyclic peptides 16, 245, 258, 260, 265, 268, 269, 273–275

D

Data analysis.. 98, 208–209
Dead end elimination (DEE)................................... 174, 301
Decoys 53, 128, 130, 147, 151–153, 300
DEE. *See* Dead end elimination (DEE)
Degrees of freedom 28, 57, 140, 145, 162, 177, 182, 206, 239, 280, 301
De-novo peptide folding 145–146
Docking
    ATTRACT 27–29, 38, 49–65, 96
    blind docking.....................27–30, 58, 96, 105, 165, 204
    ensemble docking 53–55, 58, 63, 111, 122
    HADDOCK............................... 57, 96, 109–137
    induced fit docking...236, 240
    Piper 241–242

E

Ensemble docking 53–55, 58, 63, 111, 122

F

Fast Fourier Transform (FFT)............................. 12, 17, 242
Flexible backbone 123, 173–184
Force field
    AMBER...............56–58, 61, 99, 114, 115, 249, 250, 297
    GROMACS...................................97, 98, 100, 101, 111, 113, 115–118, 121
    knowledge-based energy function 38–39

Ora Schueler-Furman and Nir London (eds.), *Modeling Peptide-Protein Interactions: Methods and Protocols*, Methods in Molecular Biology, vol. 1561, DOI 10.1007/978-1-4939-6798-8, © Springer Science+Business Media LLC 2017

Fpocket..297

Fragments........................ 21, 64, 81, 88, 92, 100, 104,
128, 140, 145, 147, 151–153, 164, 165, 177, 178,
249, 256, 270, 297, 302

Funnel 238, 239, 280, 285–288

G

GalaxyPepDock...37–45

Generalized-ensemble...204

Glide.............................236–241, 243, 245, 247, 248

Global minimum 174–176, 215, 301

GROMACS...............................97, 98, 100, 101, 111,
113, 115–118, 121

H

HADDOCK................................. 57, 96, 109–137

Helix mimetics 294

Holo structure 295

Hot spot residues............... 256, 260, 272–274, 292, 294, 295

Hydrogen bond surrogate

I

ILP. *See* Integer linear programming (ILP)

Induced fit docking (IFD)................................ 236, 240–241

Input......................4–7, 12–14, 38–40, 43, 44, 59, 60, 71–73,
80, 81, 88, 89, 91, 96, 98–100, 105, 126, 131, 133,
134, 141, 146, 147, 150, 151, 153, 158, 163, 174,
178–180, 215–219, 221, 222, 224, 226, 229,
238–241, 259–268, 271, 272, 281, 285, 287

Integer linear programming (ILP)................... 191, 196, 197,
216–218, 221–225, 227–229

InterProSurf 291, 297, 302

Intrinsically disordered protein..................................50, 299

K

Kinematic closure (KIC) 177, 178

Knowledge-based energy function................................38–39

L

Large-scale modelling 49

Library design 213–229

Local minimum...178, 281

Loopfinder 255–276

M

MD. *See* Molecular dynamics (MD)

Minimization 28, 53, 55, 56, 98, 99, 110, 112,
116–118, 123, 129, 141, 145, 148, 150, 151, 154,
155, 158, 159, 177, 178, 182, 207, 243, 244, 247,
294, 298, 301

MM-GBSA. *See* Molecular Mechanics-Generalized
Born Solvation Approximation Method
(MM-GBSA)

Molecular dynamics (MD)............................. 54, 56–58, 61,
63, 64, 97–99, 106, 110–118, 121–124, 126–130,
135–137, 140, 146, 197, 236, 243, 245, 248–250,
294, 295, 301

Molecular Mechanics-Generalized Born Solvation
Approximation Method
(MM-GBSA)................................. 240, 243–244

Monte Carlo..................................38, 71, 76, 139–166, 178,
180, 203–207, 215, 242, 294, 299, 301

MSA. *See* Multiple sequence alignment (MSA)

Multi-chains...13

Multiple sequence alignment (MSA)............ 215, 216, 226–228

O

OMEGA..294

P

PDZ domain 96, 97, 102–105, 162, 179–184,
195, 202–204, 208, 209, 239, 242, 245–247

pepATTRACT 51, 54–64, 96

PepBind..39

PepCrawler..................................11, 38, 96, 279–289

PEP-SiteFinder........................ 26–32, 38, 45, 145

PeptiDB..................... 22, 24, 29, 30, 38, 39, 45, 56, 57

Peptide

binding site 3–8, 11–19, 26, 50, 51, 54,
58, 63, 80, 124, 202, 256

blind peptide docking............ 50, 51, 54, 58, 85–106, 241

design 77, 80, 265, 269, 274, 288

docking protocols

AnchorDock...95–106

CABS-dock.................................... 38, 69–93, 96

GalaxyPepDock................................... 37–45, 96

pepATTRACT..................................... 51, 54–64, 96

PepCrawler 11, 38, 96, 279–289

Rosetta FlexPepDock..................38, 80, 96, 139–166,
192, 193, 195

flexibility.................................... 54, 96, 101, 182

folding 99, 105, 145–146

length.............................137, 147, 165, 180, 190

libraries...213–229

peptide-mediated interactions 11, 50, 139–140

peptide-protein interactions 32, 49–65, 105, 1
45, 159, 201, 235–252, 273

peptide-recognition domains.............................189–196

peptidomimetics 61, 235, 245, 291–303

QSAR... 236, 249–252

specificity... 143, 173–184

therapeutics..3, 80

PeptiMap.. 11–19, 106, 145

PinaColada..279–289

Piper..241–242

PocketQuery.. 256, 295, 302

Pocket V.3 297

Position specific scoring matrix (PSSM) 218–220, 225–227
PPI. *See* Protein-protein interactions (PPI)
Prime 236, 240, 242–243
Protein flexibility 51, 63, 242, 244
Protein-protein interactions (PPI)
    inhibitors 257, 258, 291, 294
    peptide-protein interactions 32, 49–65, 105, 145, 159, 201, 235–252, 273
    protein-protein interfaces 53, 89, 249, 266, 270, 291, 293, 299, 302
PSSM. *See* Position specific scoring matrix (PSSM)

R

Ranking of models 164
Rapidly exploring Random Tree (RRT) 280, 283
Receptor backbone 154, 165
Refinement 40, 49–65, 79, 80, 96, 103, 106, 110, 111, 123, 129, 130, 140, 141, 143, 145–154, 158–160, 163–166, 226, 242–243
Results analysis 61, 64, 70, 73–76, 103, 147–154
Rosetta
    backrub 176
    Rosetta FlexPepBind 142, 154–156, 167
    Rosetta FlexPepDock 38, 80, 96, 139–166, 192, 193, 195
    Rosetta FlexPepDock ab-initio 141, 145–146
    Rosetta PeptiDerive 257
Rotamers 145, 150, 151, 164, 174, 182, 216, 229, 249, 293, 298, 299, 301
RRT. *See* Rapidly exploring Random Tree (RRT)

S

Sampling parameters 163–164
Schrödinger 98, 235–252
Scoring function 4, 110, 123, 124, 155, 193, 238, 239, 241, 293, 299, 301–303
Secondary structure
    DSSP 297
    STRIDE 297
Second-site suppressor approach 181–182
Sequence
    sequence-based design 91, 156, 157, 189–199
    space 190–192, 196, 197, 202, 210, 213, 214, 280, 293
    tolerance 175–178
Side chain flexibility 55, 147, 164, 165
Simulated annealing 98, 99, 106, 123, 243, 249, 301
Simulation techniques
    ant colony optimization 280, 281, 288

dead end elimination 174, 301
Fast Fourier Transform 12, 17, 242
integer linear programming 191, 196, 197, 216–218, 221–225, 227–229
kinematic closure 177, 178
minimization 28, 53, 55, 56, 98, 99, 110, 112, 116–118, 123, 129, 141, 145, 148, 150, 151, 154, 155, 158, 159, 177, 178, 182, 207, 243, 244, 247, 294, 298, 301
MM-GBSA 240, 243–244
molecular dynamics 54, 56–58, 61, 63, 64, 97–99, 106, 110–118, 121–124, 126–130, 135–137, 140, 146, 197, 236, 243, 245, 248–250, 294, 295, 301
Monte Carlo 38, 71, 76, 139–166, 178, 180, 203–207, 215, 242, 294, 299, 301
rapidly expanding random trees 280, 283
refinement 40, 49–65, 79, 80, 96, 103, 106, 110, 111, 123, 129, 130, 140, 141, 143, 145–154, 158–160, 163–166, 226, 242–243
rotamers 145, 150, 151, 164, 174, 182, 216, 229, 249, 293, 298, 299, 301
Schrödinger 98, 235–252
second-site suppressor approach 181–182
side chain felixibility 55, 147, 164, 165
simulated annealing 98, 99, 106, 123, 243, 249, 301
WaterMap 236, 245, 246, 248
Solvent-accessible surface area 256, 293
Solvent mapping 140
Starting pose 298–299, 301
STATIUM 216, 218, 220
Surface mapping,

T

Template-based modelling 37–45, 80
Template selection 39, 43
Test set 45, 158, 195–196
Training set 155, 157, 158, 193–196

U

Unbound 6, 8, 23, 50, 53, 54, 58, 61, 72, 88, 89, 96, 97, 99–101, 106, 124, 126, 135–137, 147, 150, 151, 153, 154, 158, 164, 166, 204–206, 208, 293, 299

W

WaterMap 236, 245, 246, 248
Web interface 51, 58–64, 81
Web server 3, 38, 62, 70, 71, 111, 134, 143, 178–179, 181, 281, 282, 284–288

Printed in the United States
By Bookmasters